Adult Activity Books Series

The Giant Book of

Masyu

1000 Easy to Hard Puzzles (10x10)

Vol. 50

Khalid Alzamili, Ph.D.

A Special Request

Your brief review could really help us.

Thank you for your support

August 2020

Copyright © 2020 Dr. Khalid Alzamili

ISBN: 9789922636771

Dr. Khalid Alzamili Pub

www.alzamili.com

Author Email : khalid@alzamili.com

CONTENTS

INTRODUCTION

Masyu is a logic puzzle with simple rules and challenging solutions. It is played on a rectangular grid of squares, some of which contain circles; each circle is either "white" (empty) or "black" (filled).

The rules of Masyu are simple:

1- Draw a single, non-intersecting loop with lines passing through the centers of cells, horizontally or vertically. The loop never crosses itself, branches off, or goes through the same cell twice.

2- The loop must go straight through the cells with white circles, with a turn in at least one of the cells immediately before/after each white circle.

3- The loop passing through black circles must make a right-angled turn in its cell, then it must go straight through the next cell (till the middle of the second cell) on both sides.

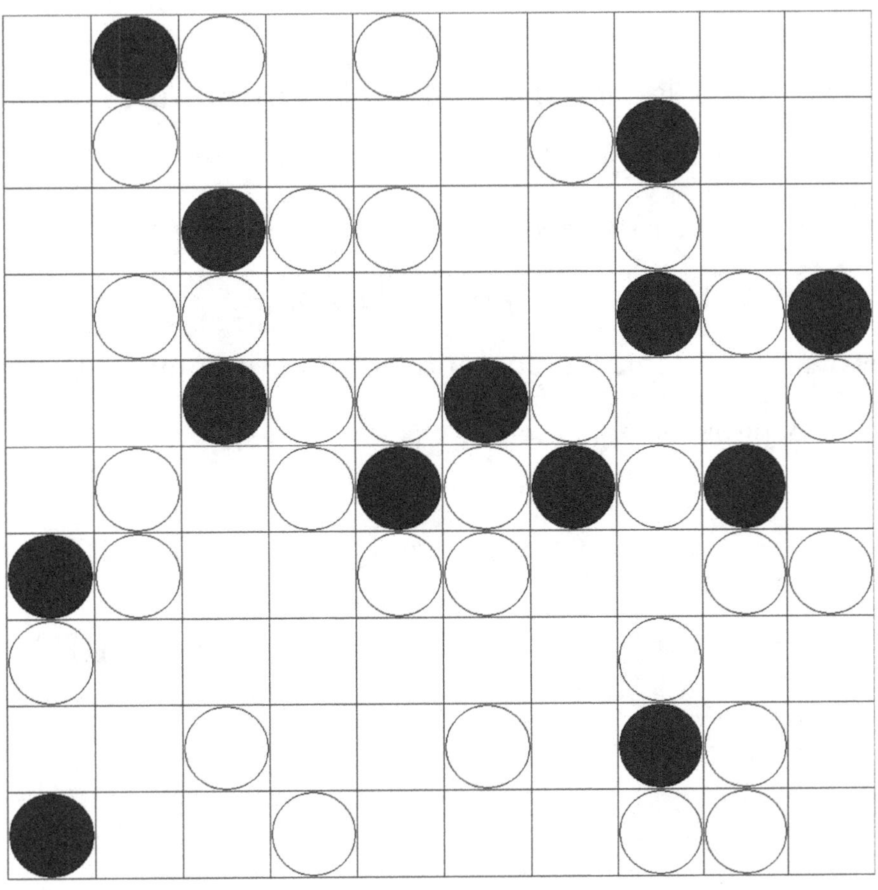

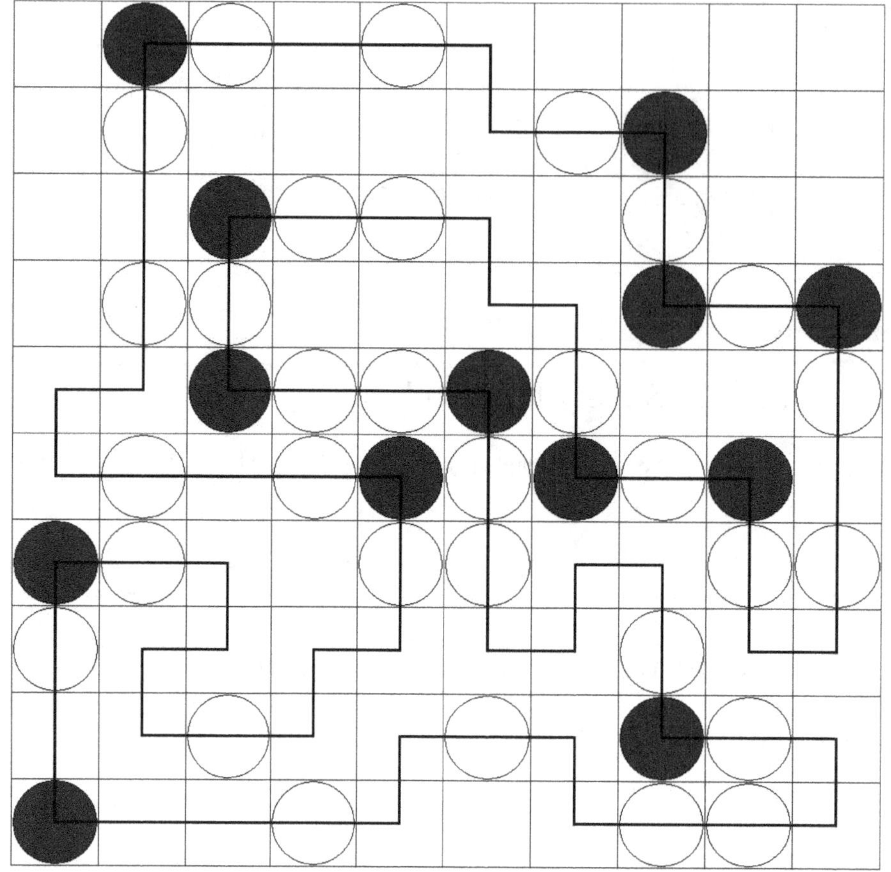

This Logic Puzzles book is packed with the following features:

- 1000 Masyu (10x10) Puzzles from Easy to Hard.

- Answers to every puzzle are provided.

- Each puzzle is guaranteed to have only one solution.

We hope this will be an entertaining and uplifting mental workout, enjoy Tons of Masyu for Adults & Seniors Book.

Khalid Alzamili, Ph.D.

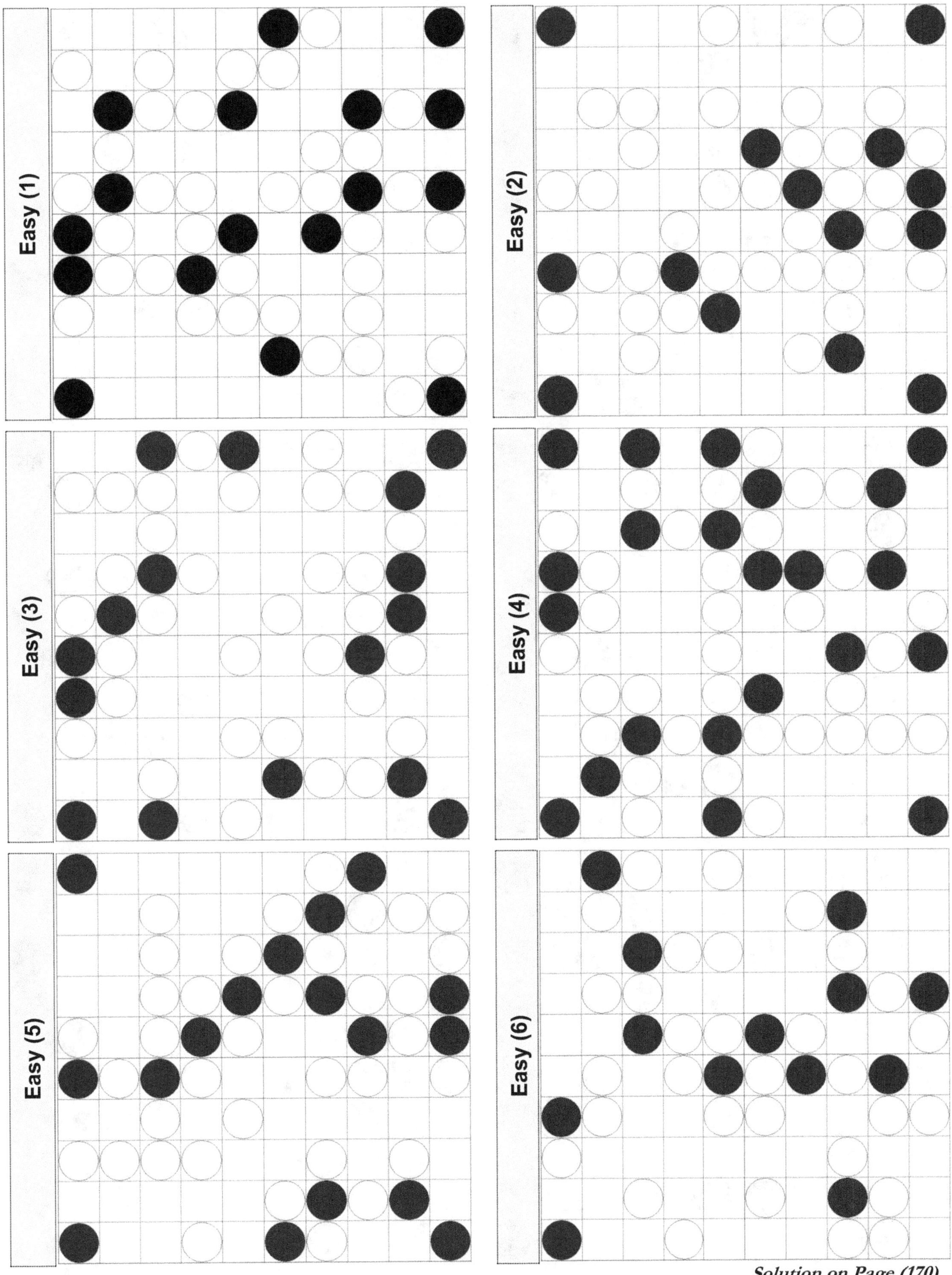

Easy (1)
Easy (2)
Easy (3)
Easy (4)
Easy (5)
Easy (6)

Solution on Page (170)

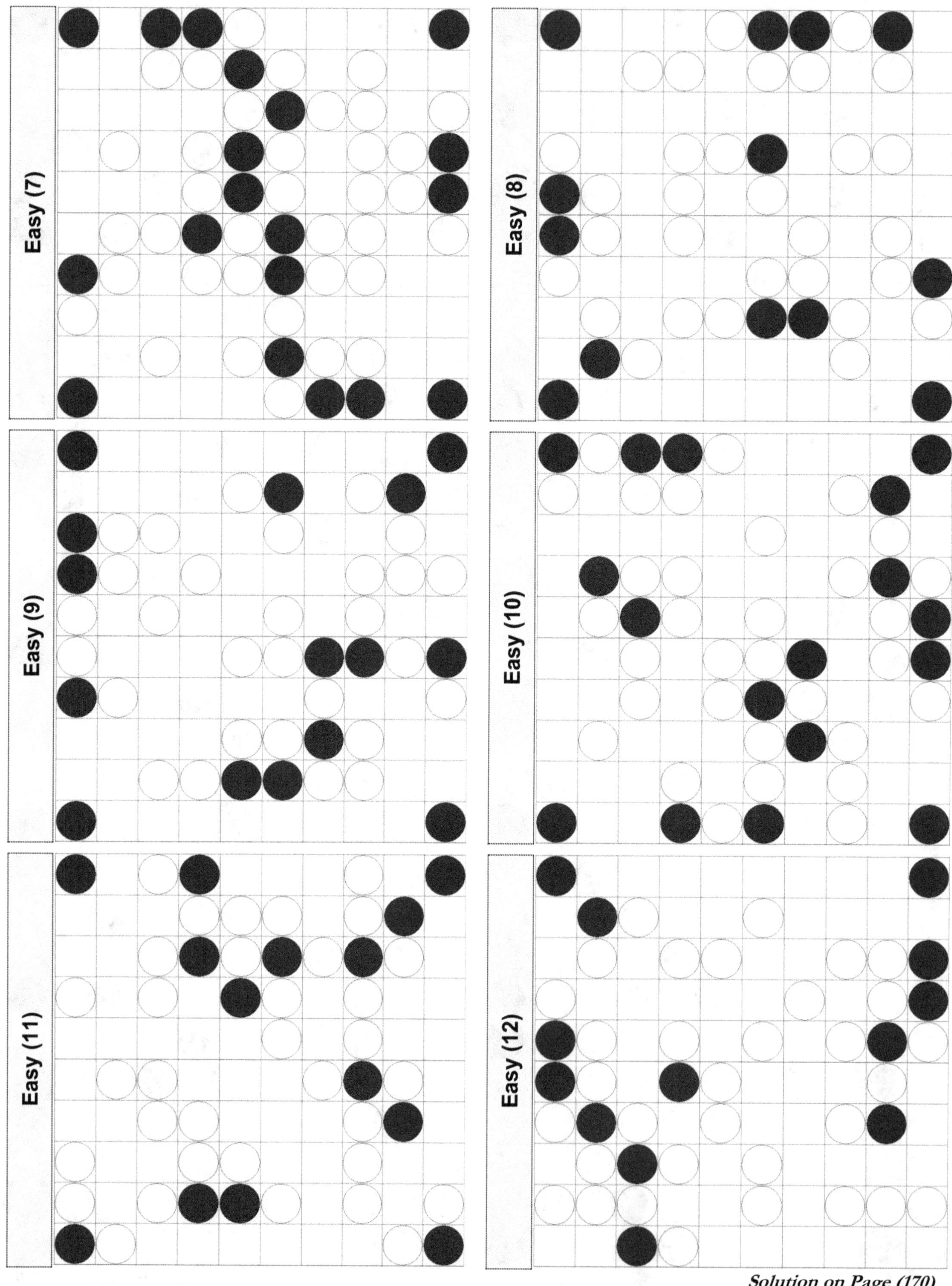

Solution on Page (170)

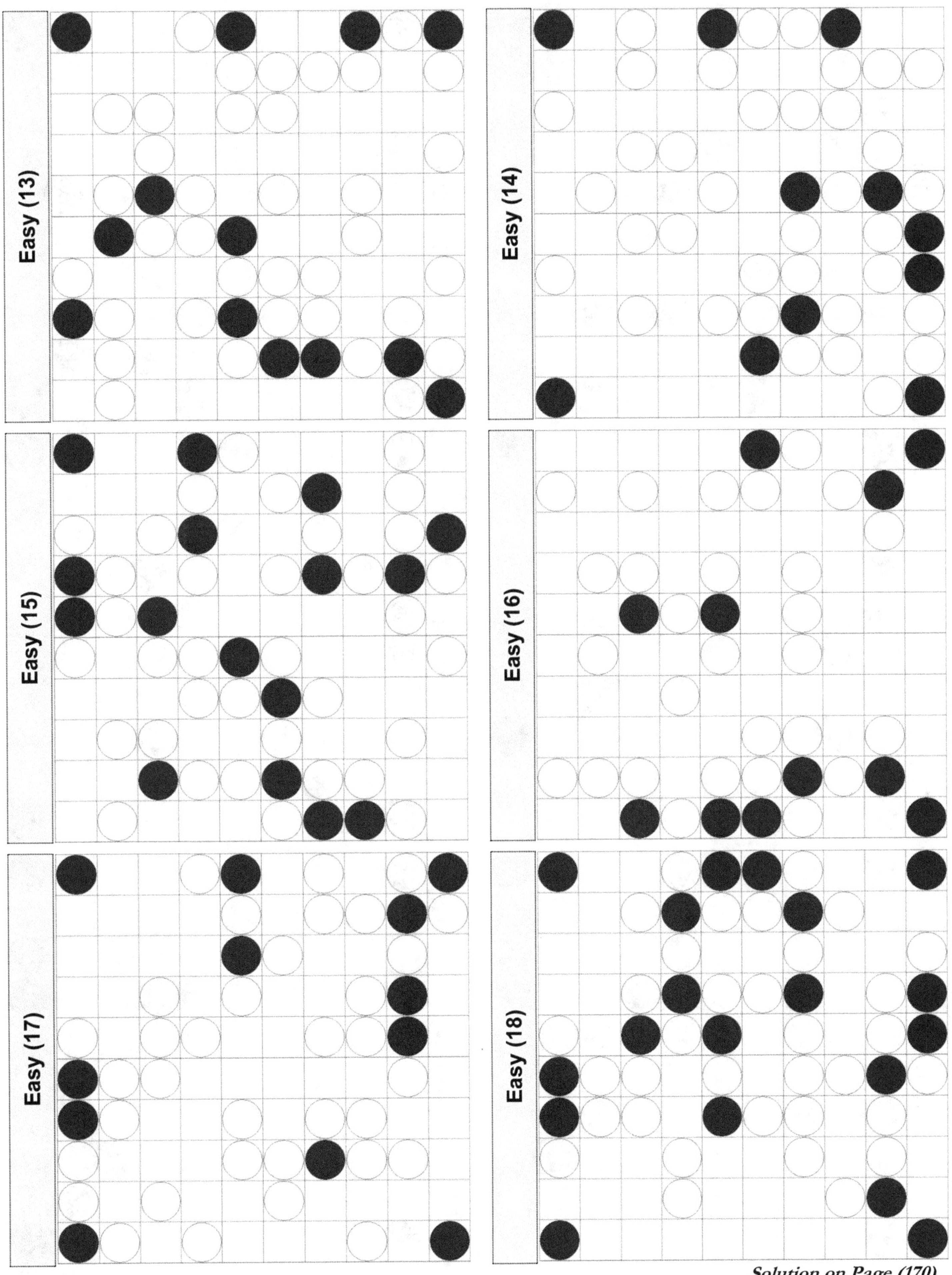

Solution on Page (170)

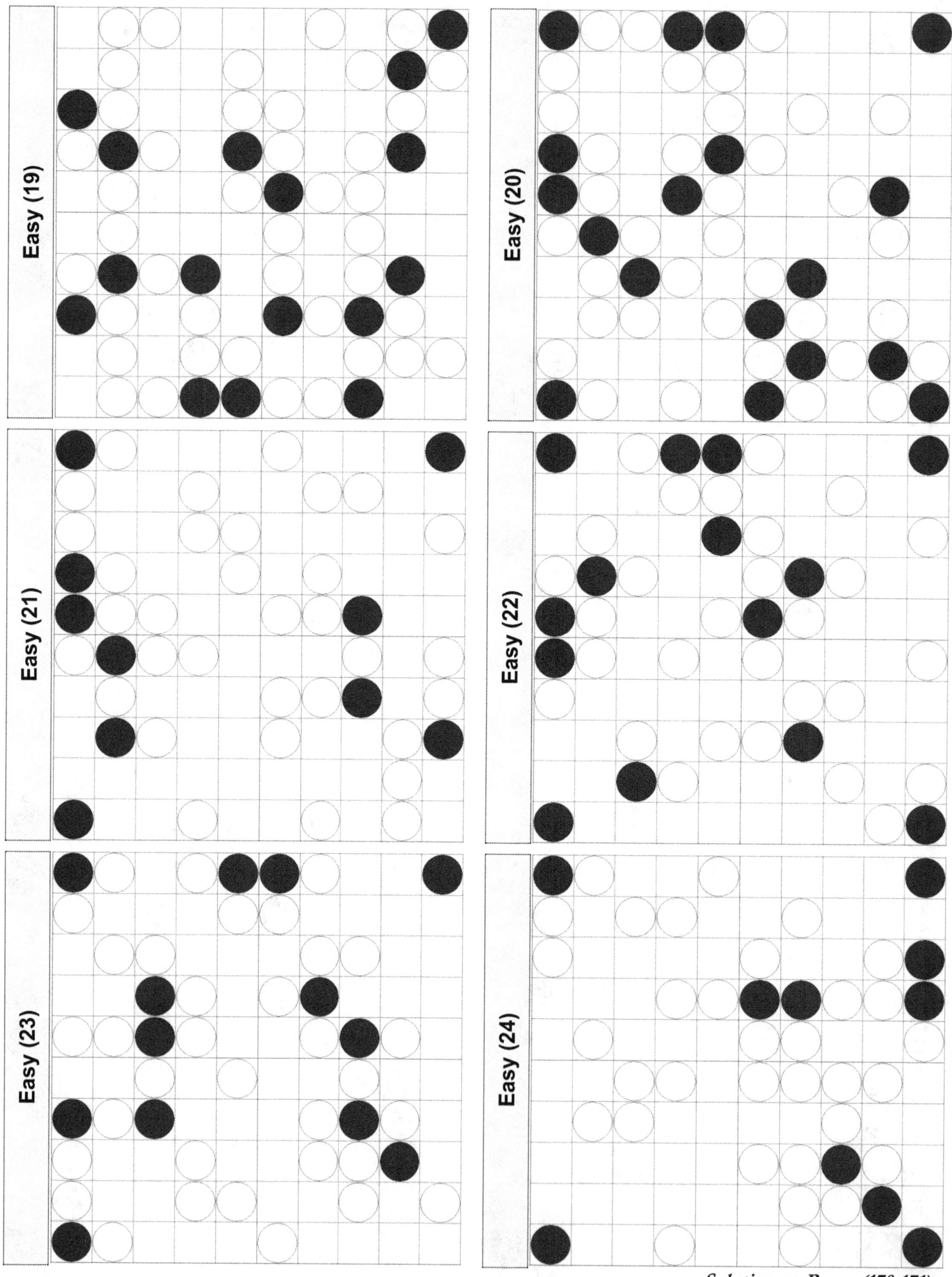

Solution on Pages (170-171)

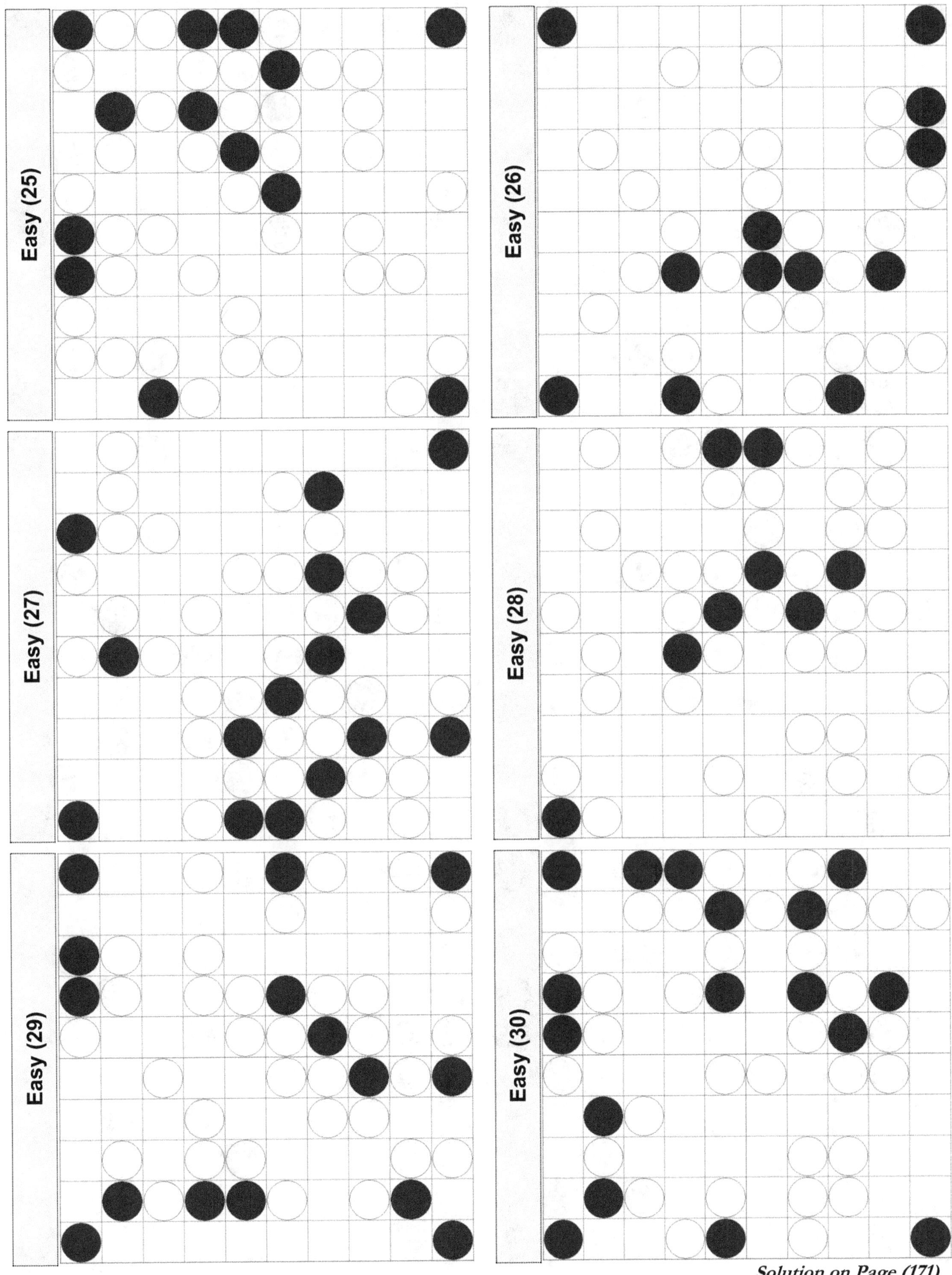

Solution on Page (171)

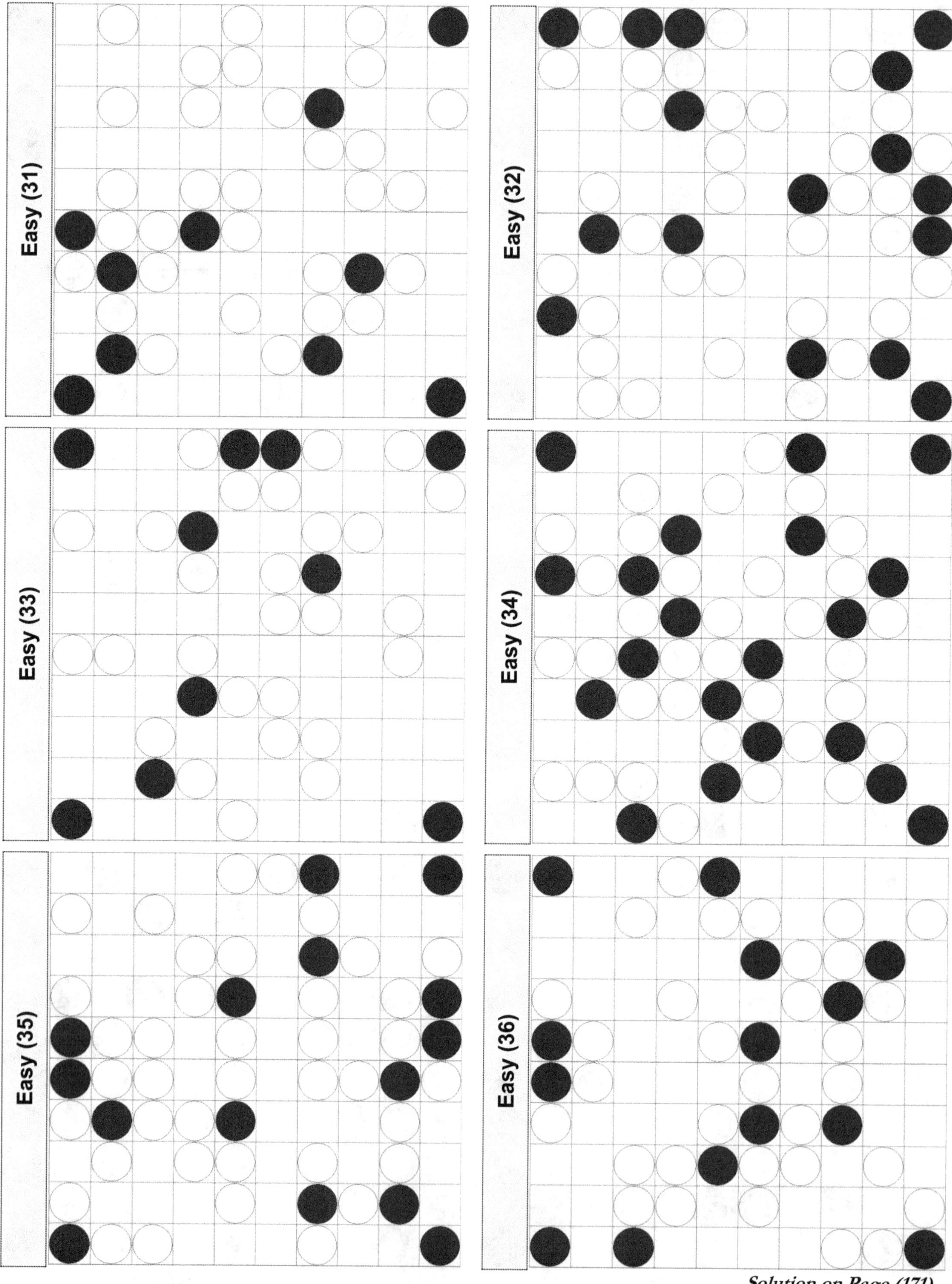

Easy (31)
Easy (32)
Easy (33)
Easy (34)
Easy (35)
Easy (36)
Solution on Page (171)

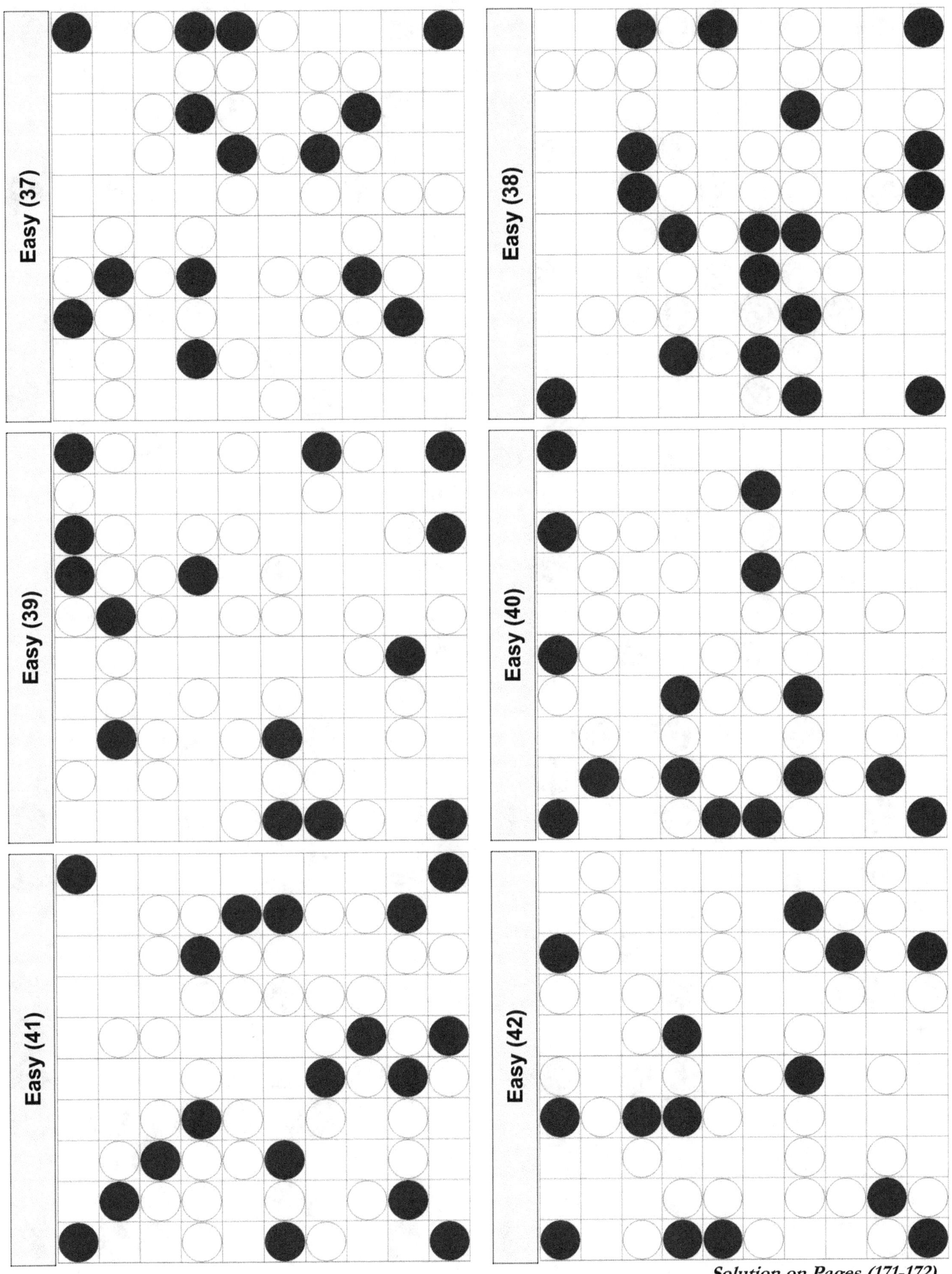

Solution on Pages (171-172)

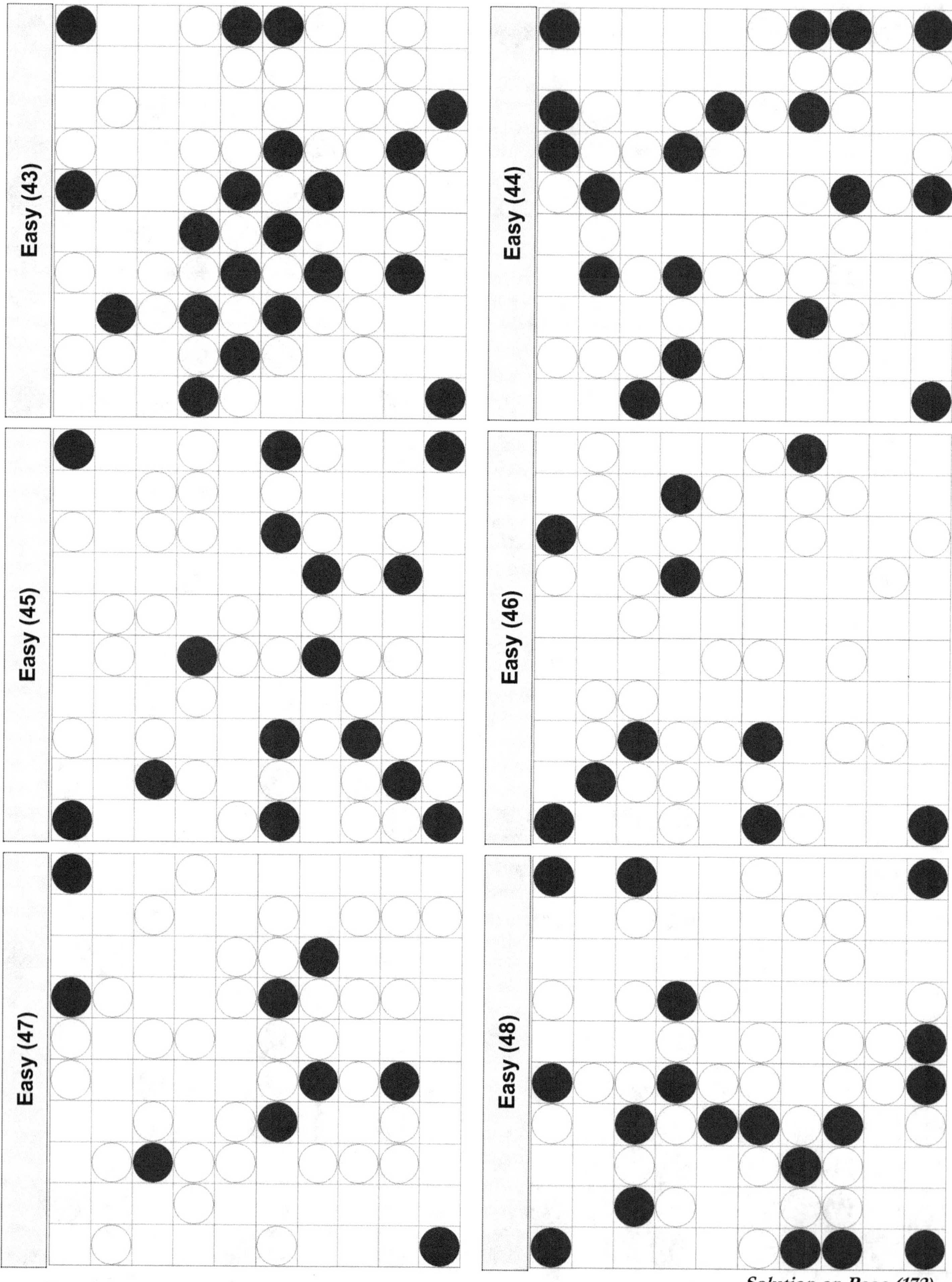

Solution on Page (172)

(10)

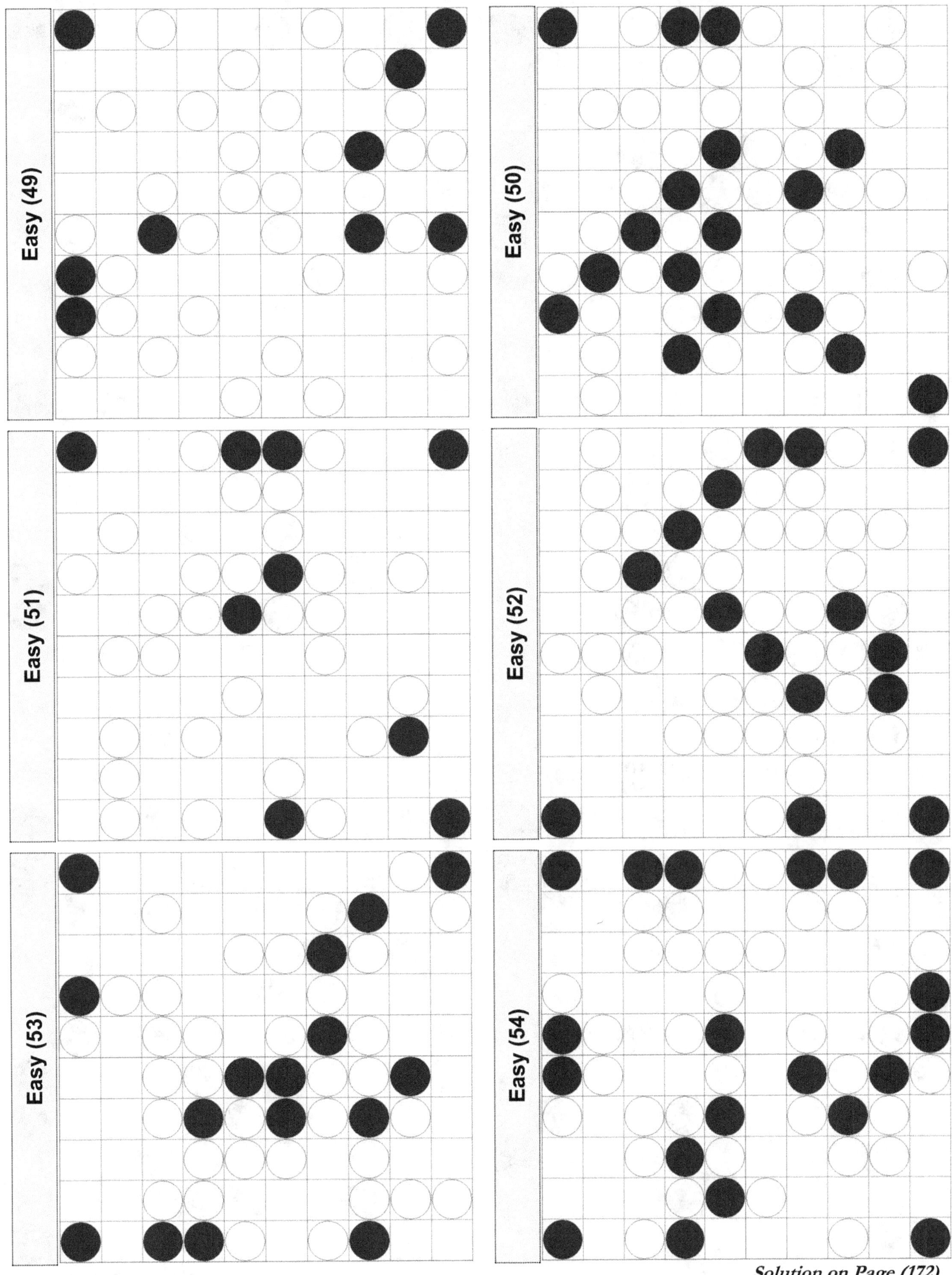

Solution on Page (172)

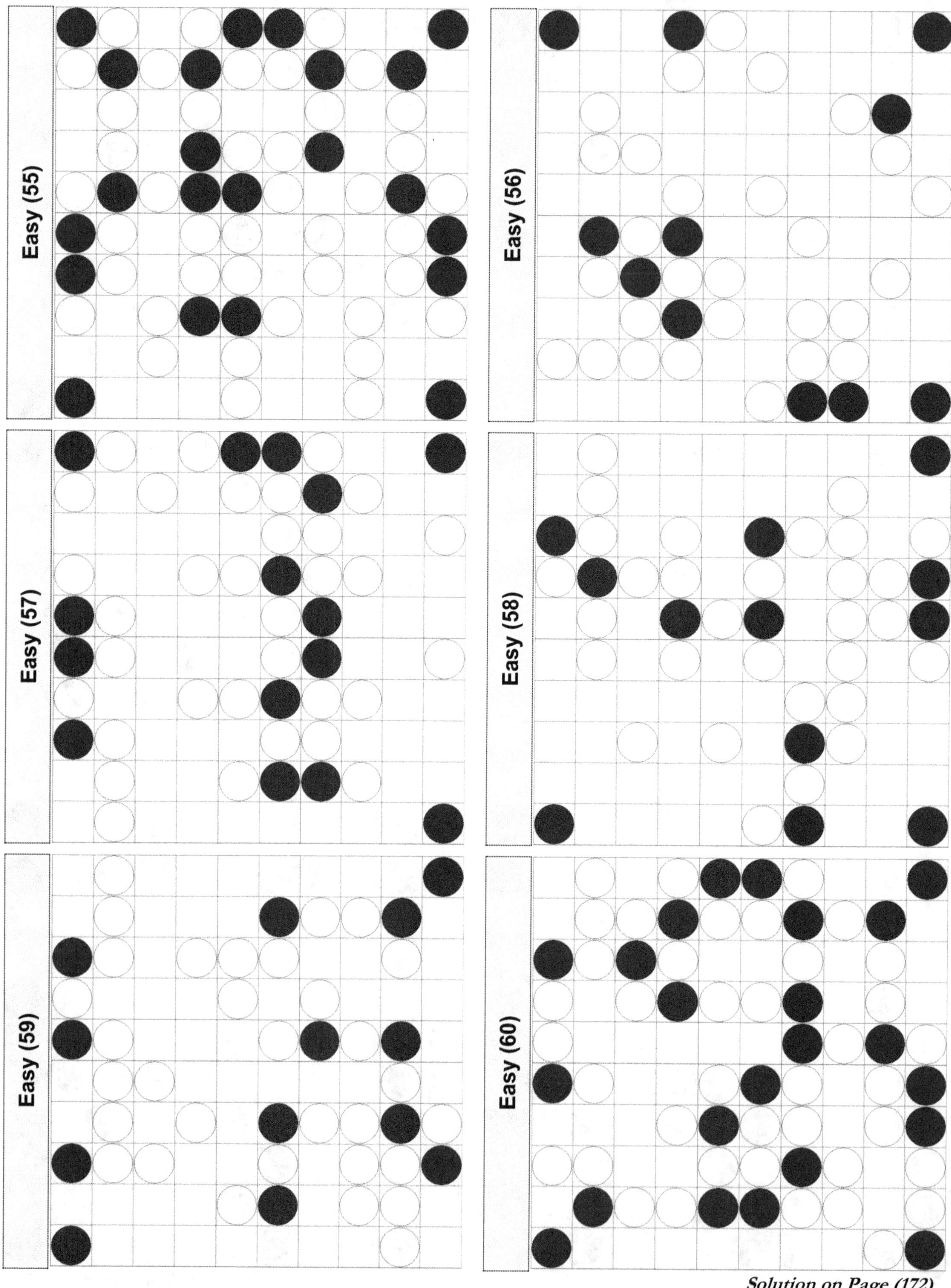

(12)

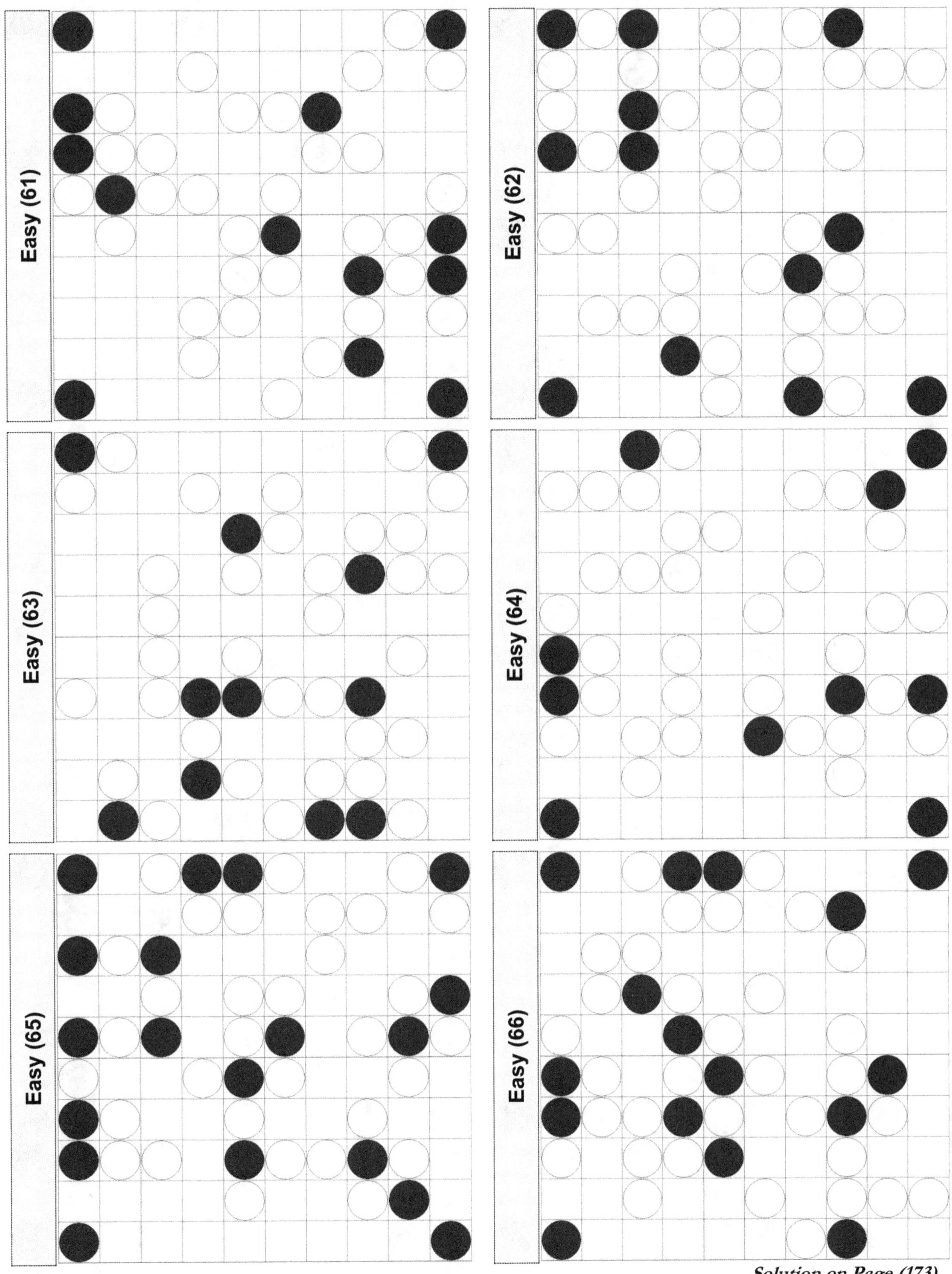

Solution on Page (173)

(13)

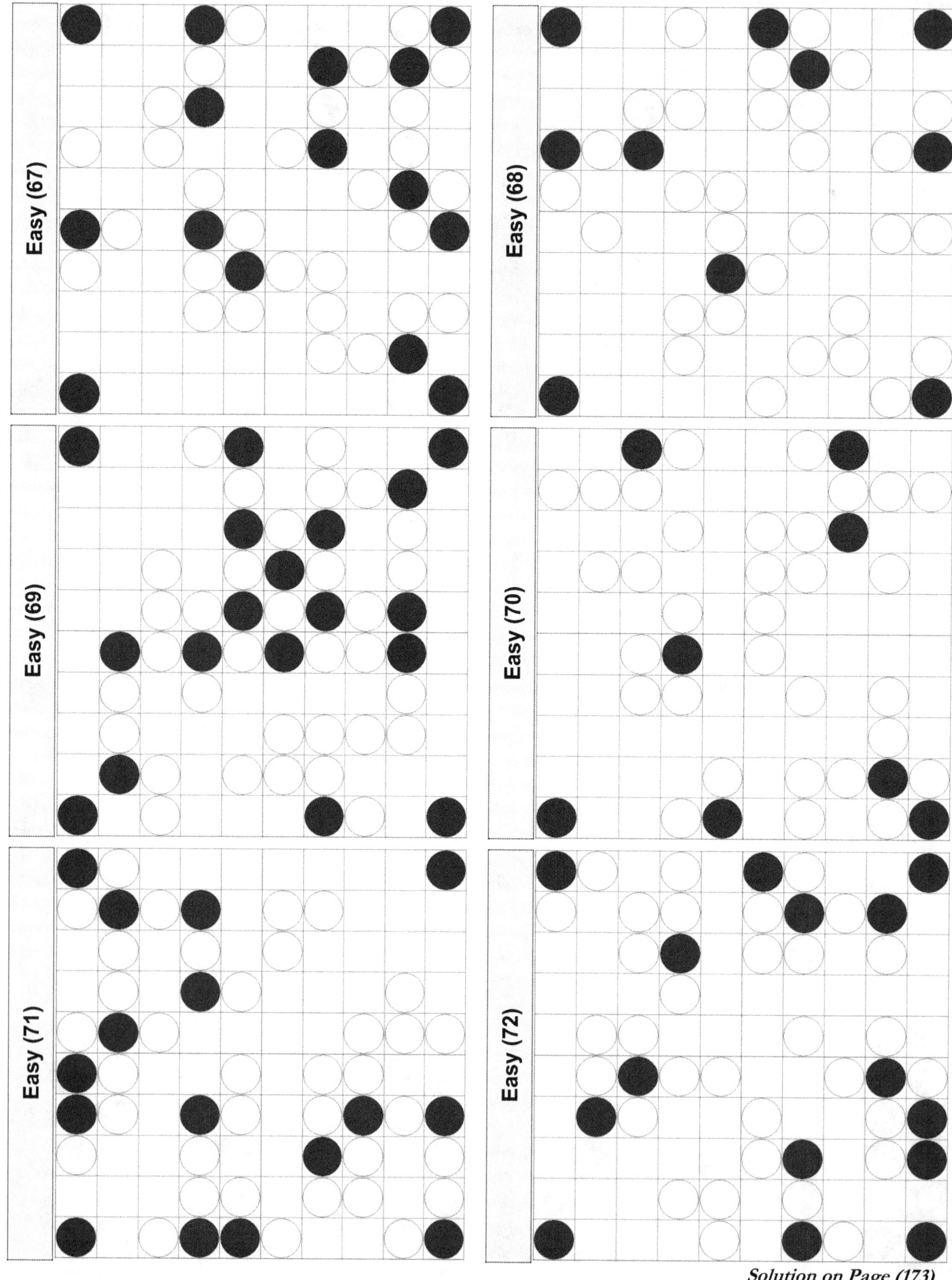

Solution on Page (173)

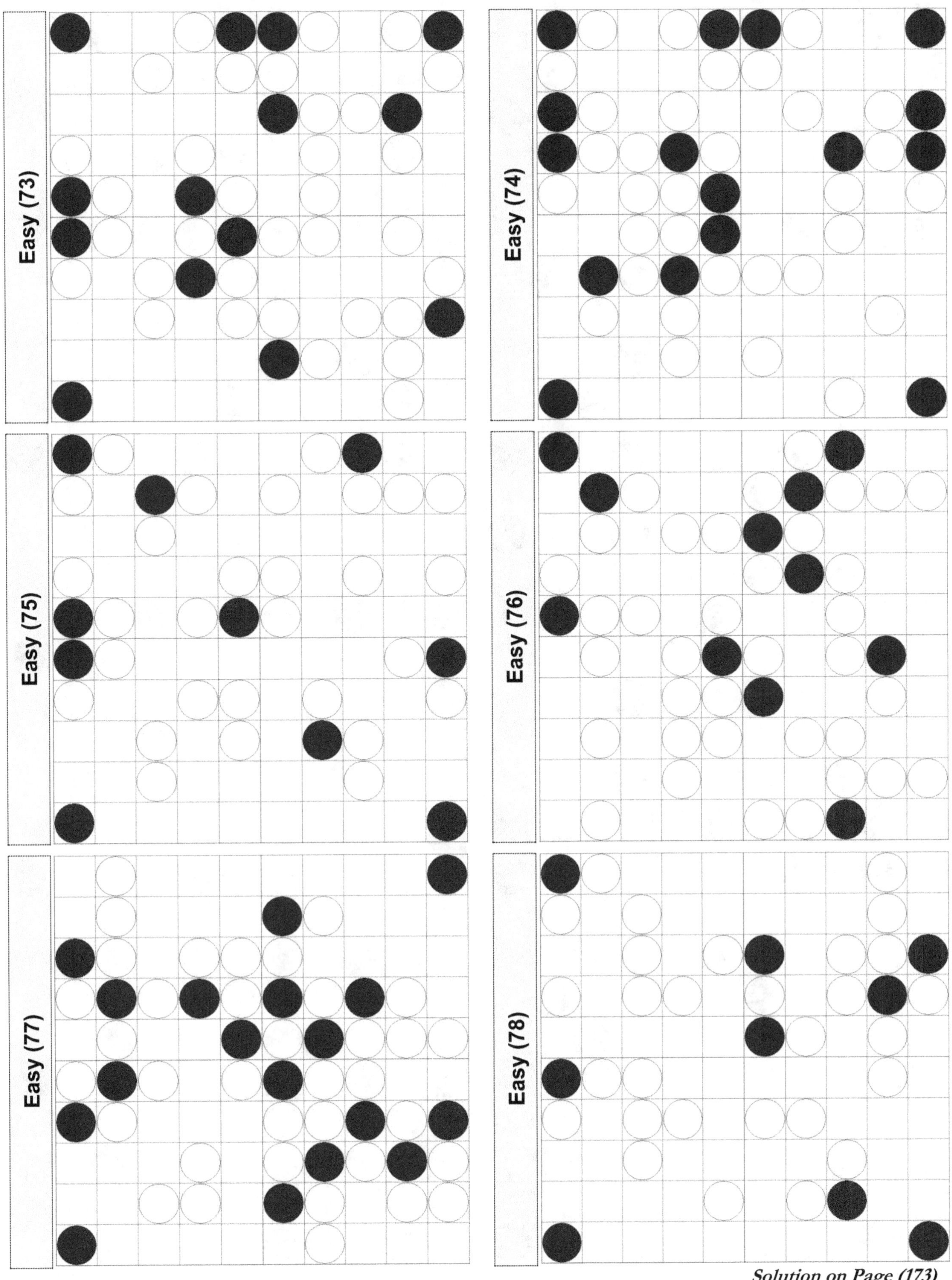

Solution on Page (173)

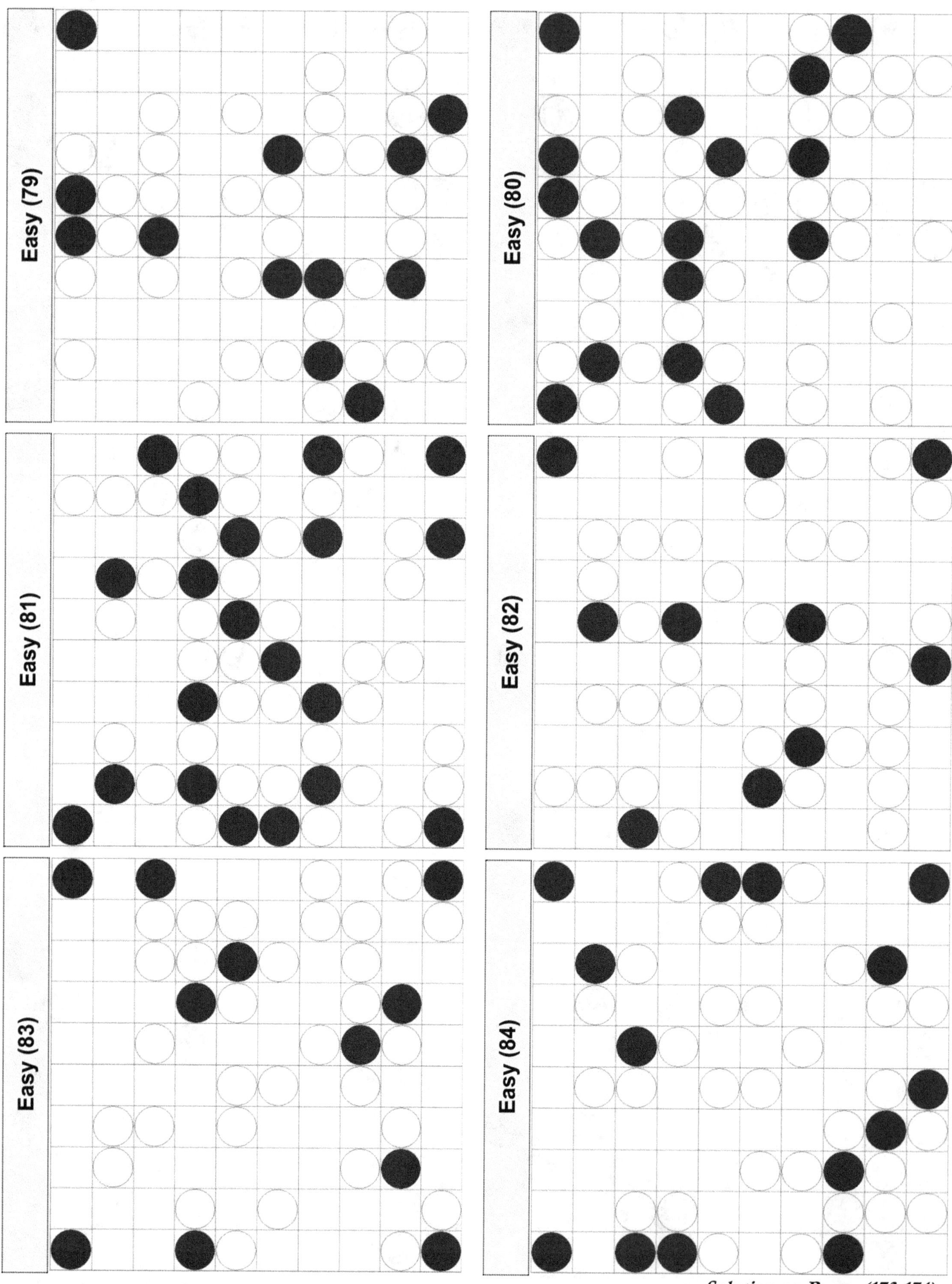

Solution on Pages (173-174)

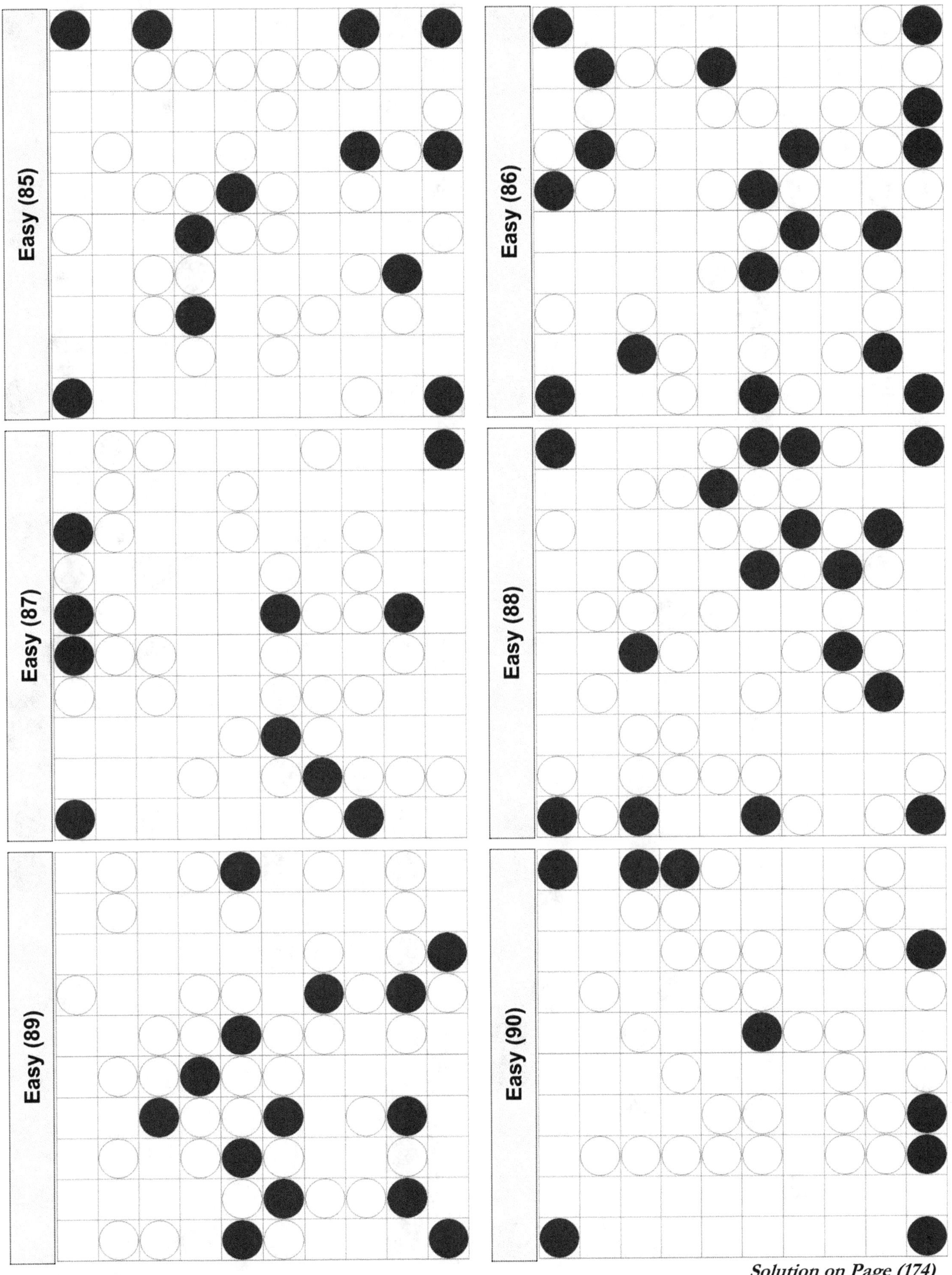

Solution on Page (174)

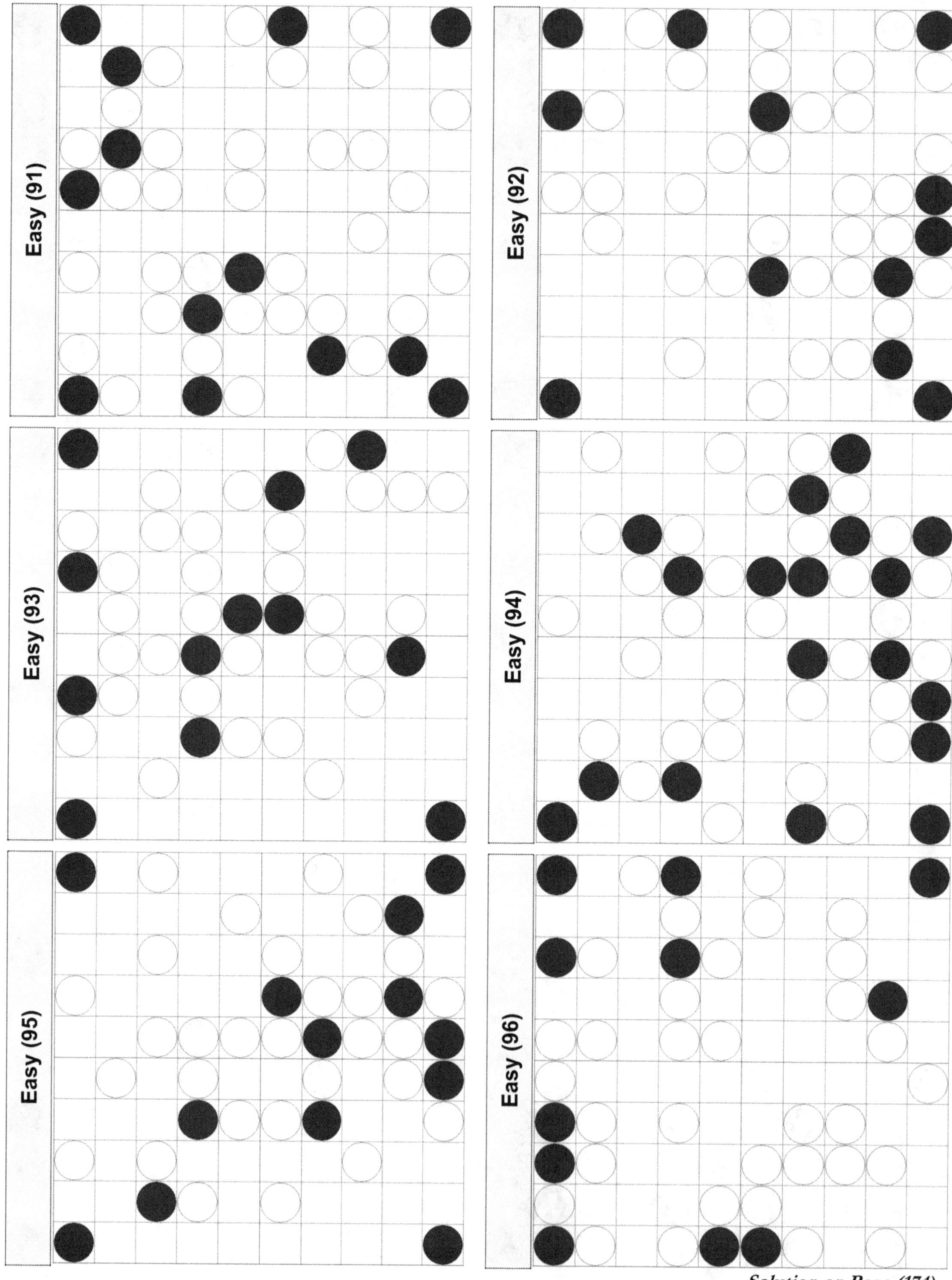

Easy (91)
Easy (92)
Easy (93)
Easy (94)
Easy (95)
Easy (96)
Solution on Page (174)

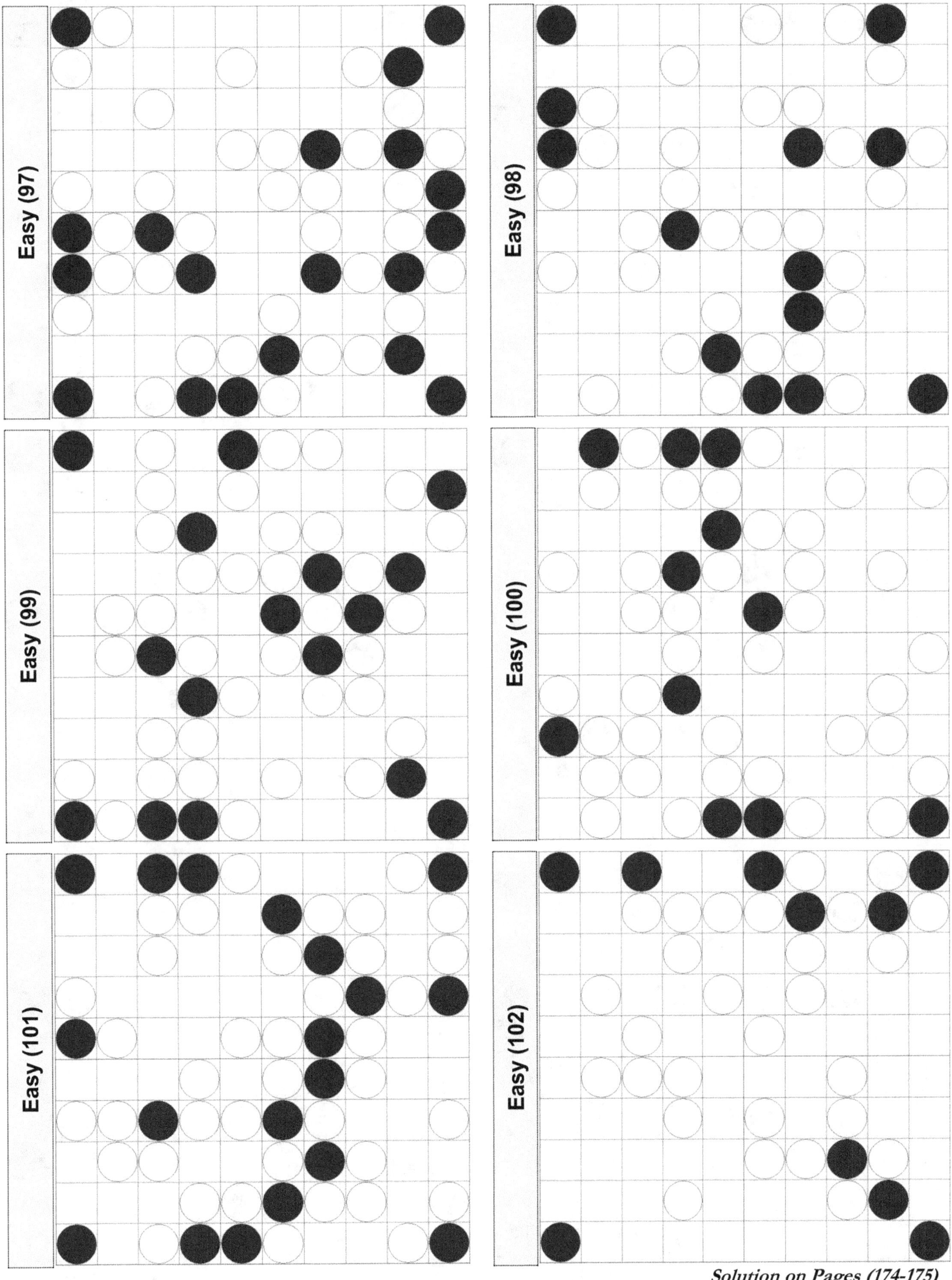

Easy (97)

Easy (98)

Easy (99)

Easy (100)

Easy (101)

Easy (102)

Solution on Pages (174-175)

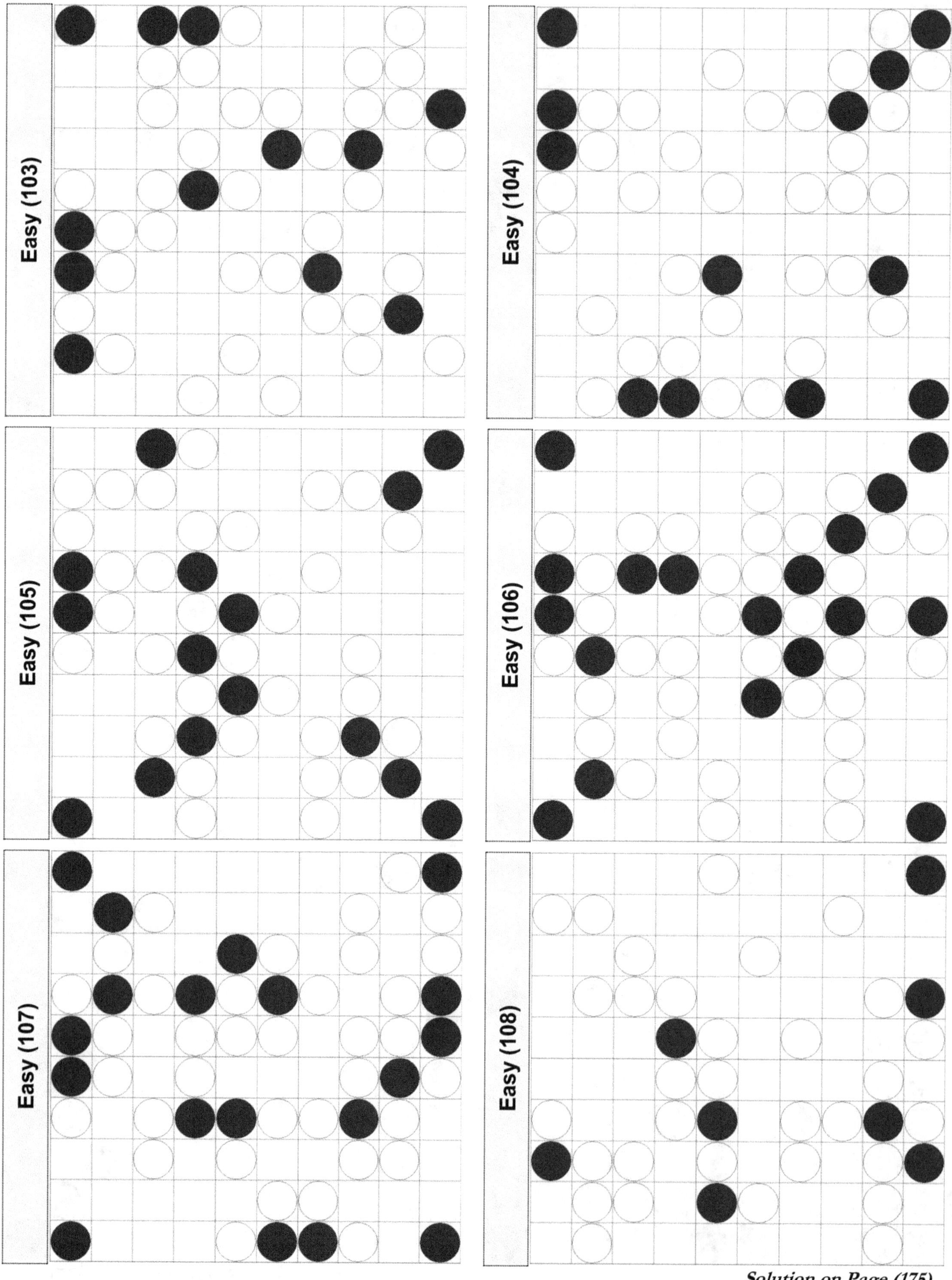

(20)

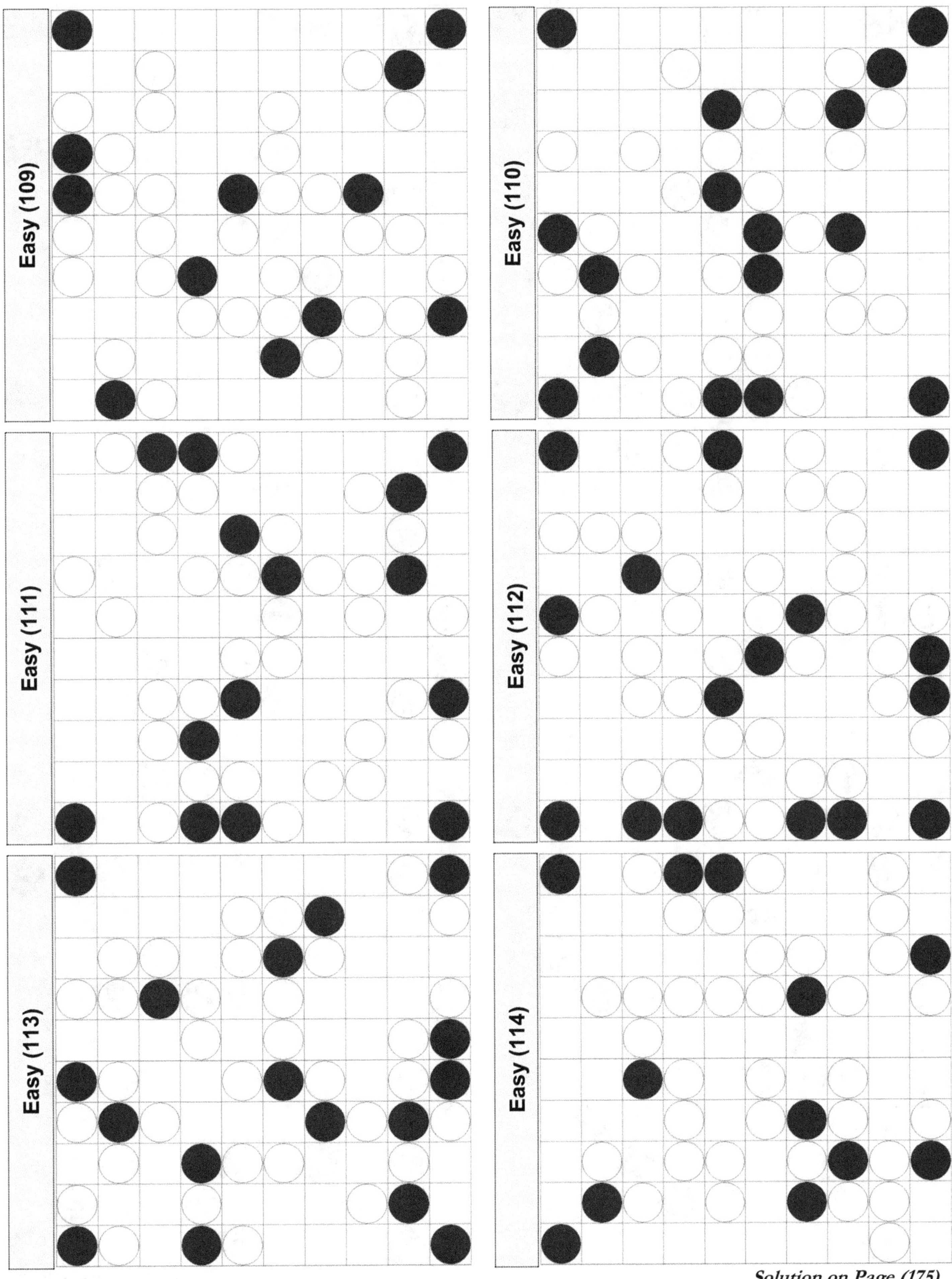

Solution on Page (175)

(21)

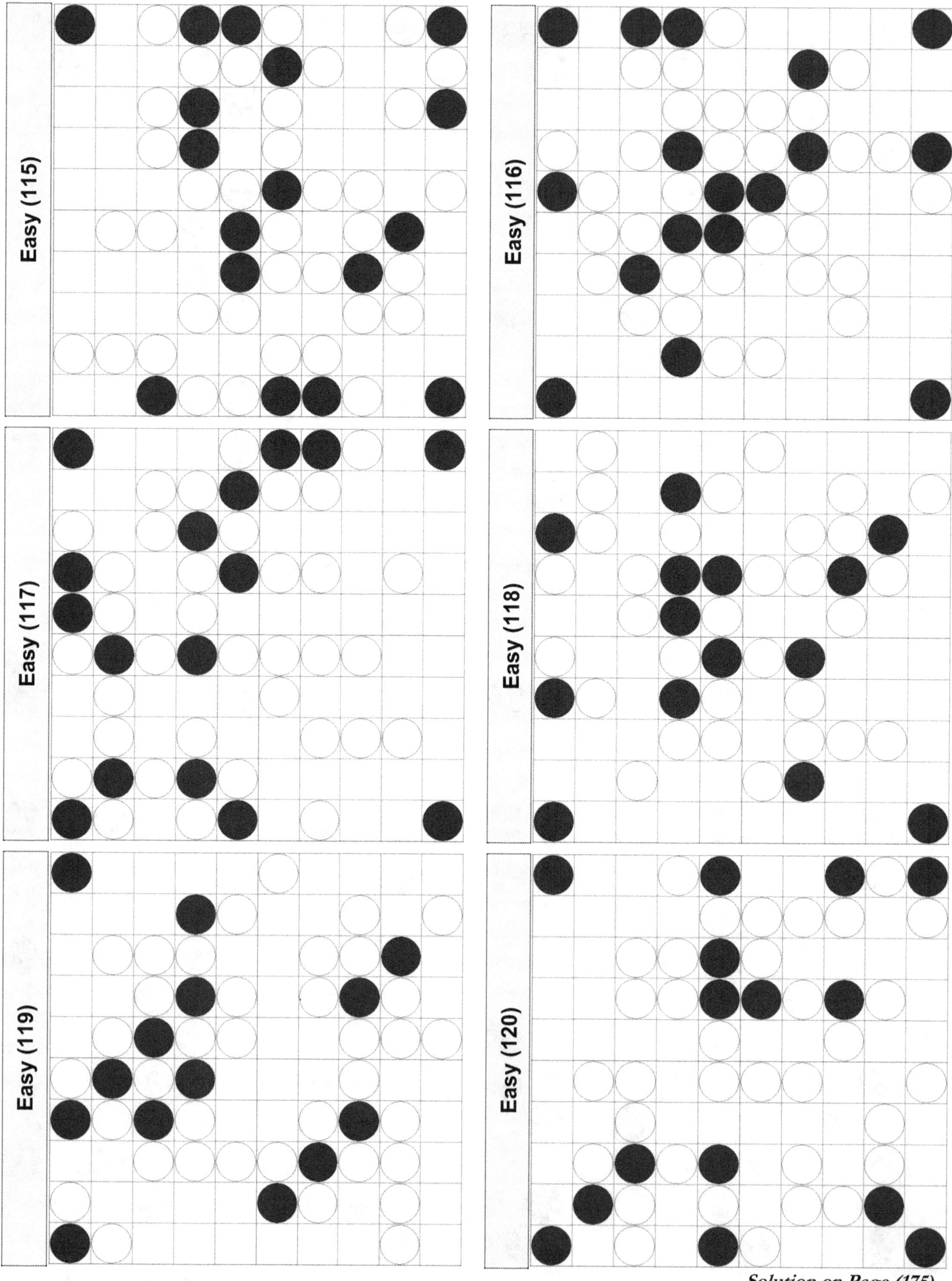

Easy (115)

Easy (116)

Easy (117)

Easy (118)

Easy (119)

Easy (120)

Solution on Page (175)

(22)

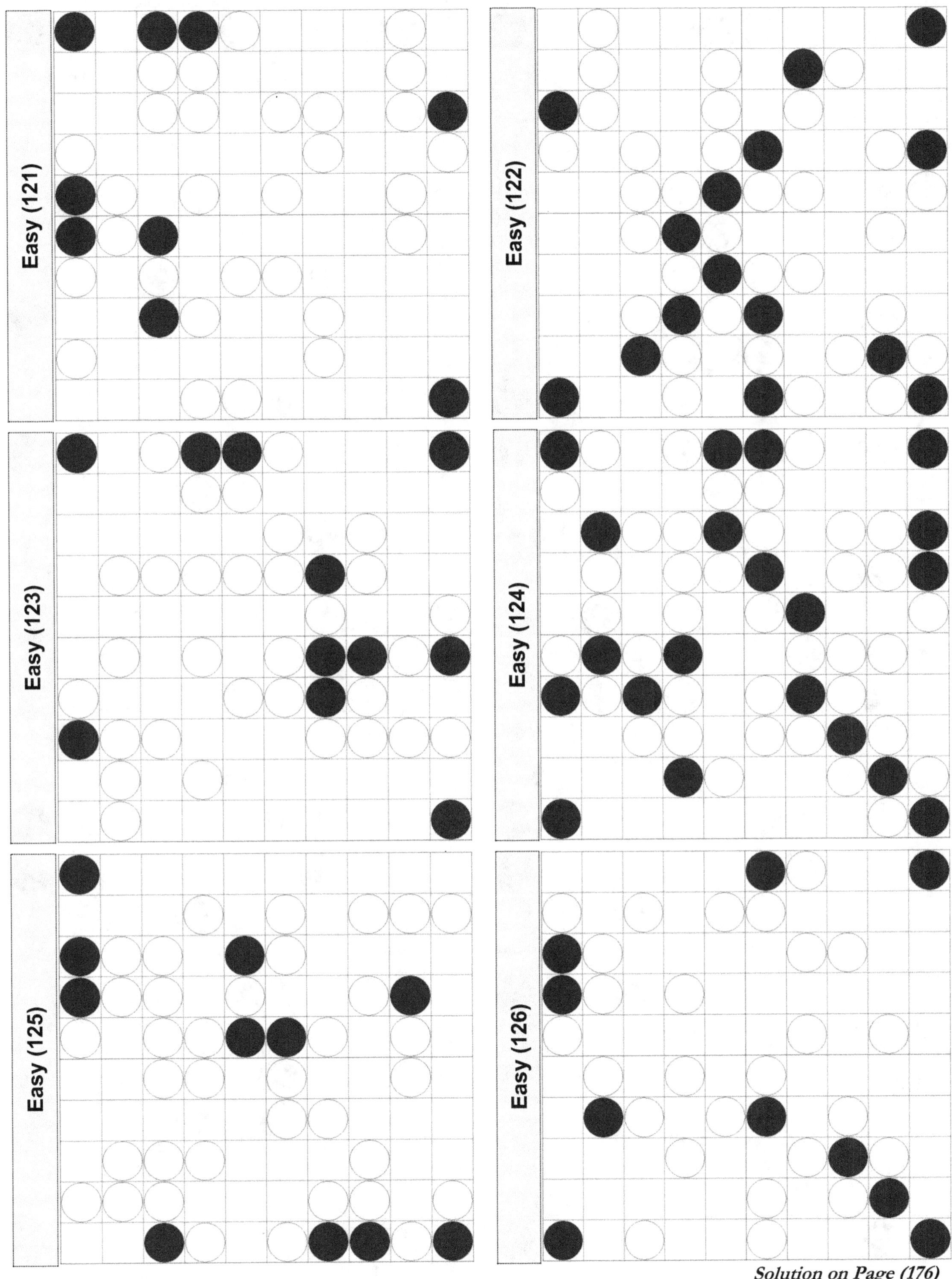

Easy (121)
Easy (122)
Easy (123)
Easy (124)
Easy (125)
Easy (126)
Solution on Page (176)

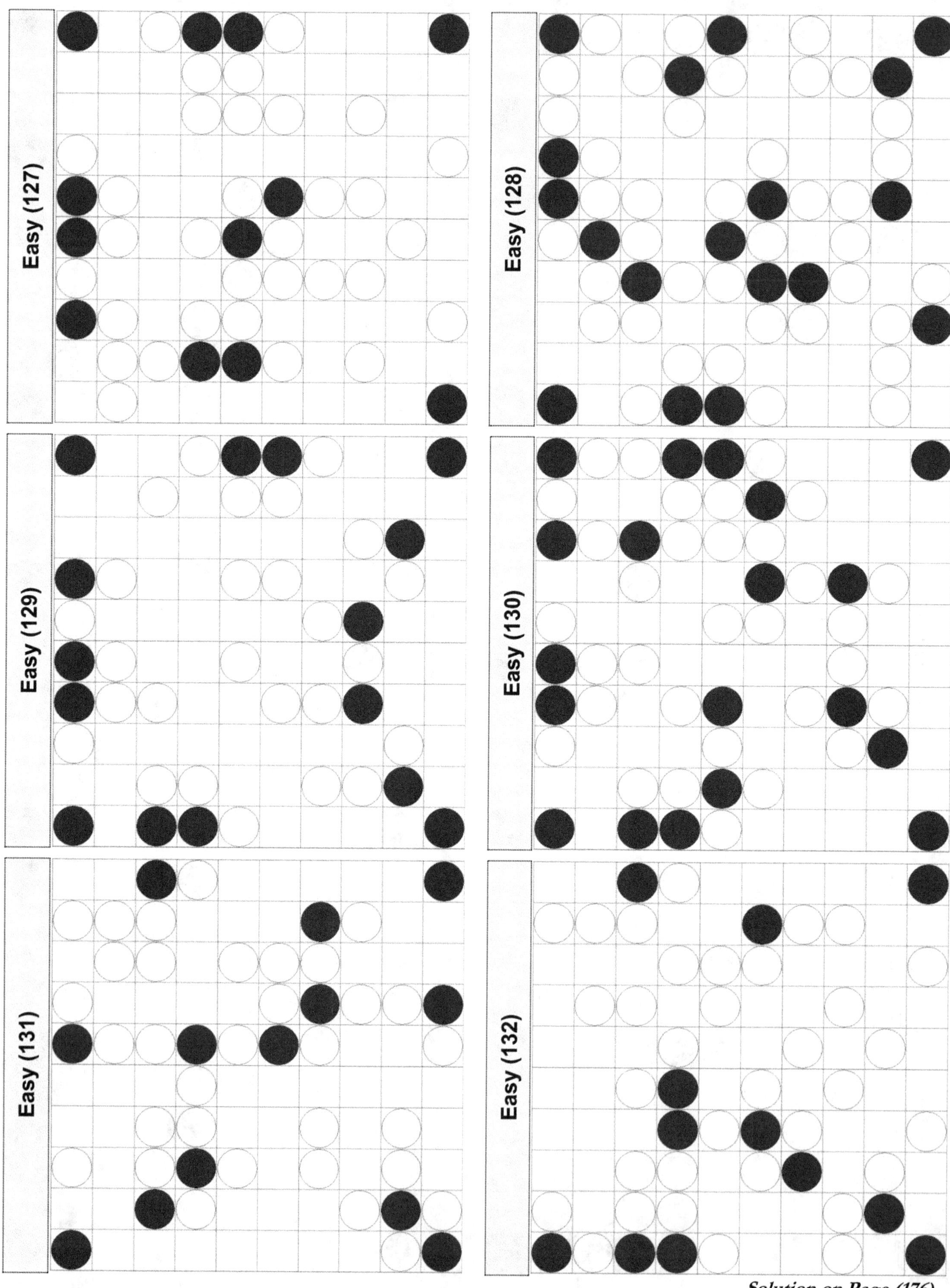

Solution on Page (176)

(24)

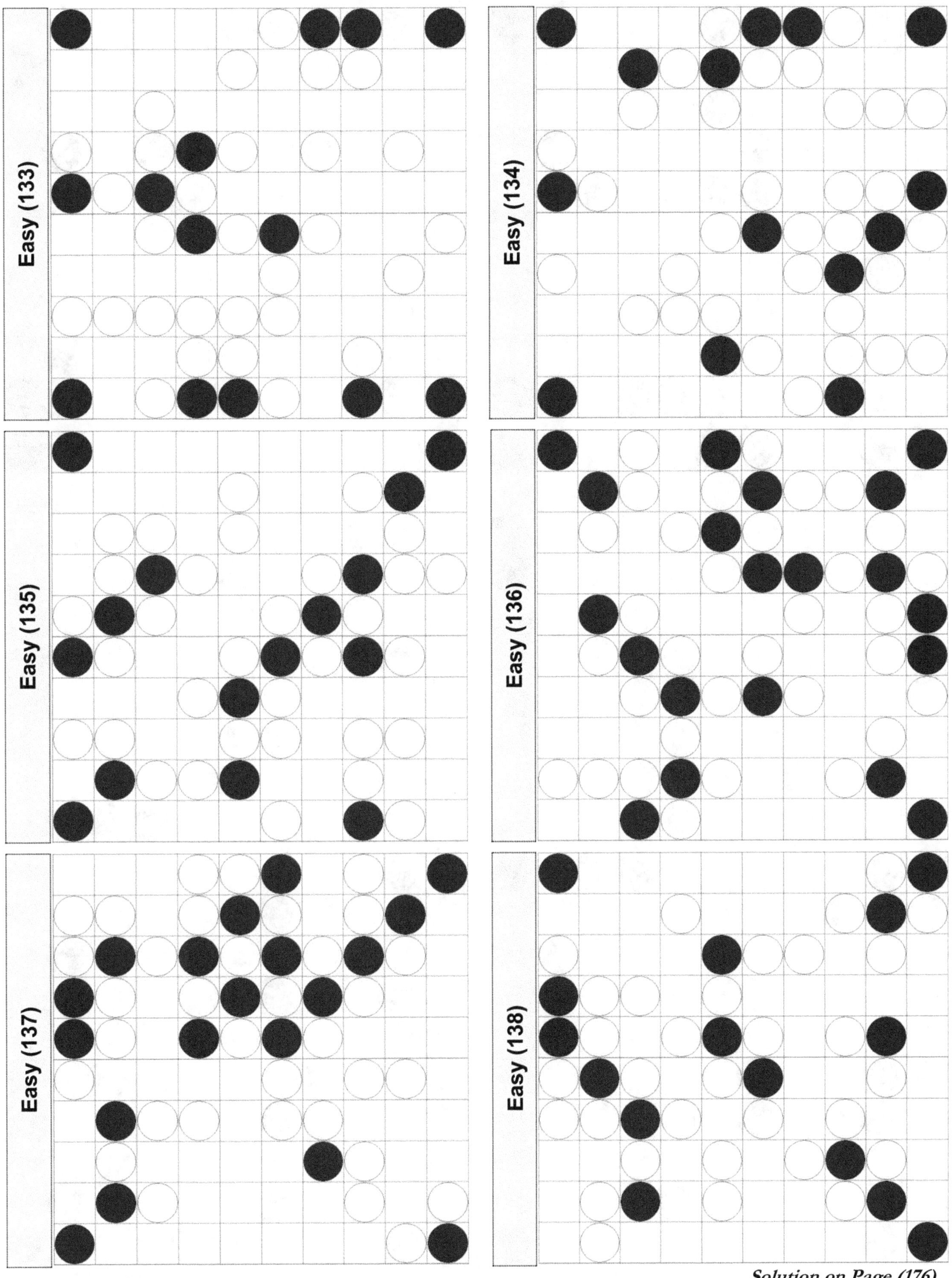

Solution on Page (176)

(25)

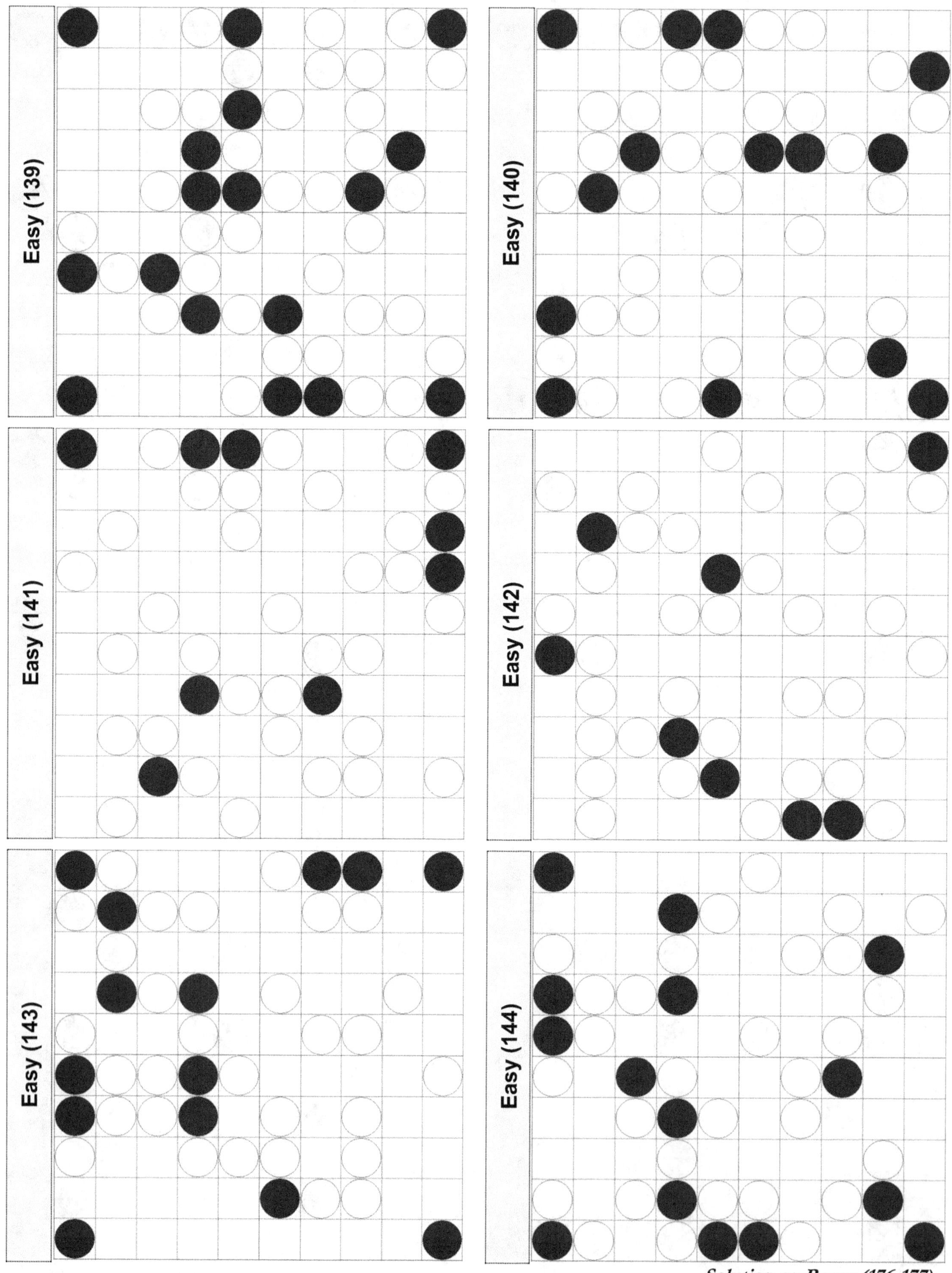

Solution on Pages (176-177)

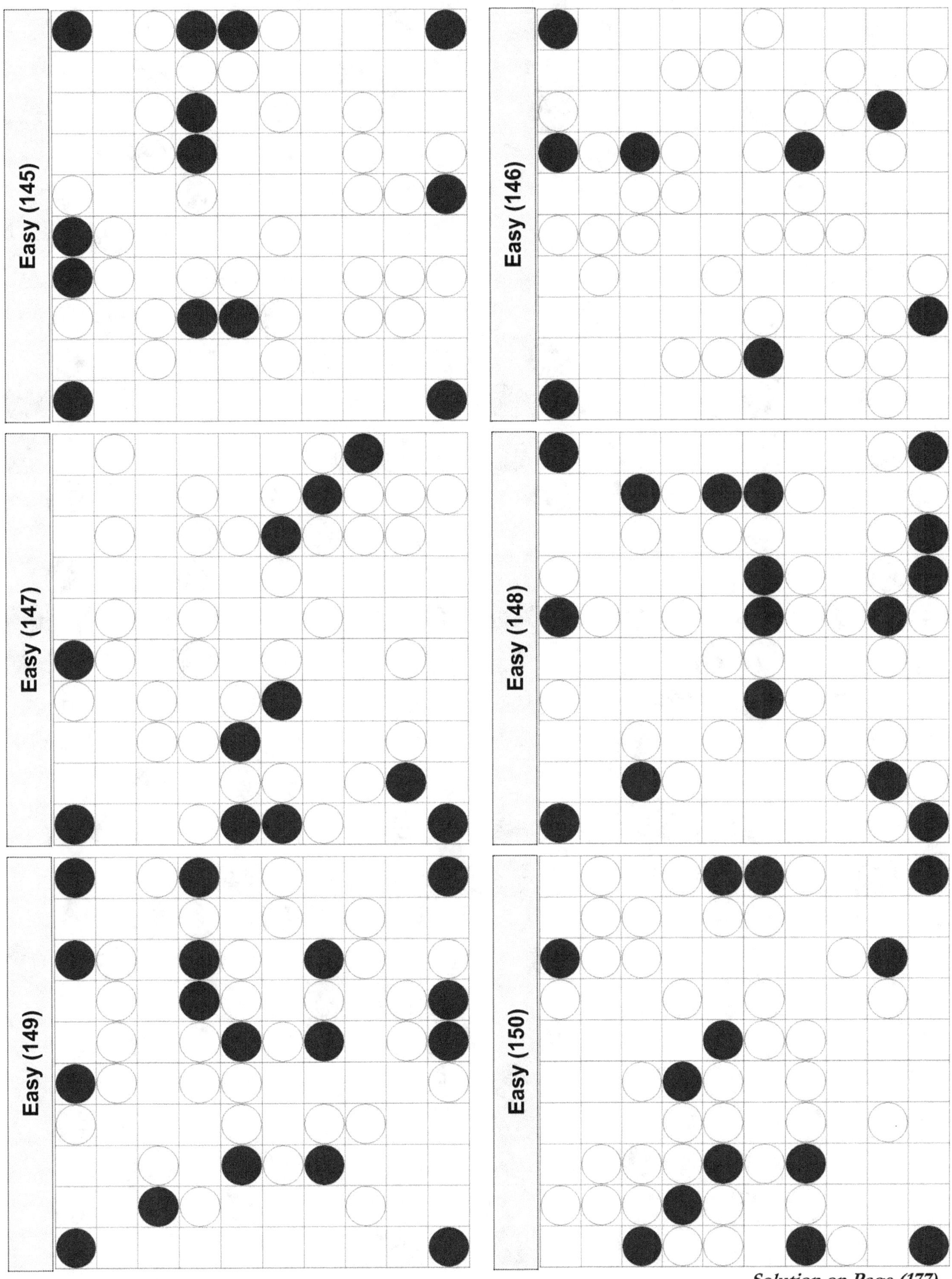

Solution on Page (177)

(27)

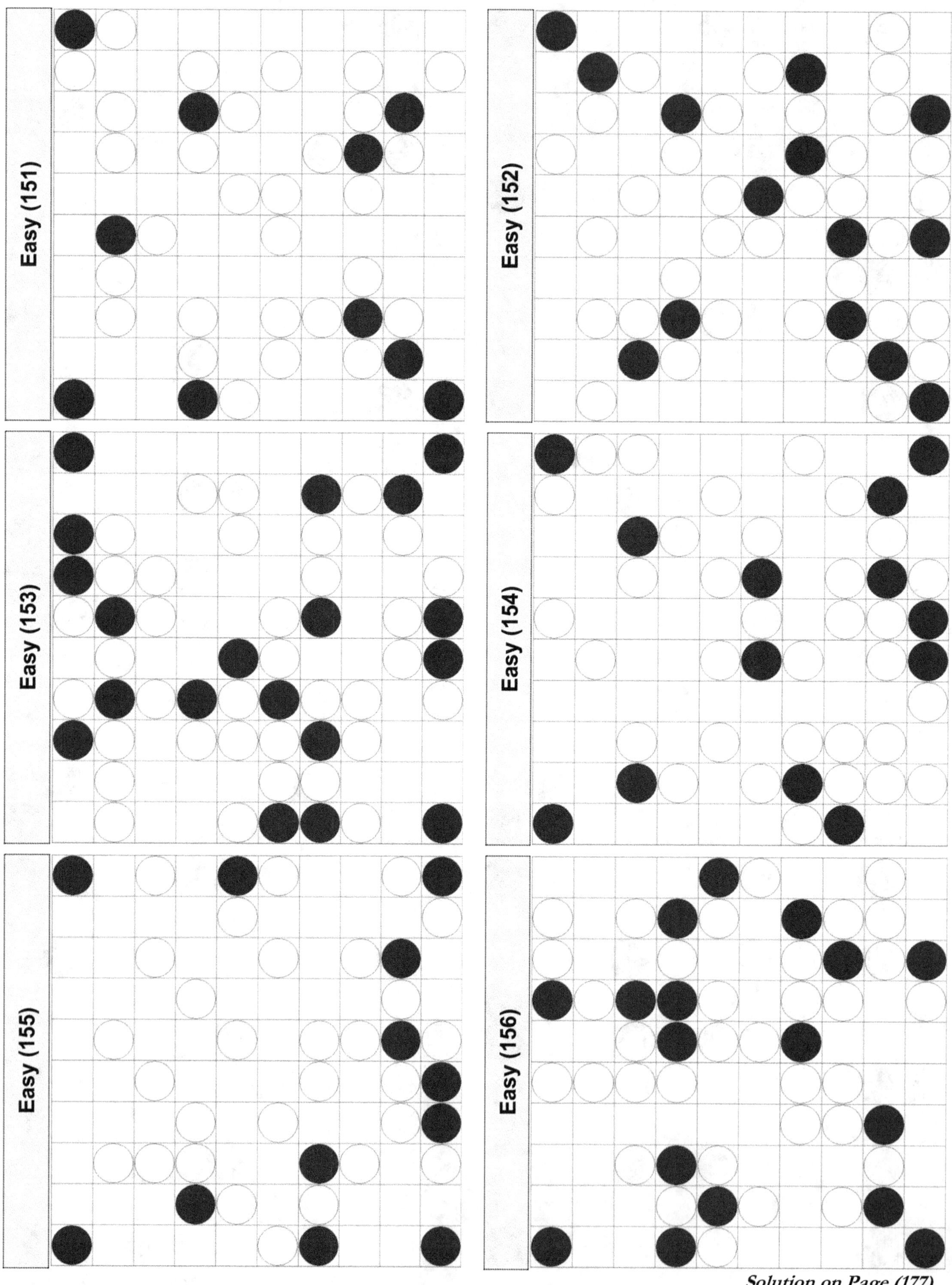

Easy (151)
Easy (152)
Easy (153)
Easy (154)
Easy (155)
Easy (156)
Solution on Page (177)

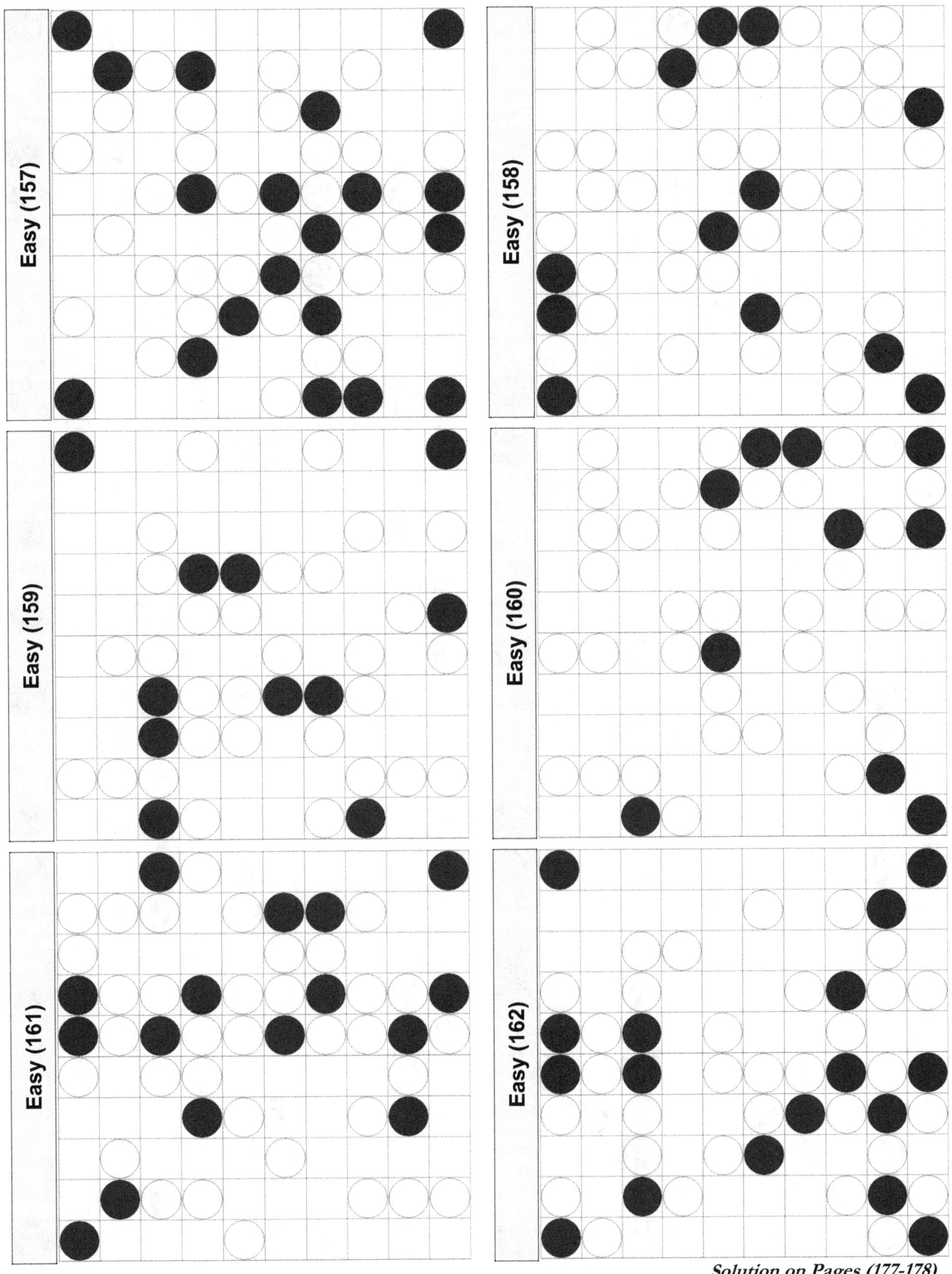

Easy (157)

Easy (158)

Easy (159)

Easy (160)

Easy (161)

Easy (162)

Solution on Pages (177-178)

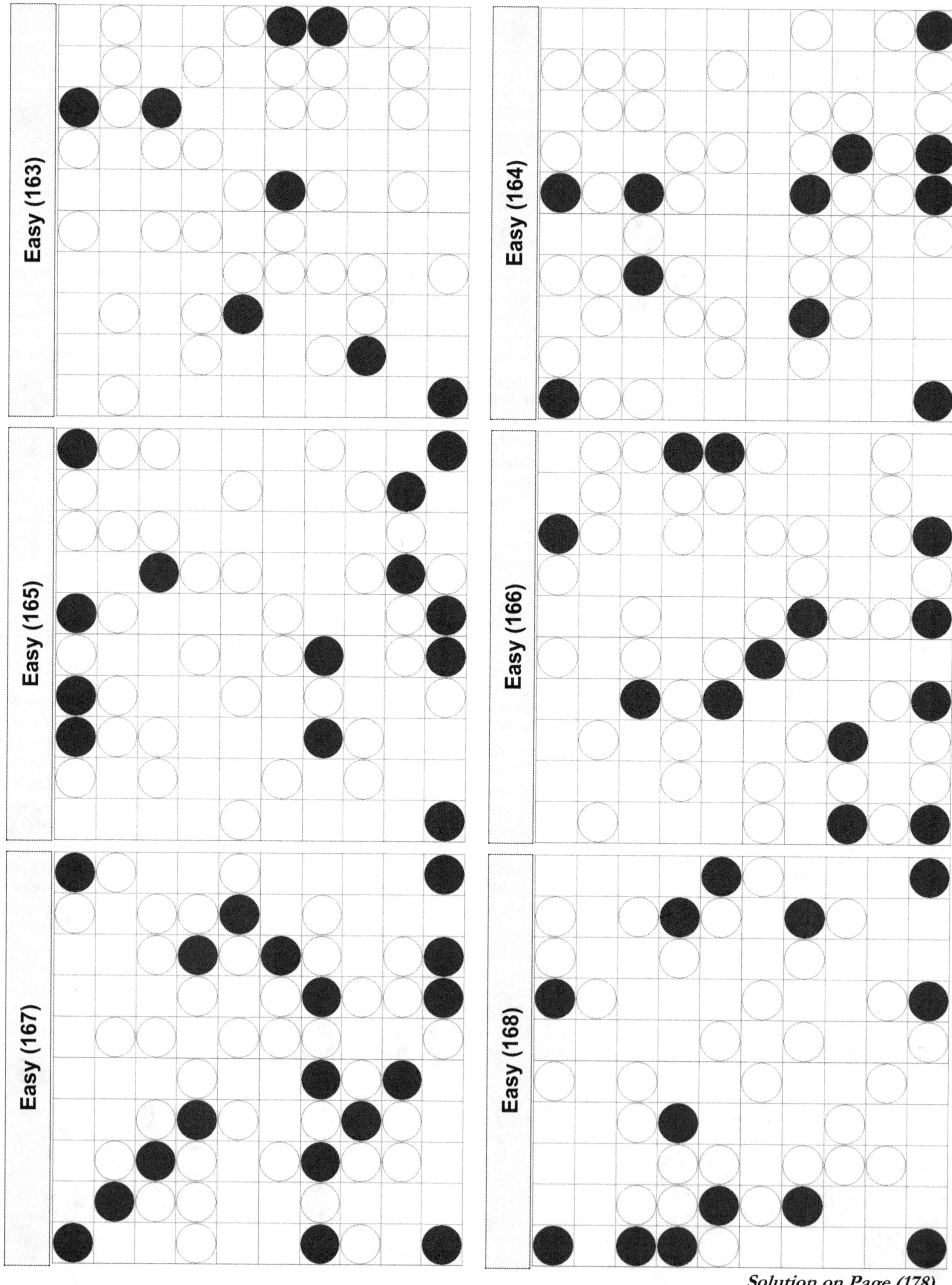

Solution on Page (178)

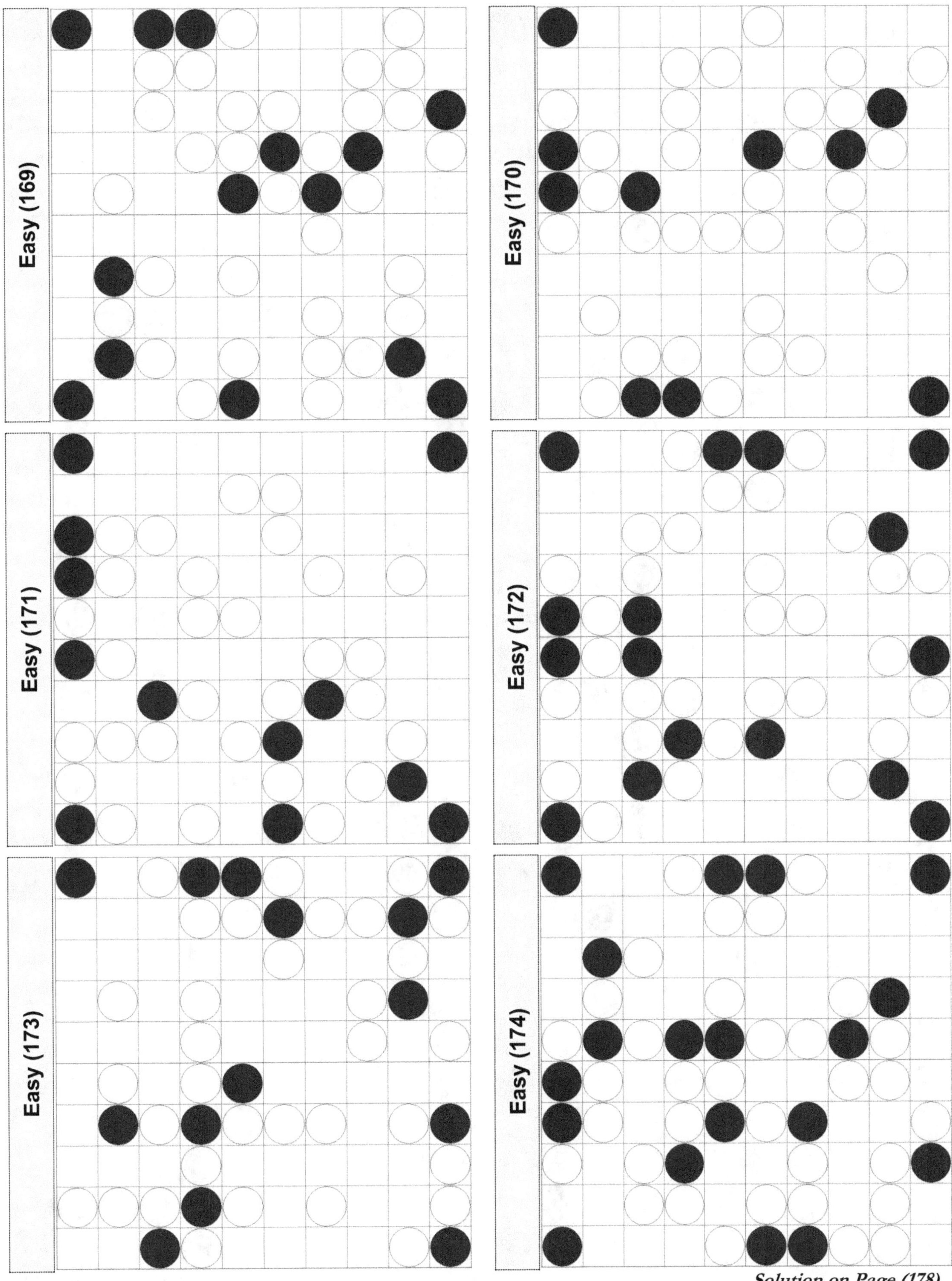

Solution on Page (178)

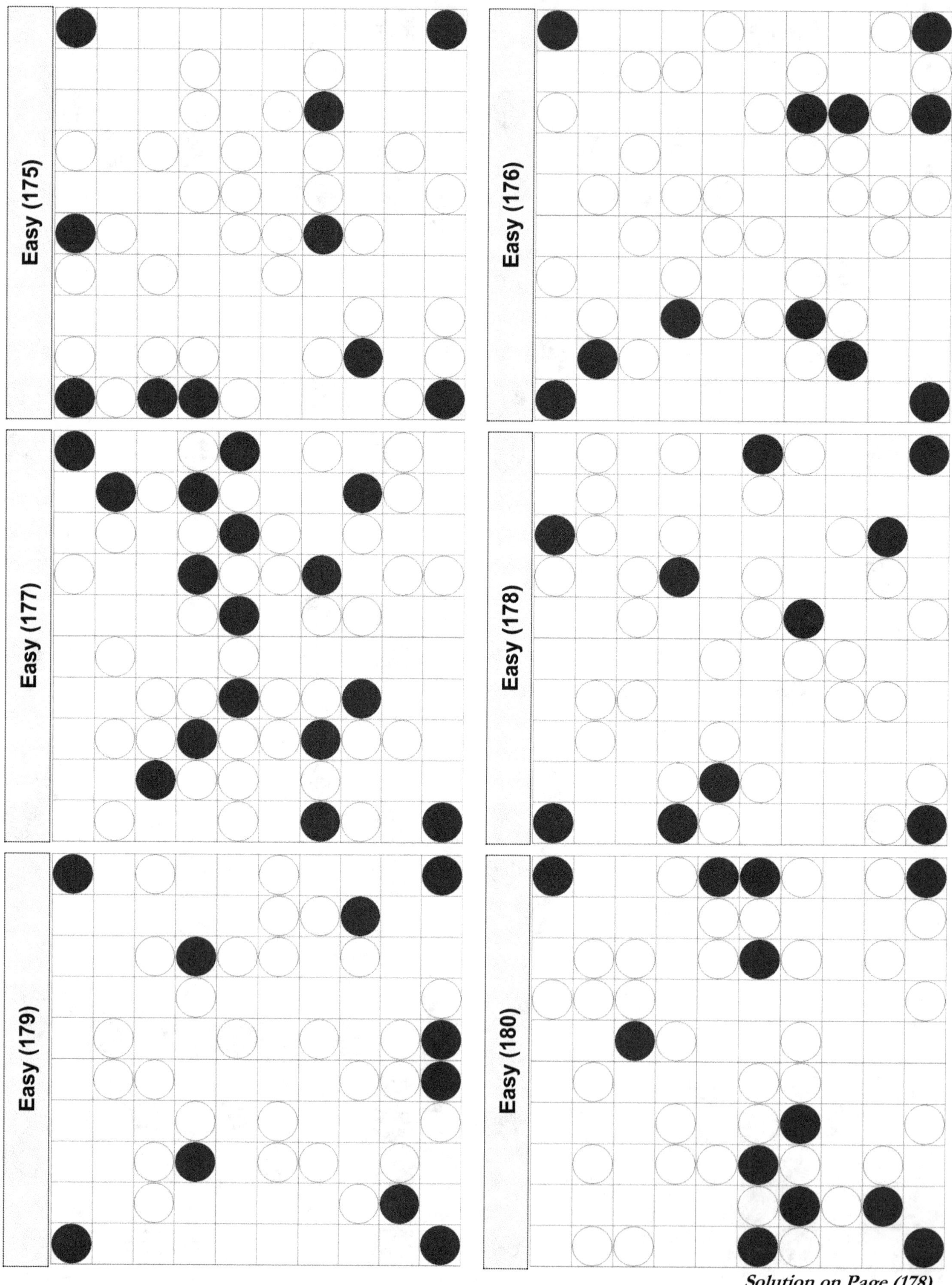

Solution on Page (178)

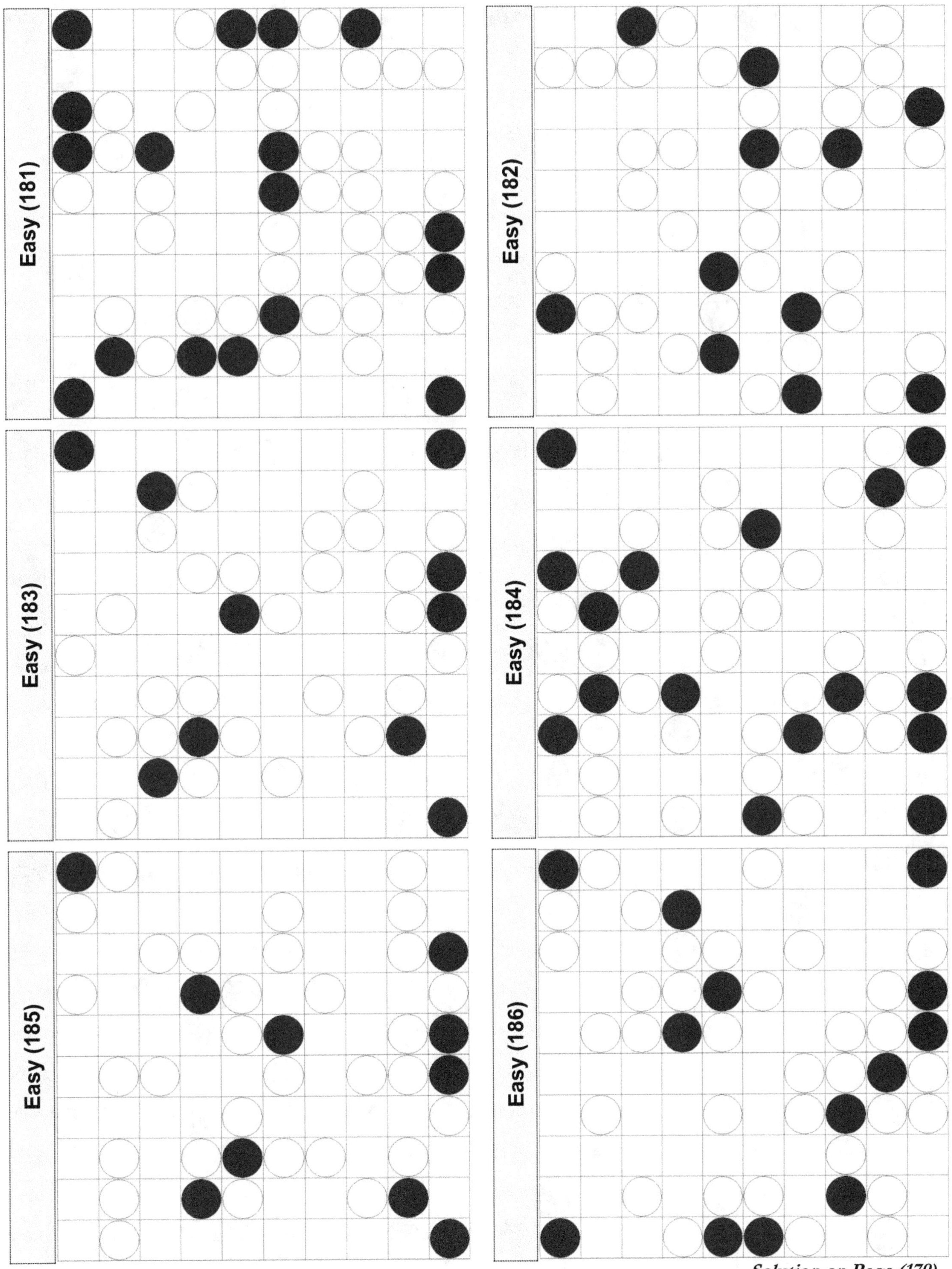

Solution on Page (179)

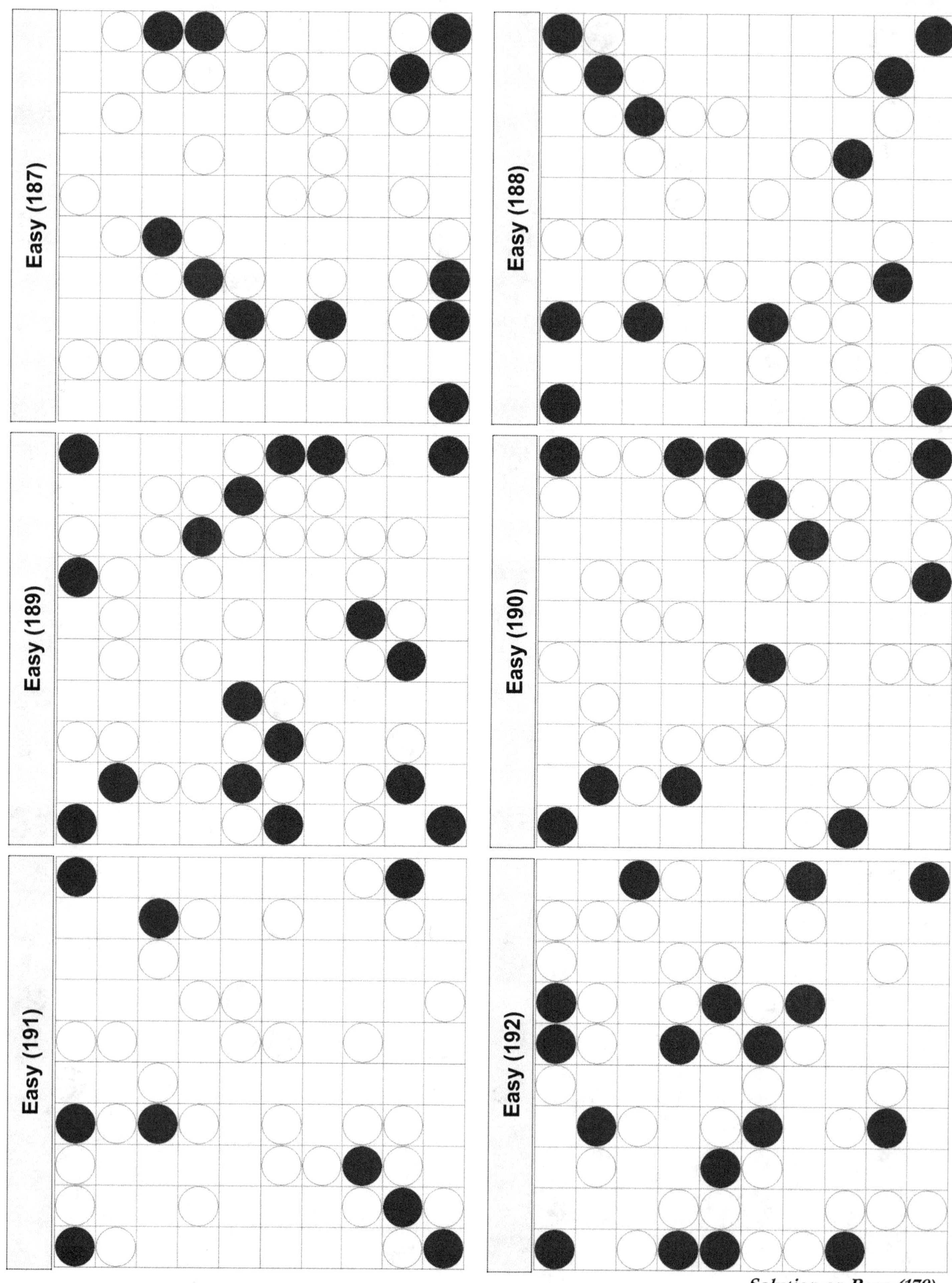

Easy (187)

Easy (188)

Easy (189)

Easy (190)

Easy (191)

Easy (192)

Solution on Page (179)

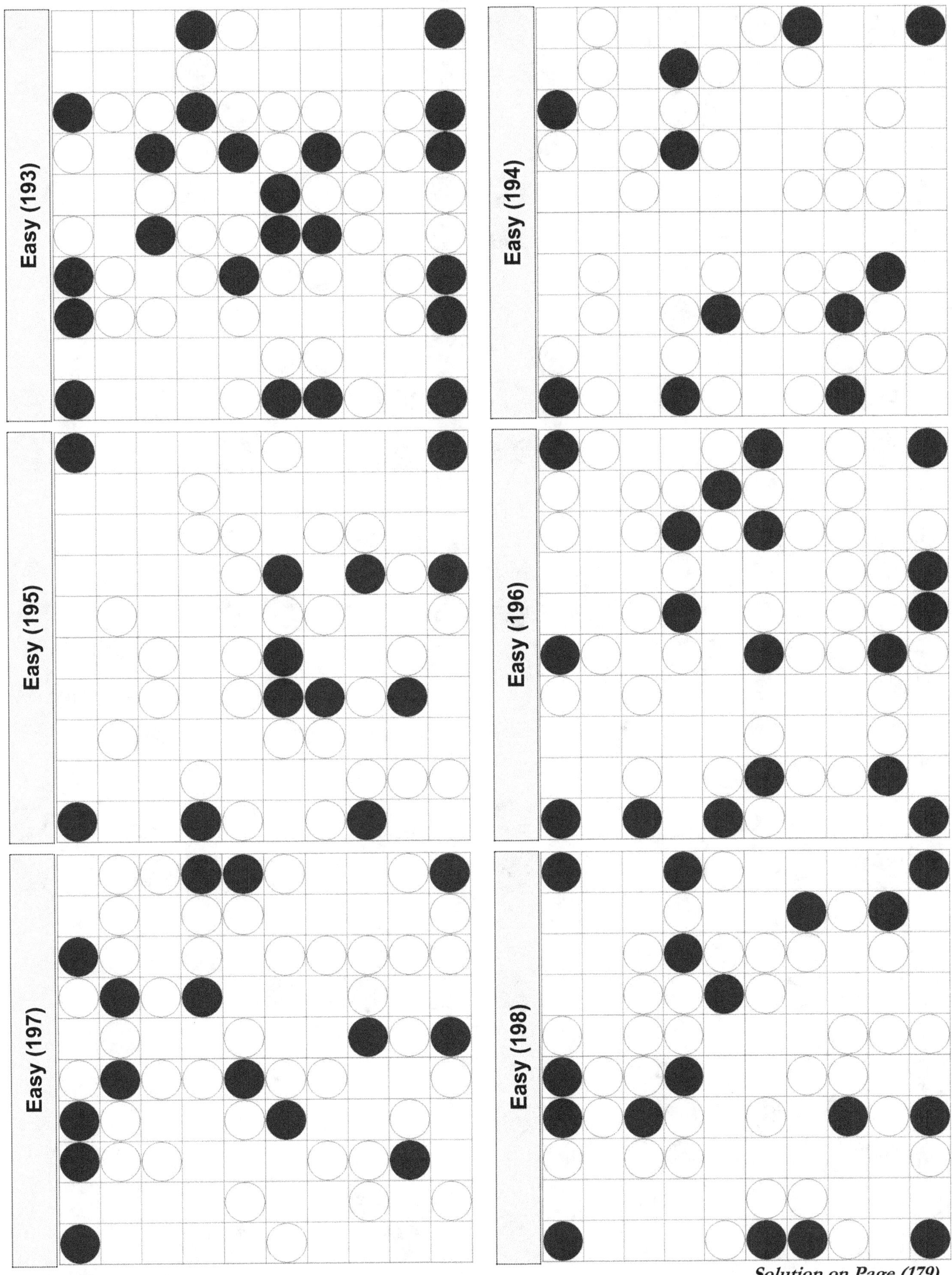

Solution on Page (179)

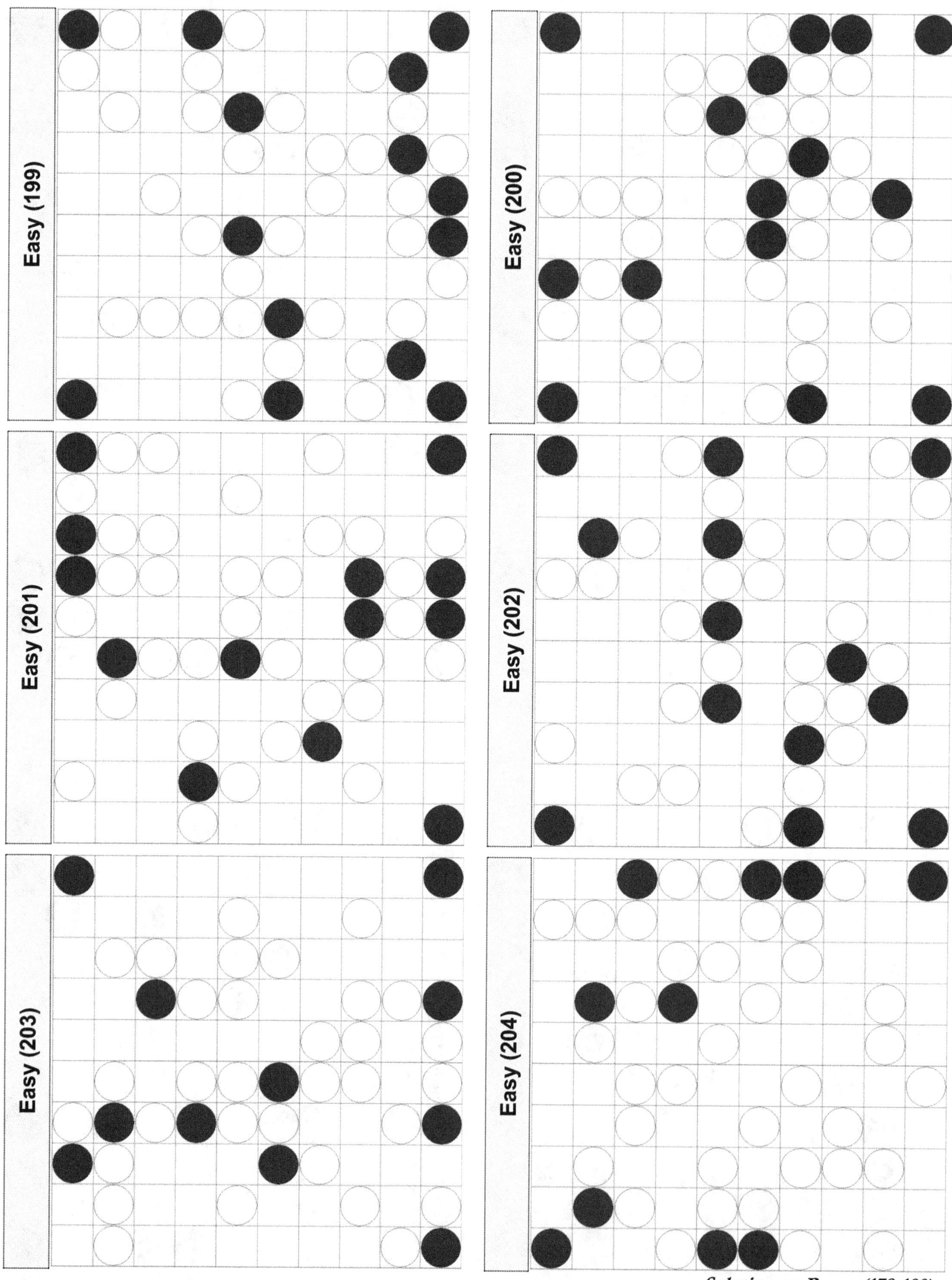

Solution on Pages (179-180)

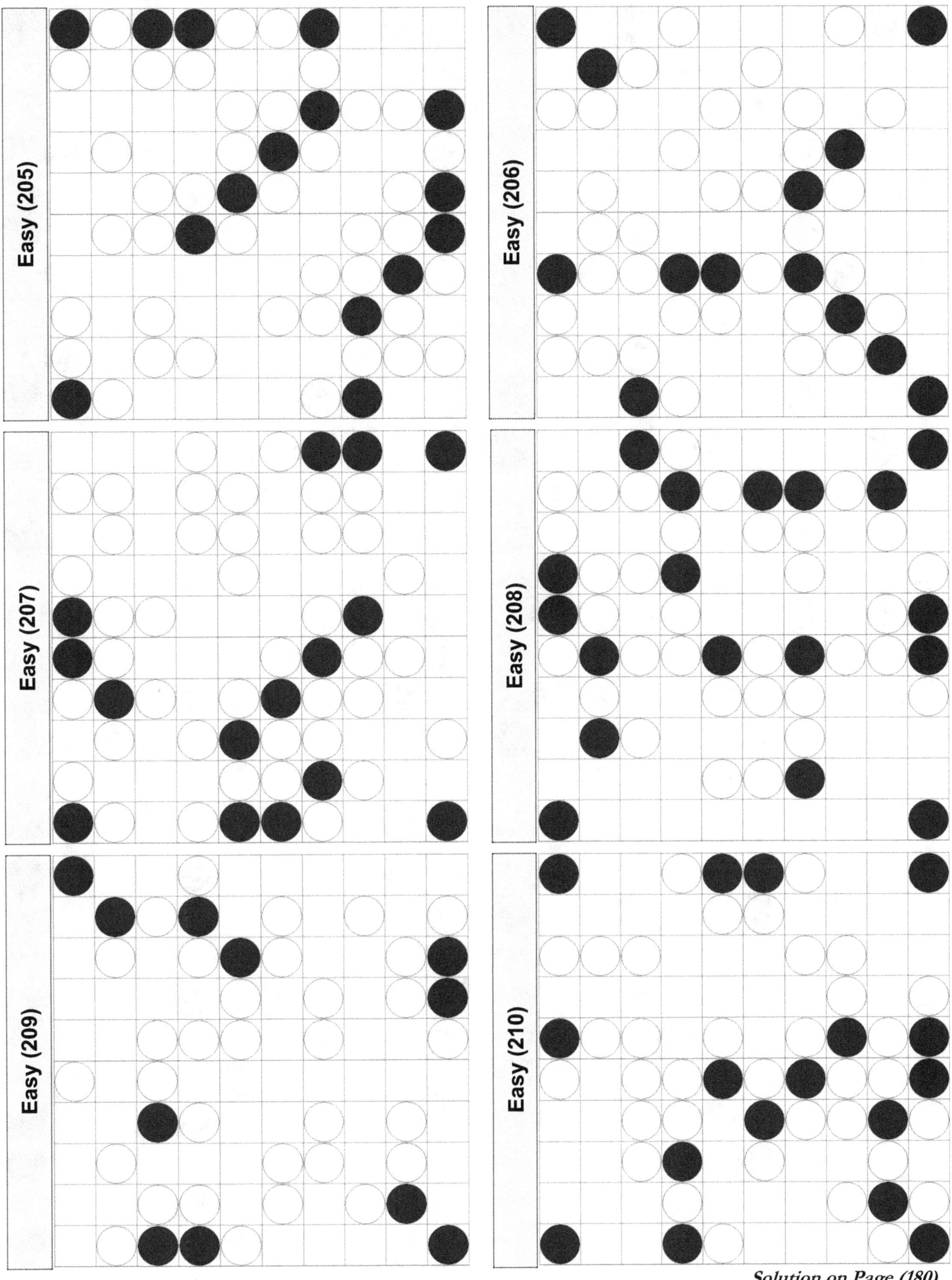

Solution on Page (180)

(37)

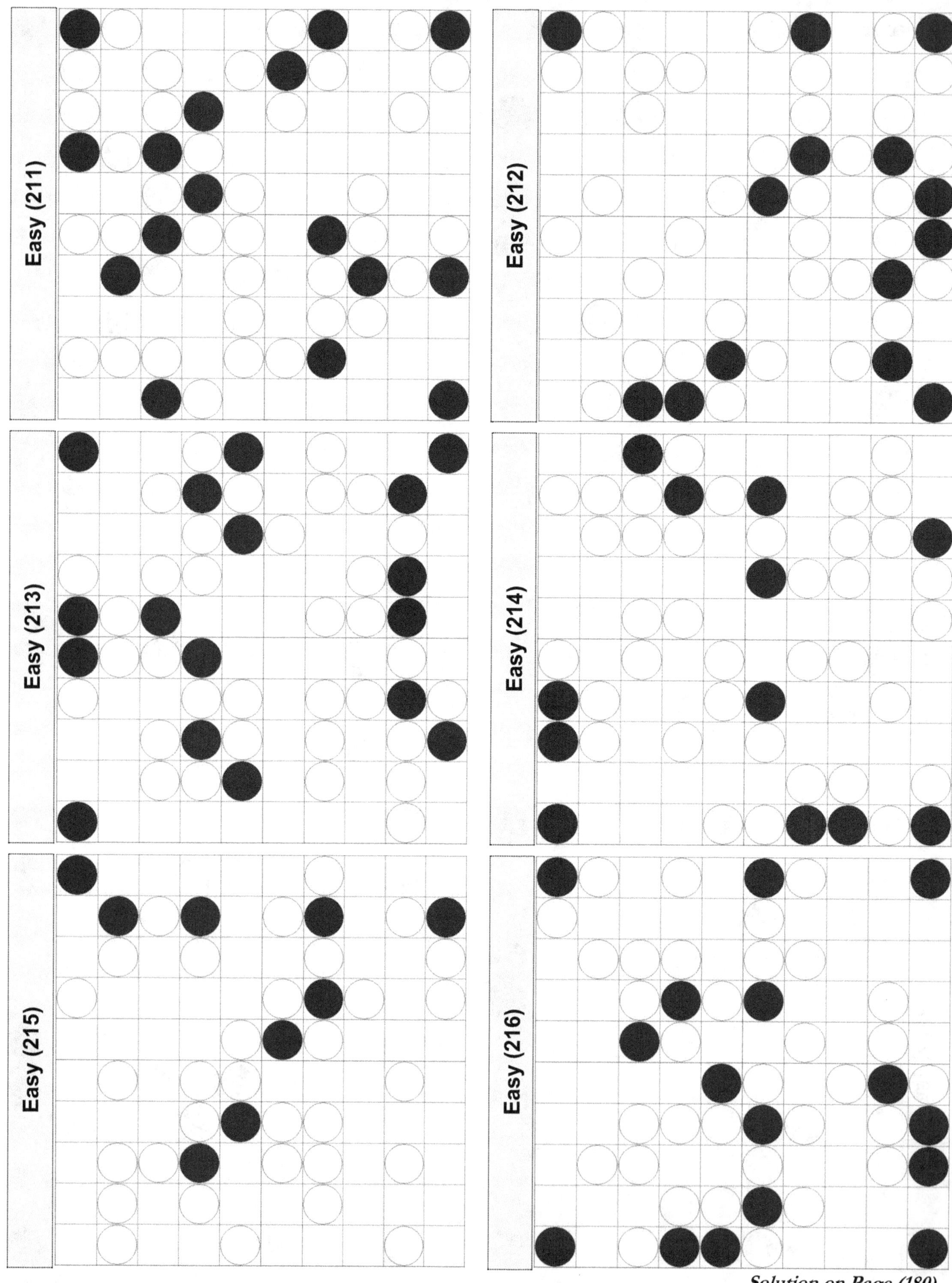

Solution on Page (180)

(38)

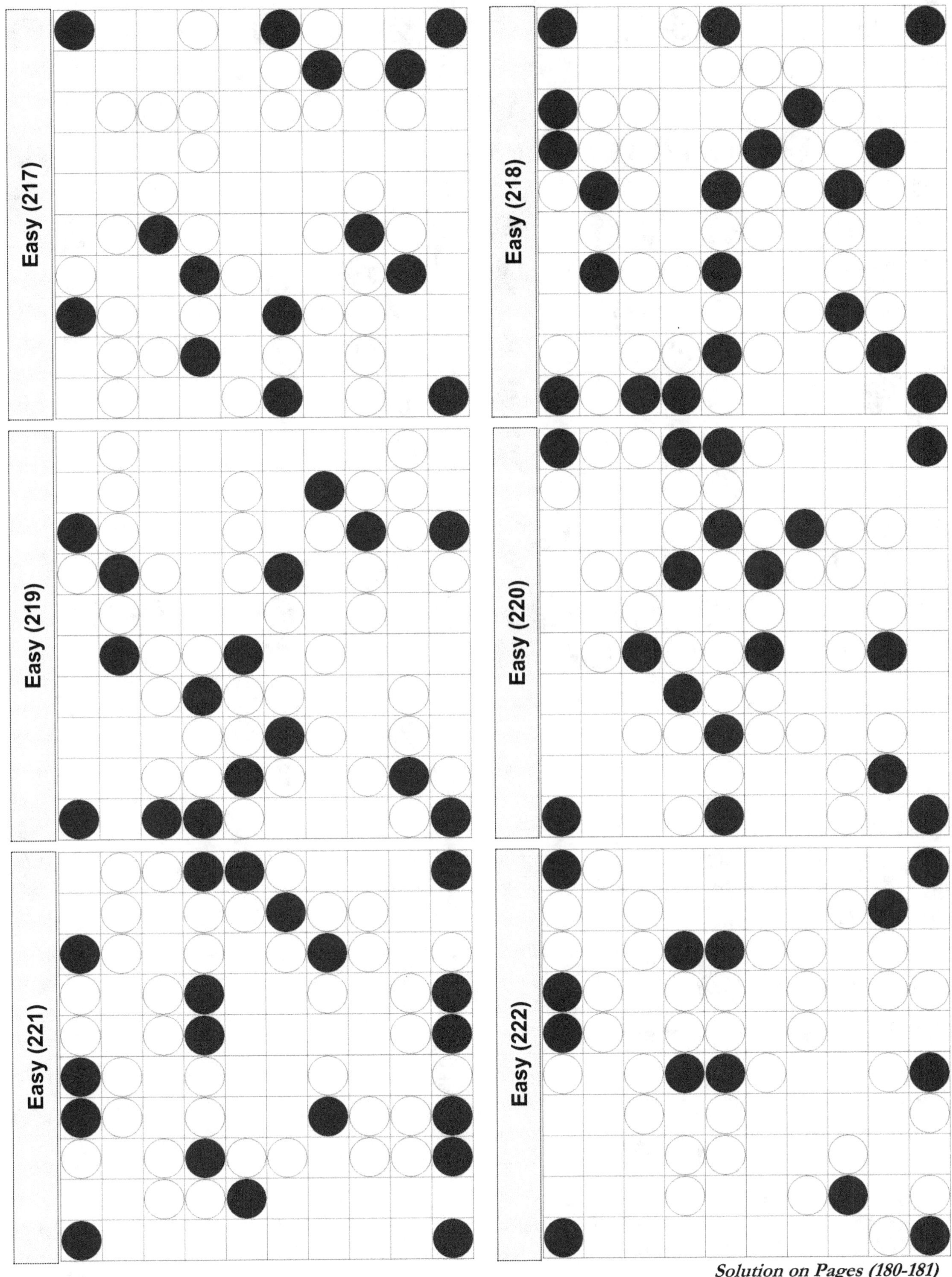

Easy (217)
Easy (218)
Easy (219)
Easy (220)
Easy (221)
Easy (222)
Solution on Pages (180-181)

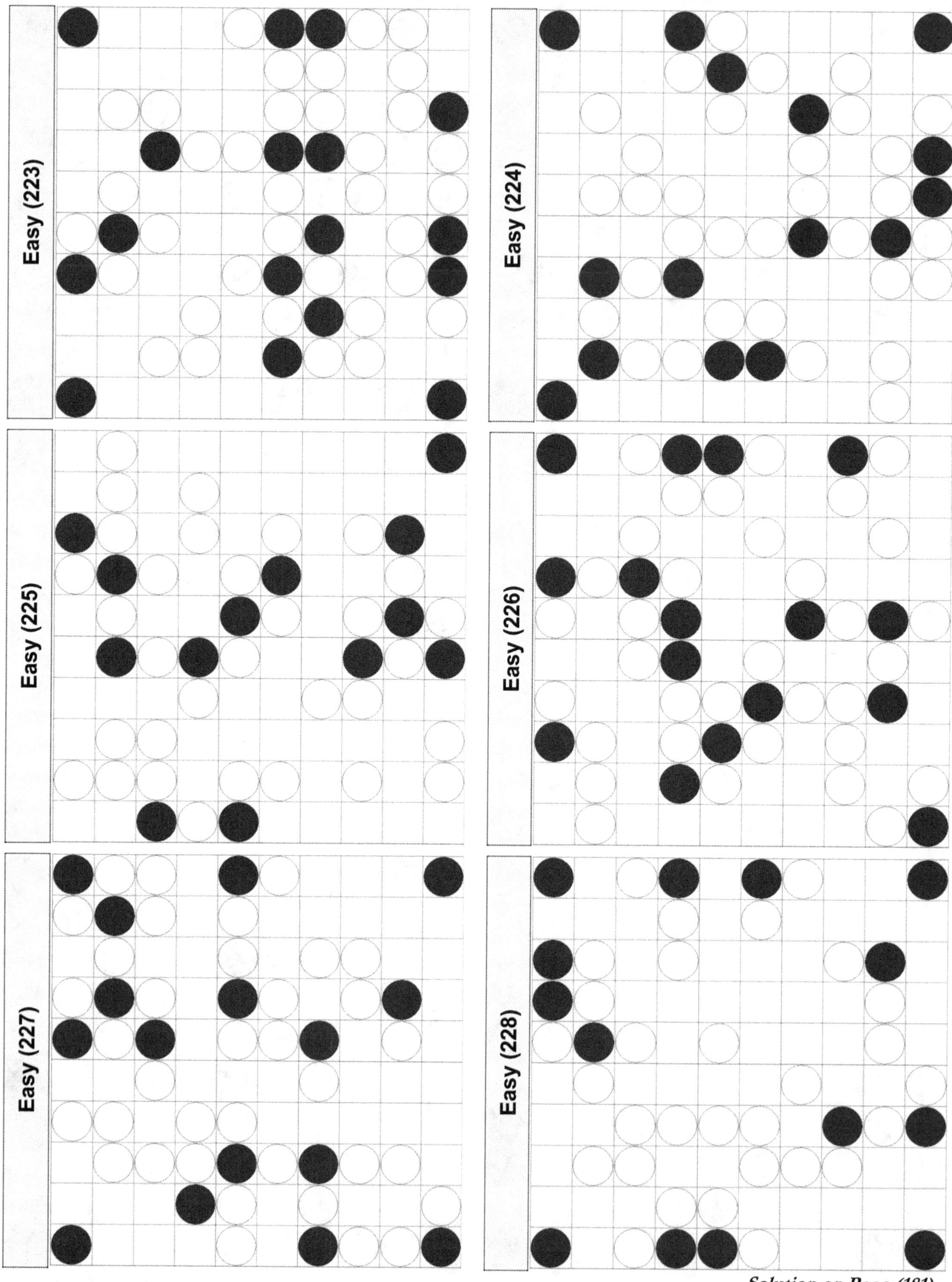

Solution on Page (181)

(40)

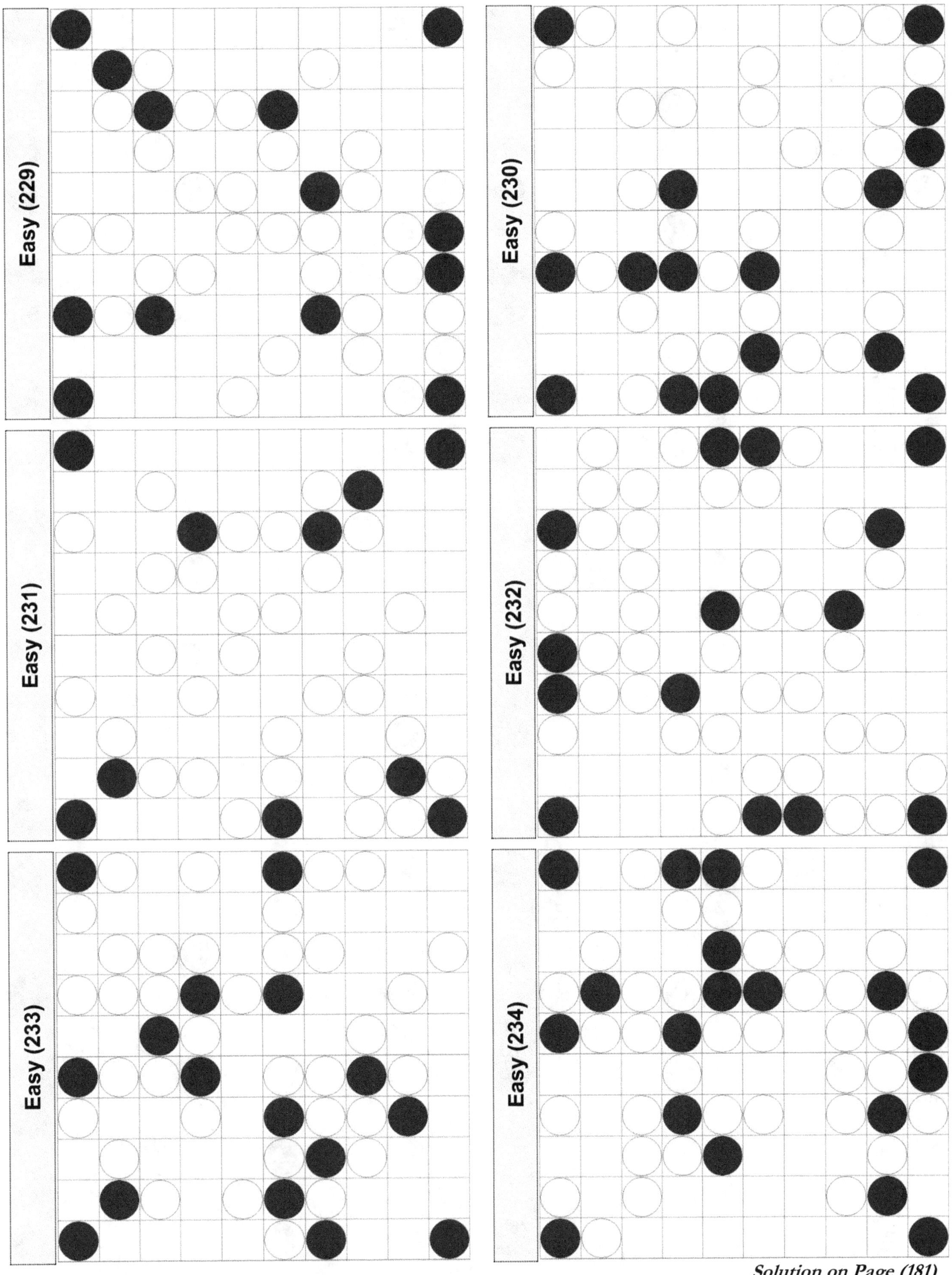

(41)

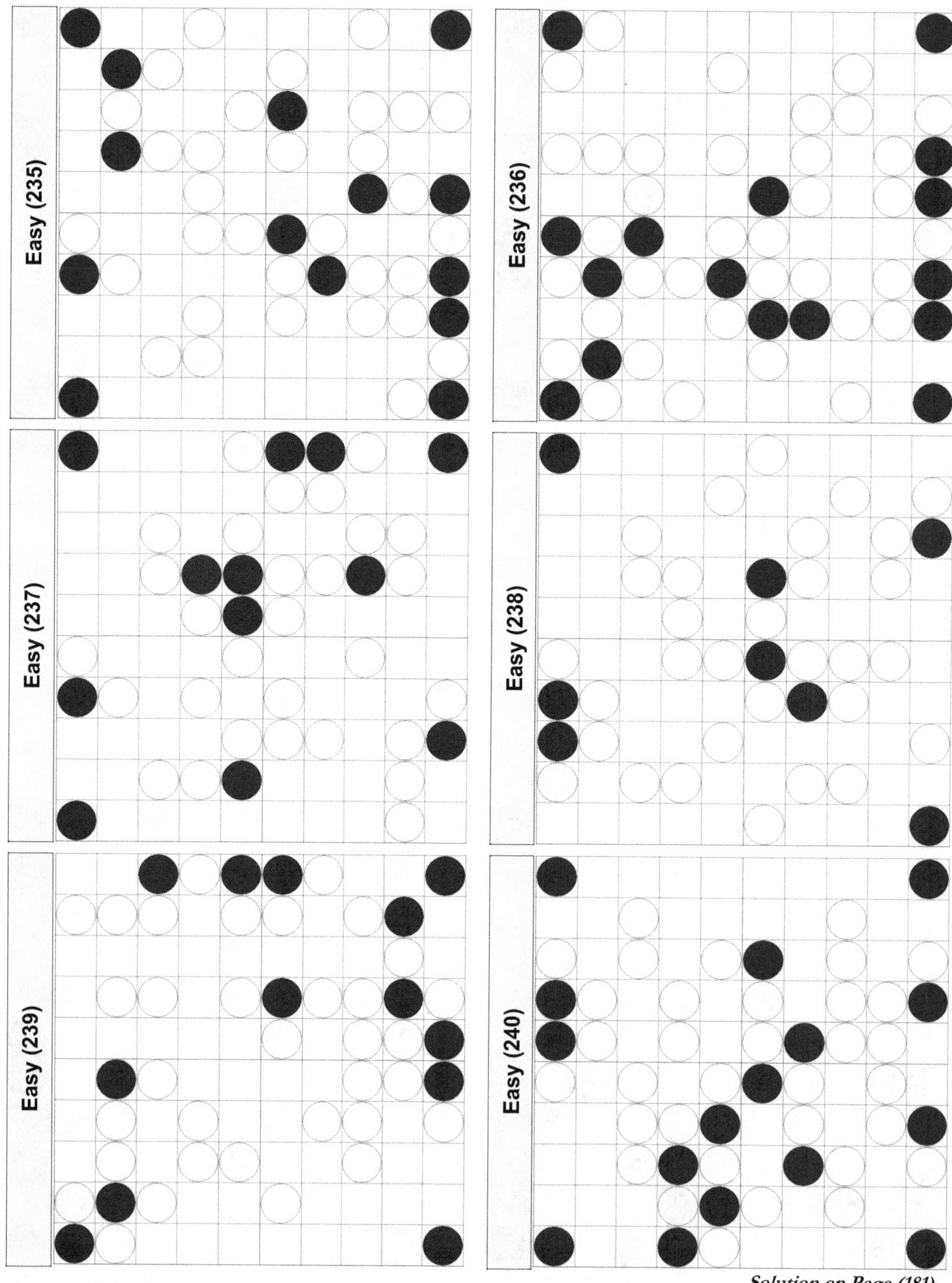

Solution on Page (181)

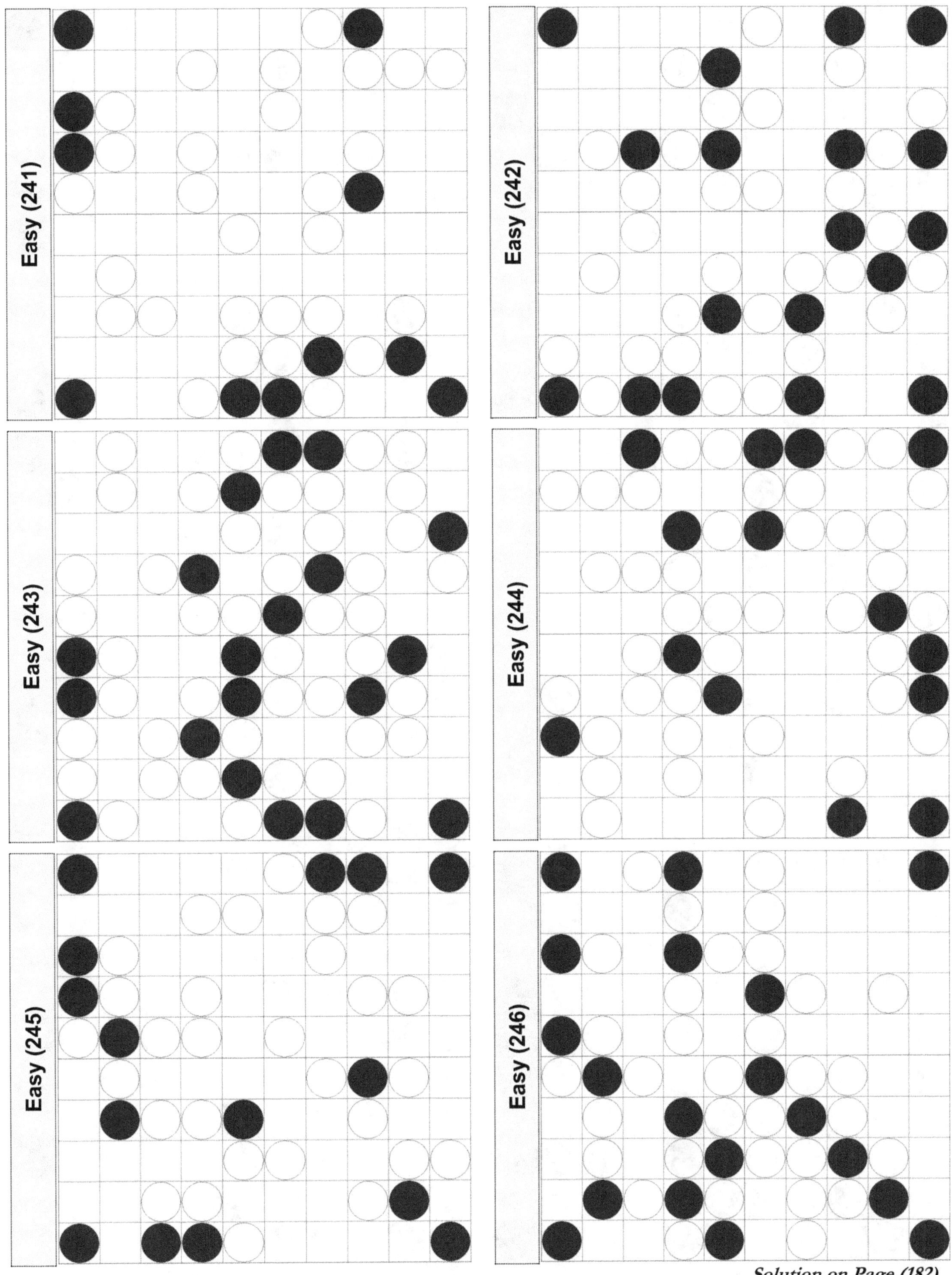

Solution on Page (182)

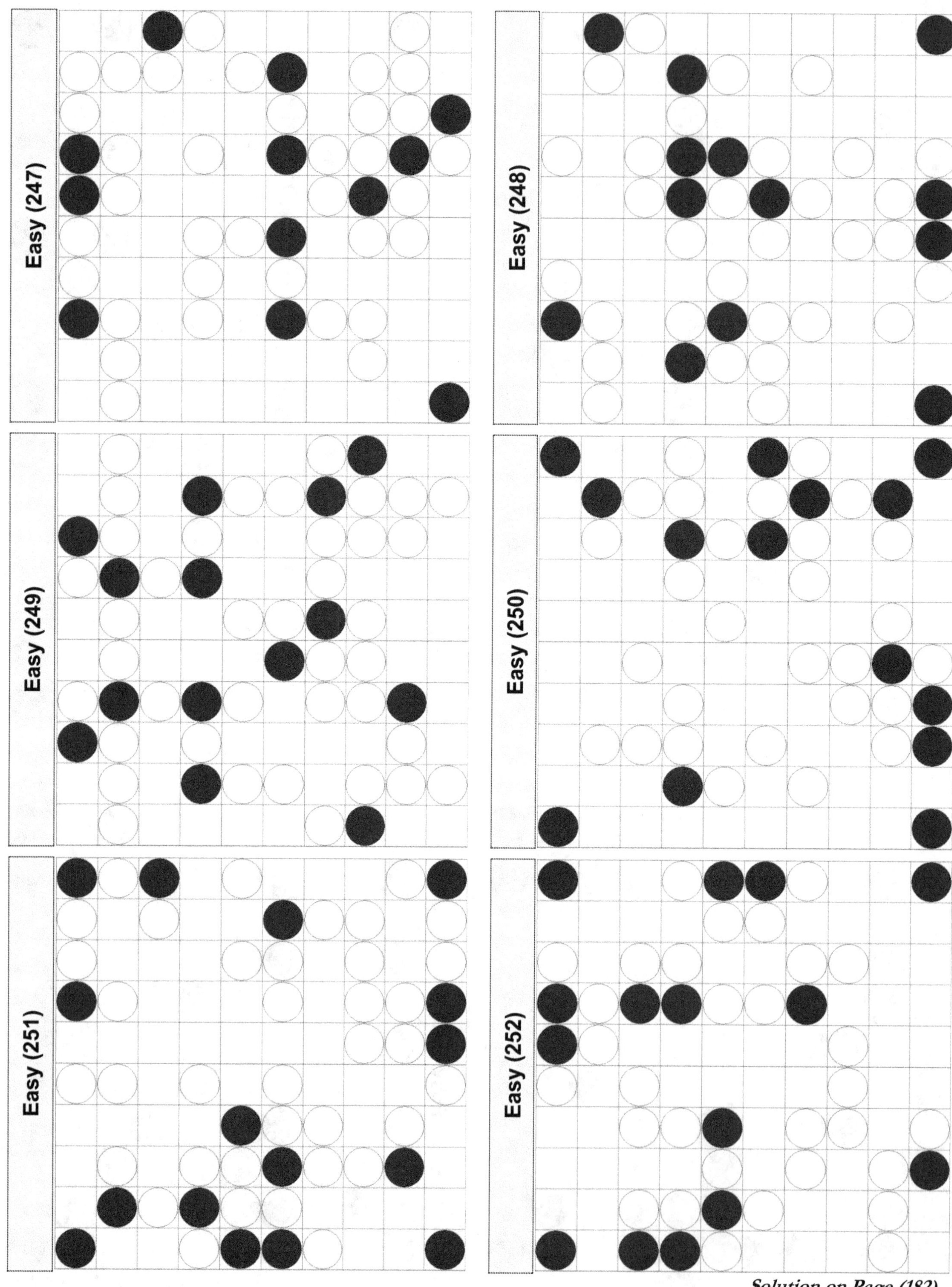

Easy (247)

Easy (248)

Easy (249)

Easy (250)

Easy (251)

Easy (252)

Solution on Page (182)

(44)

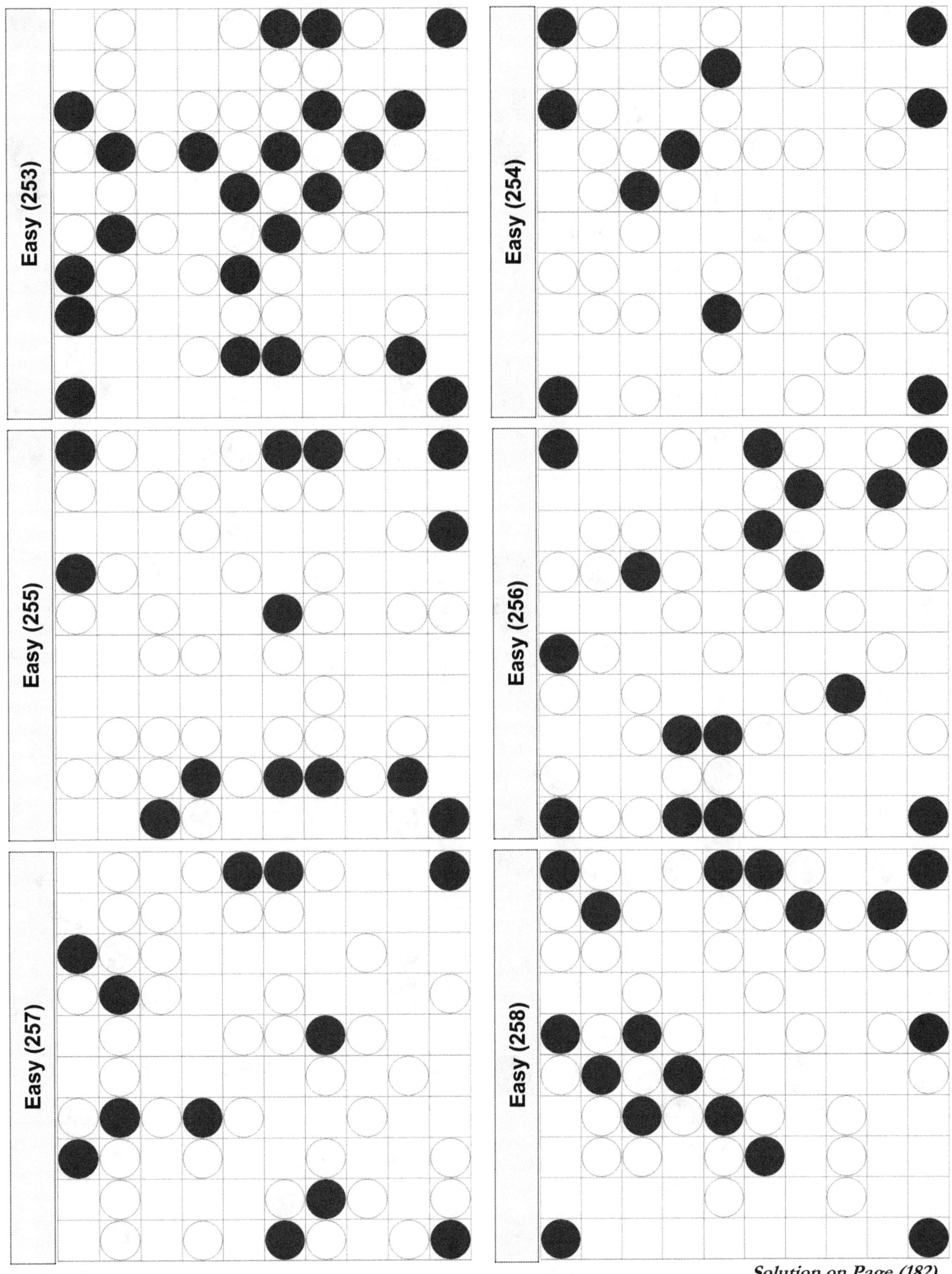

Easy (253)
Easy (254)
Easy (255)
Easy (256)
Easy (257)
Easy (258)

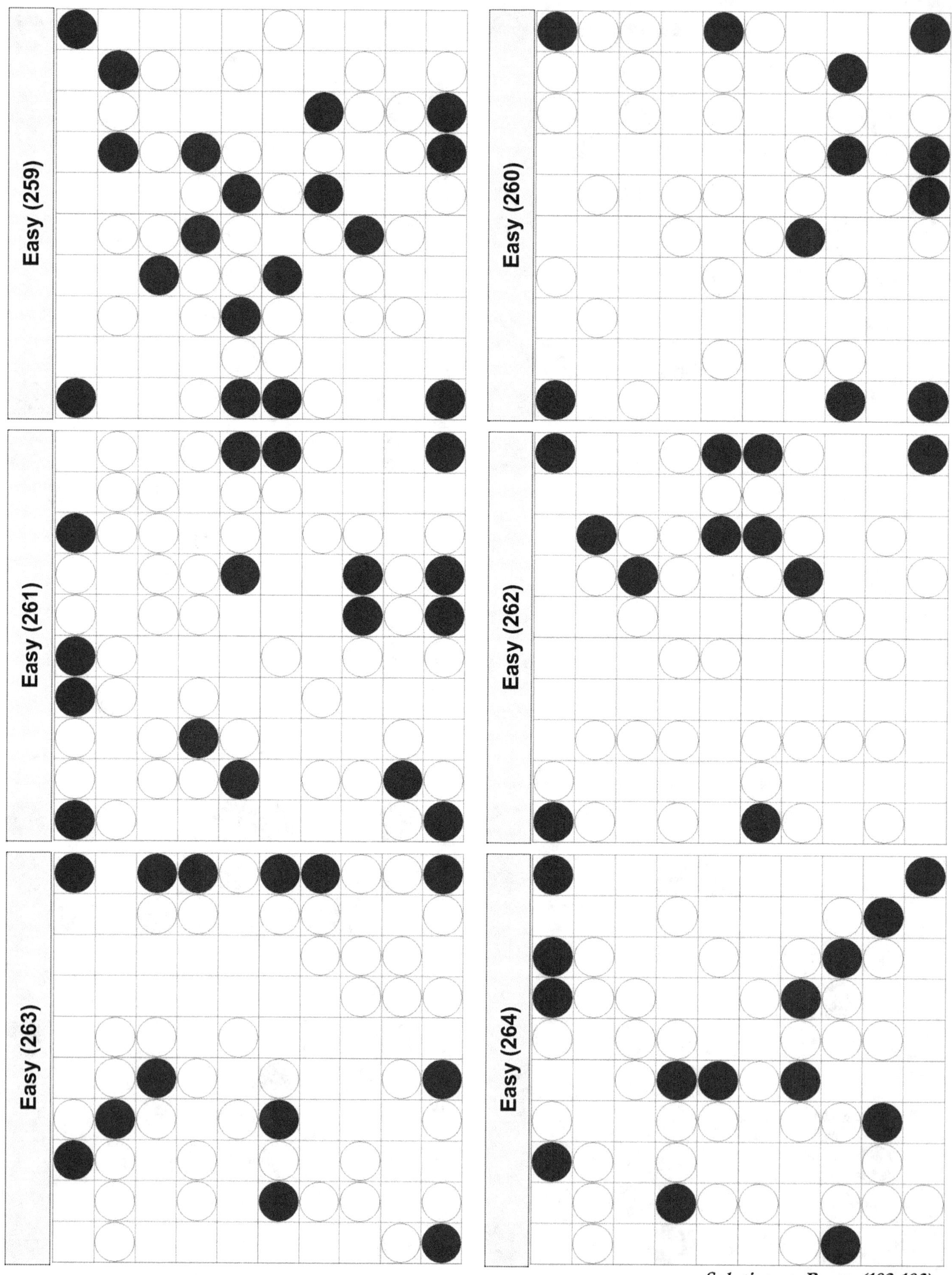

Solution on Pages (182-183)

(46)

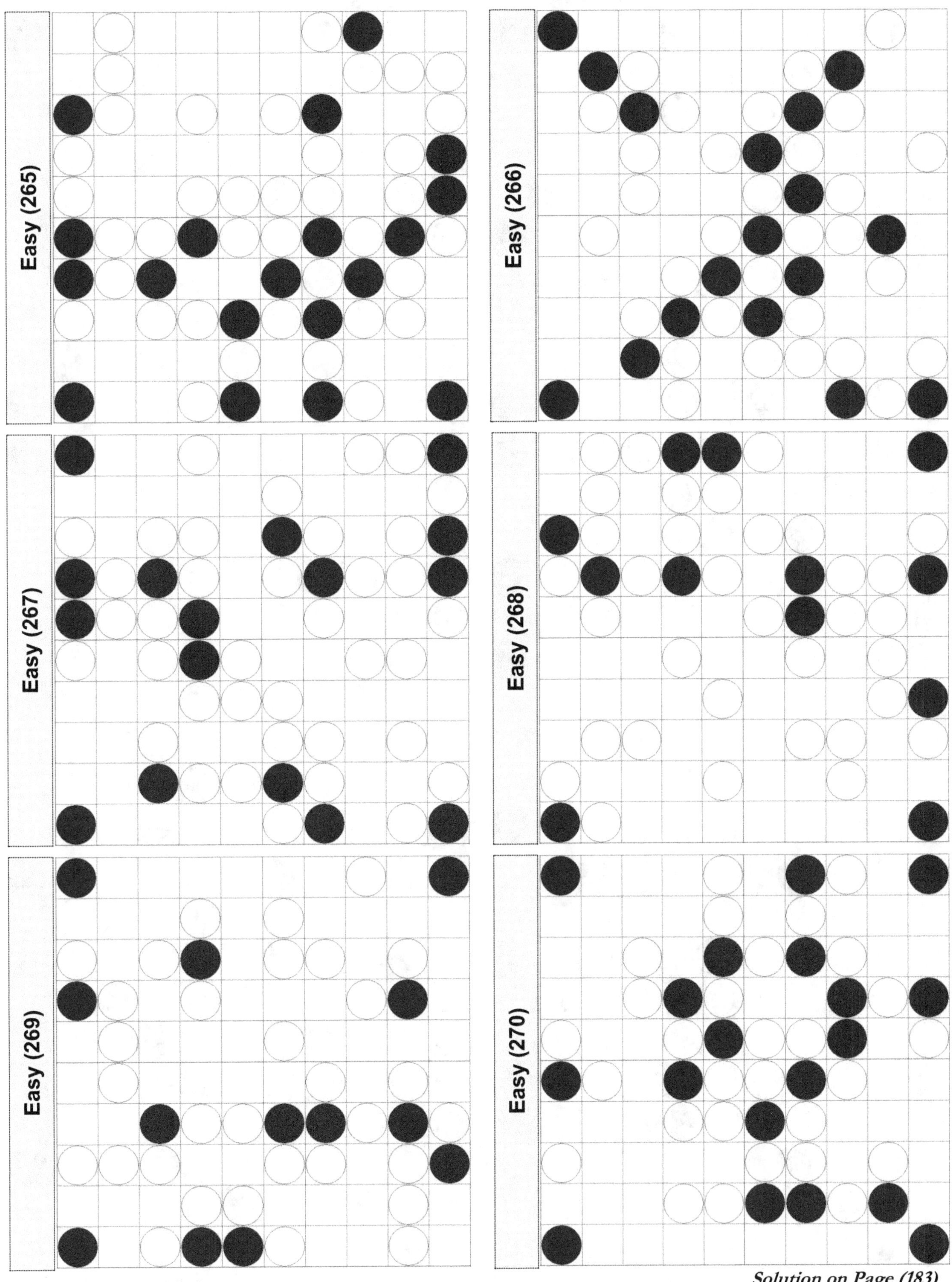

Easy (265)

Easy (266)

Easy (267)

Easy (268)

Easy (269)

Easy (270)

Solution on Page (183)

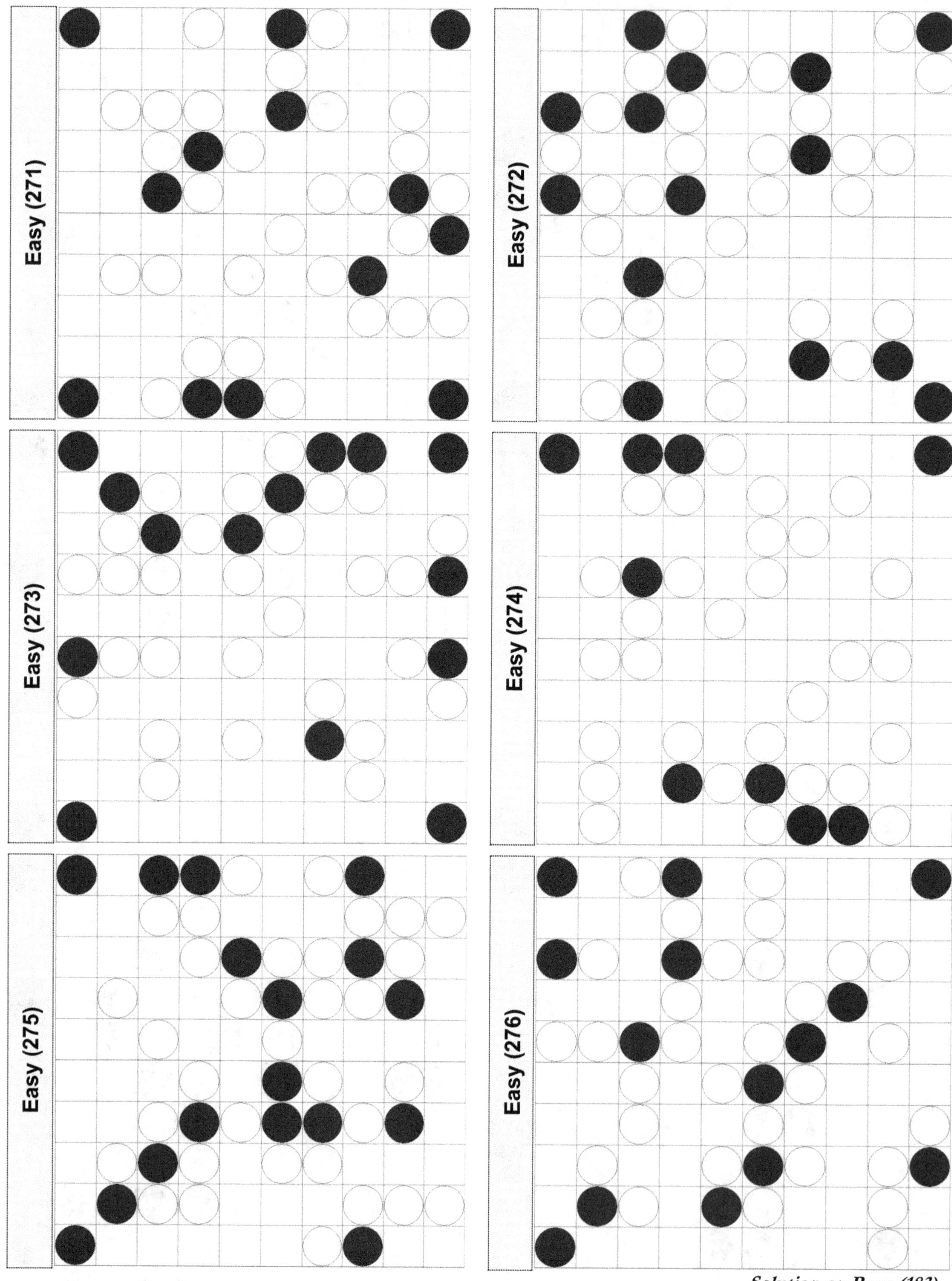

(48)

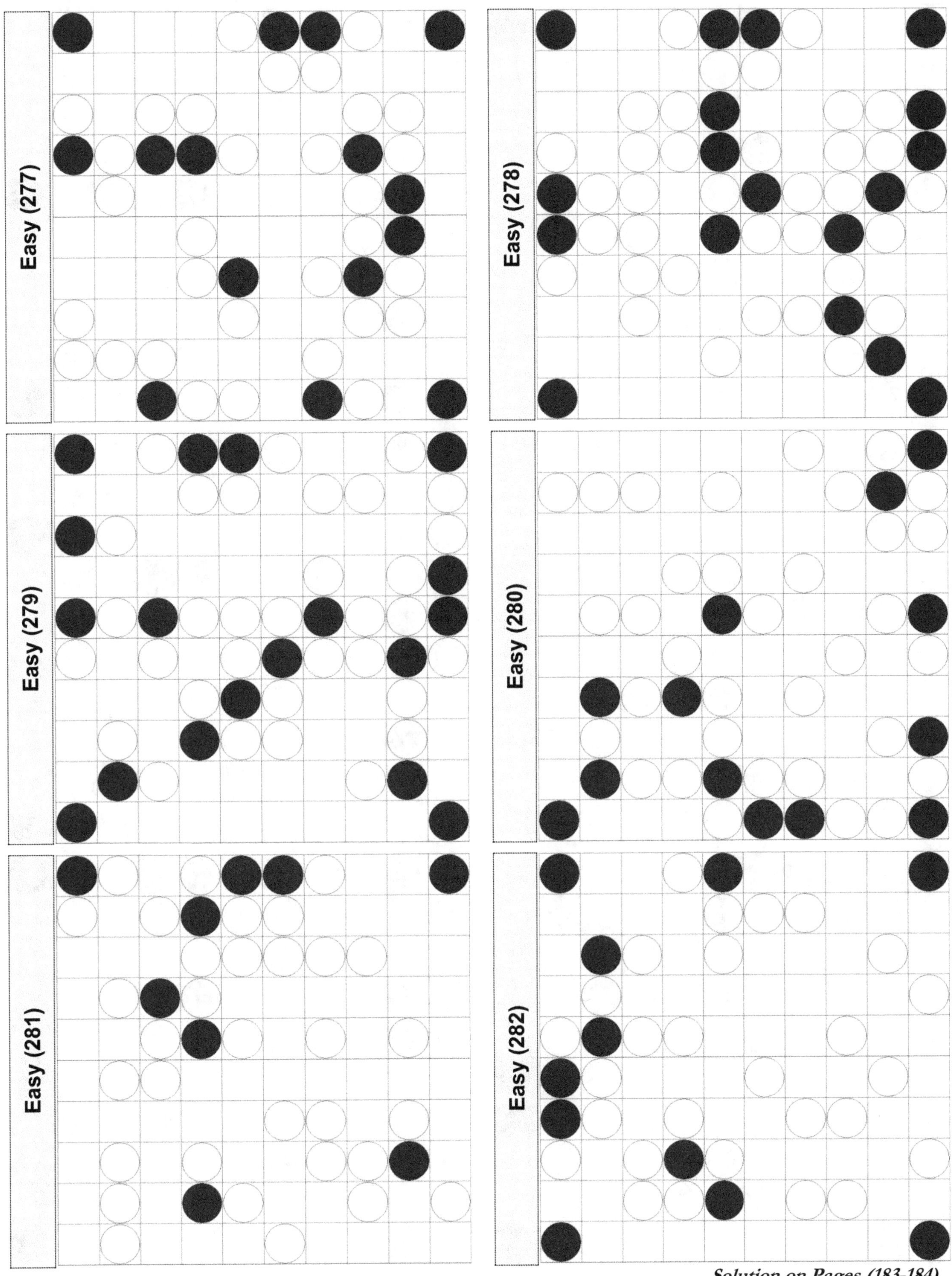

Solution on Pages (183-184)

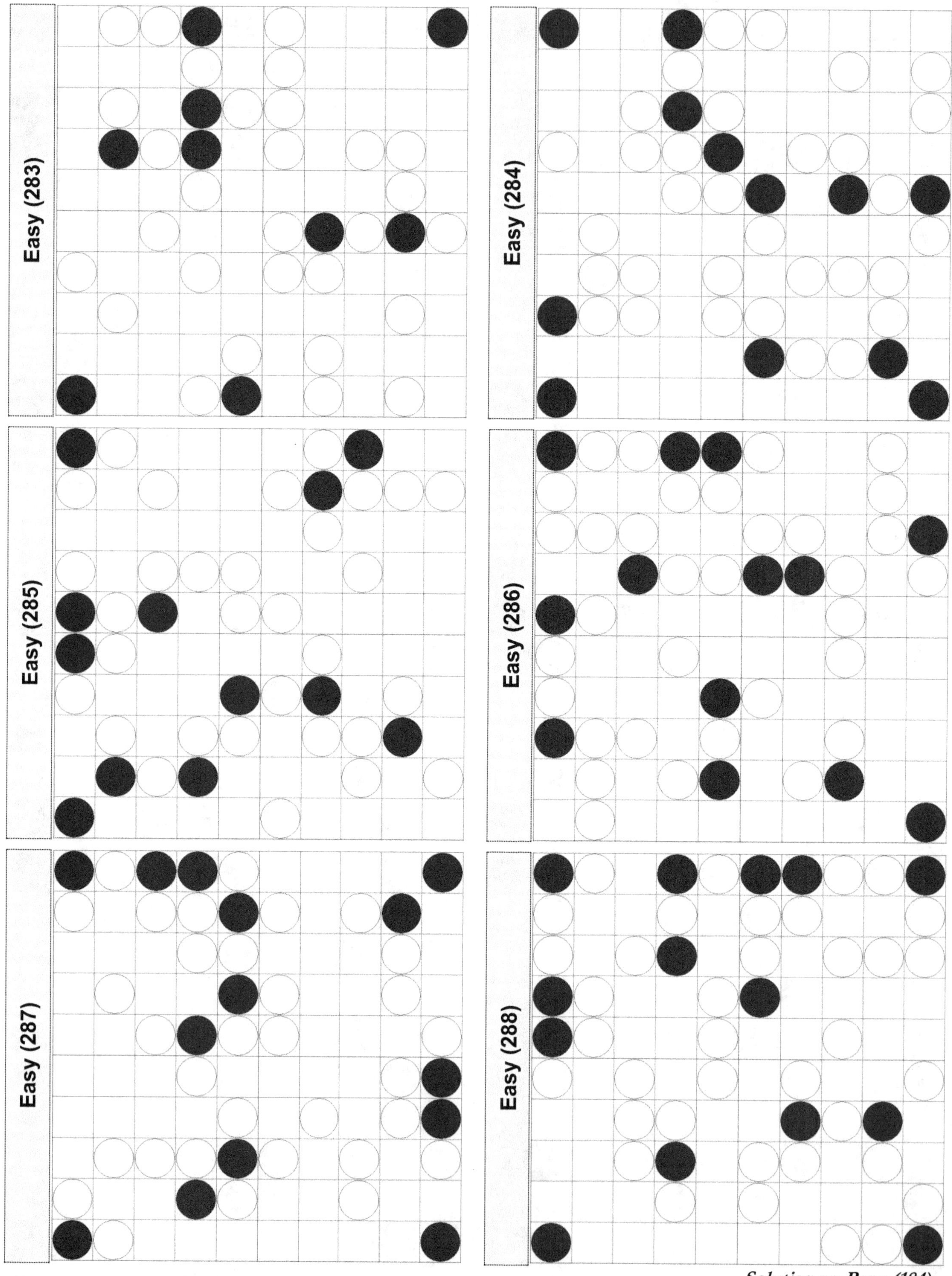

Easy (283)

Easy (284)

Easy (285)

Easy (286)

Easy (287)

Easy (288)

Solution on Page (184)

(50)

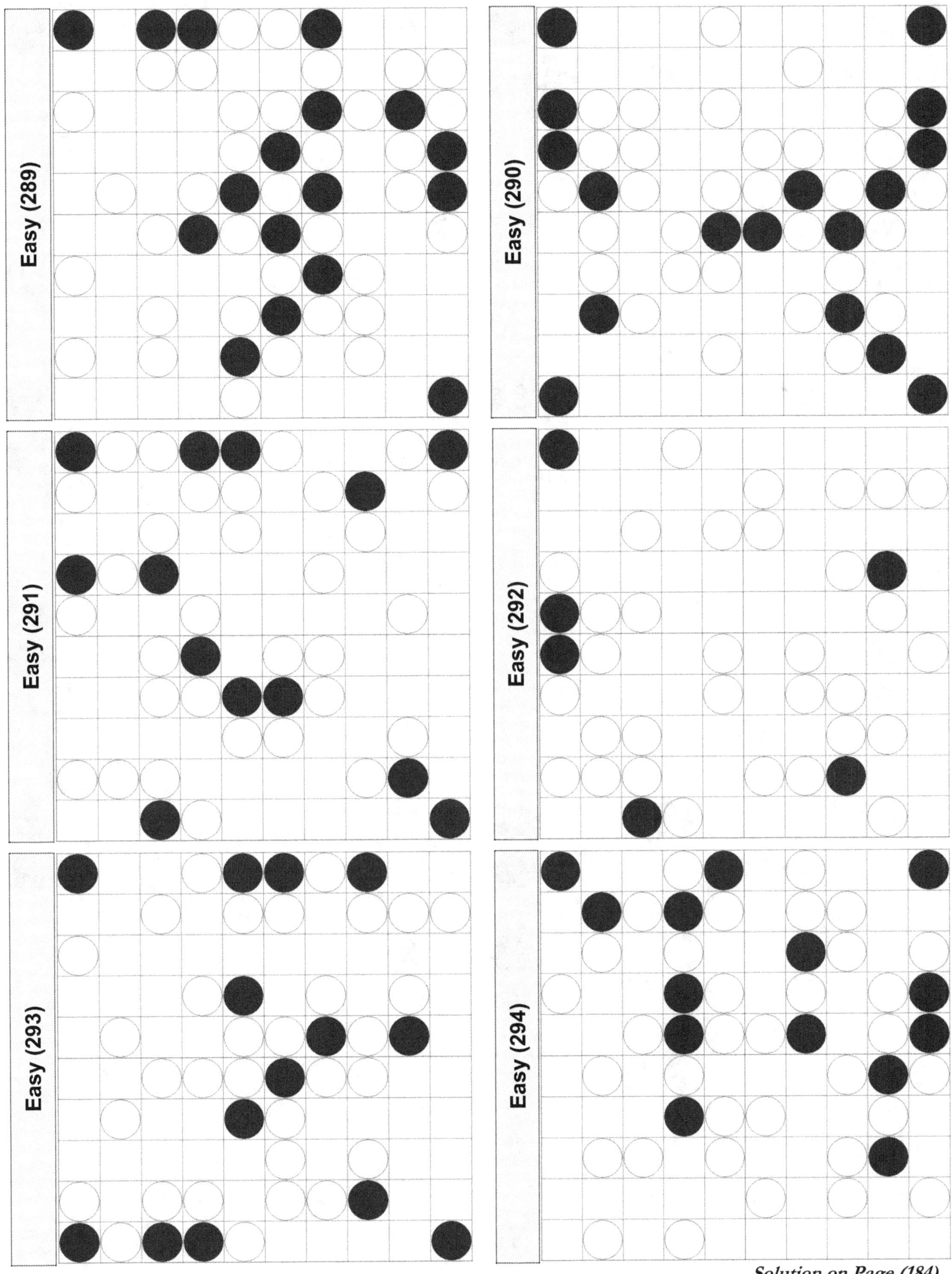

Solution on Page (184)

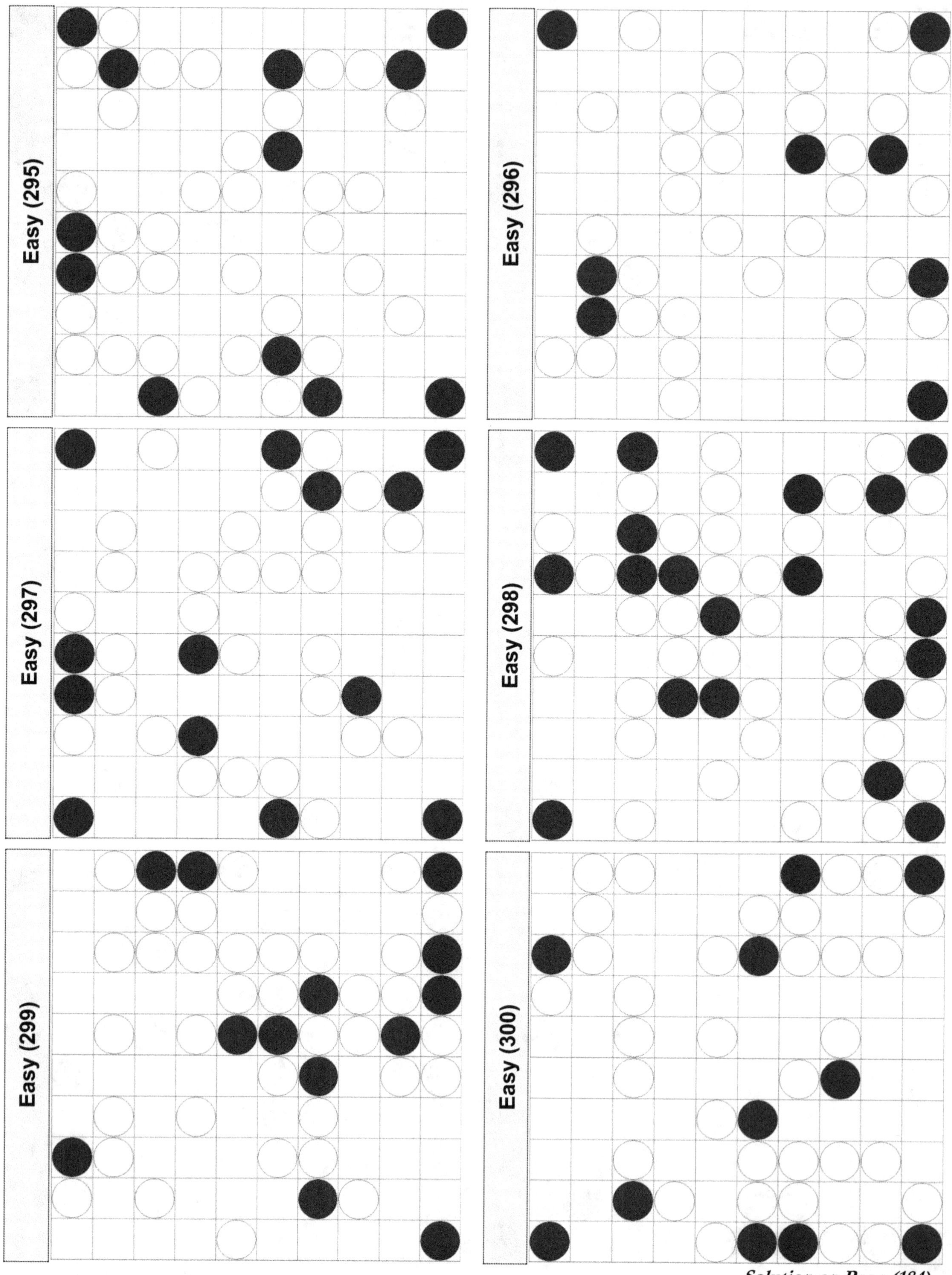

(52)

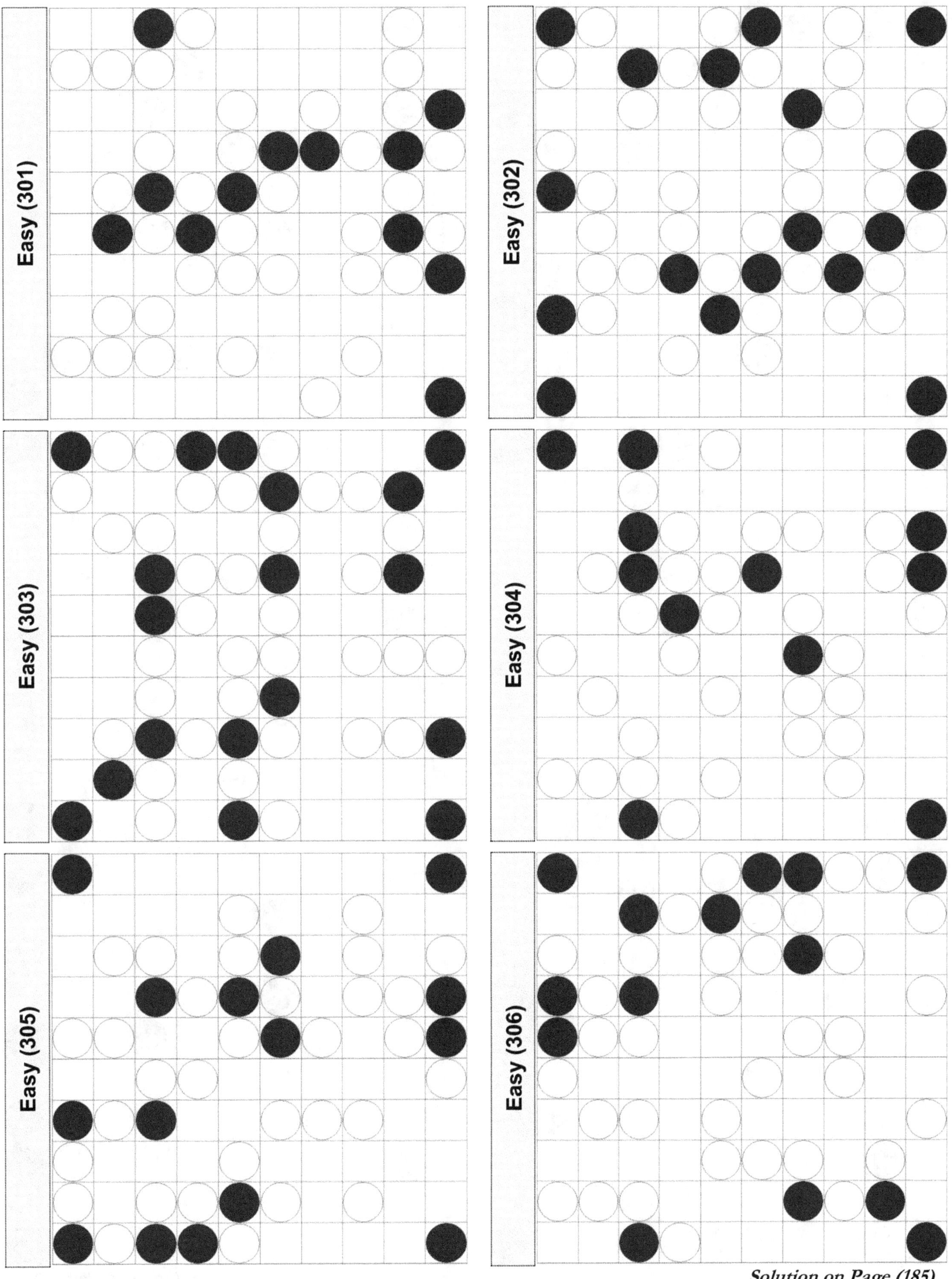

Solution on Page (185)

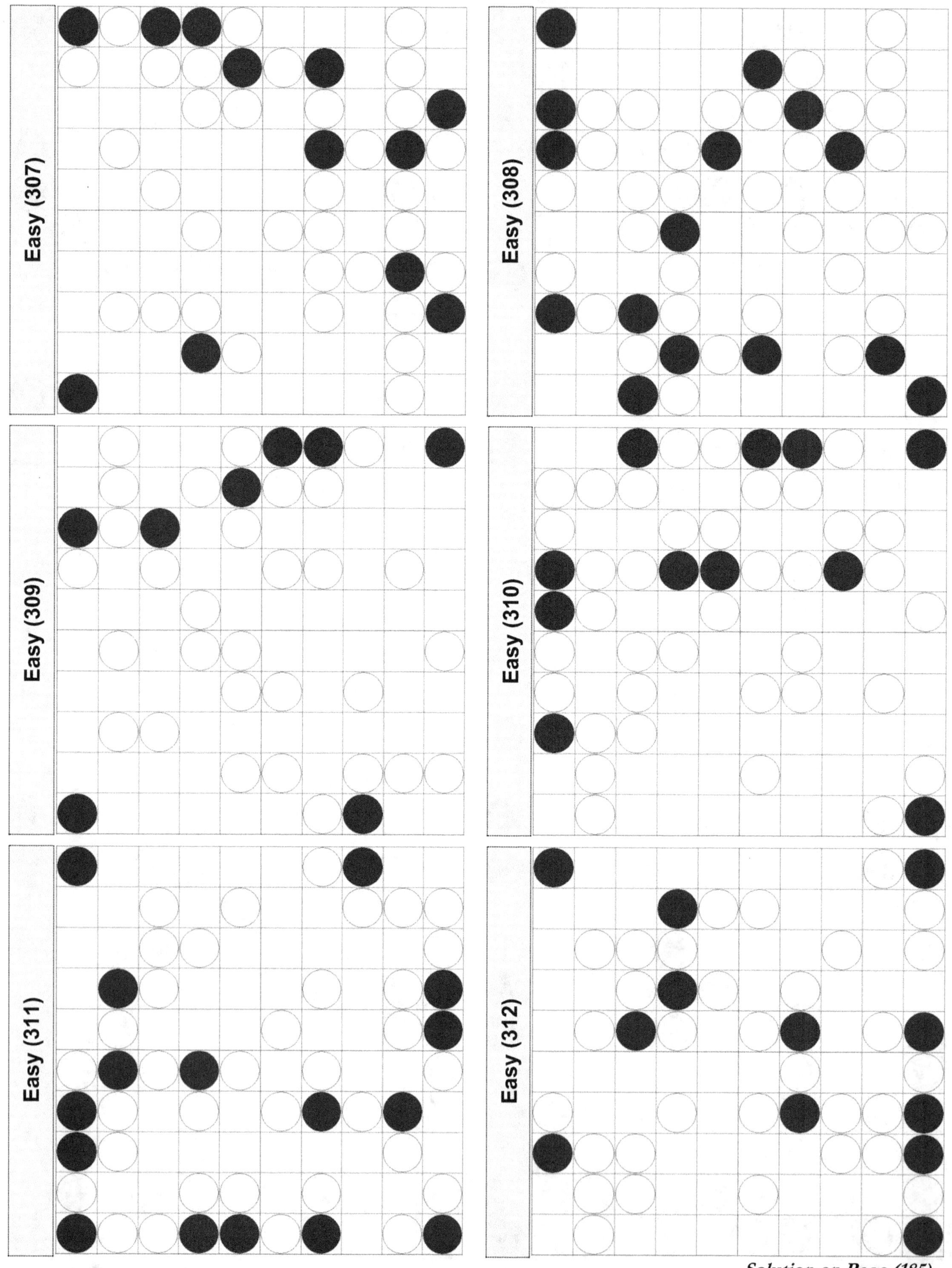

Solution on Page (185)

(54)

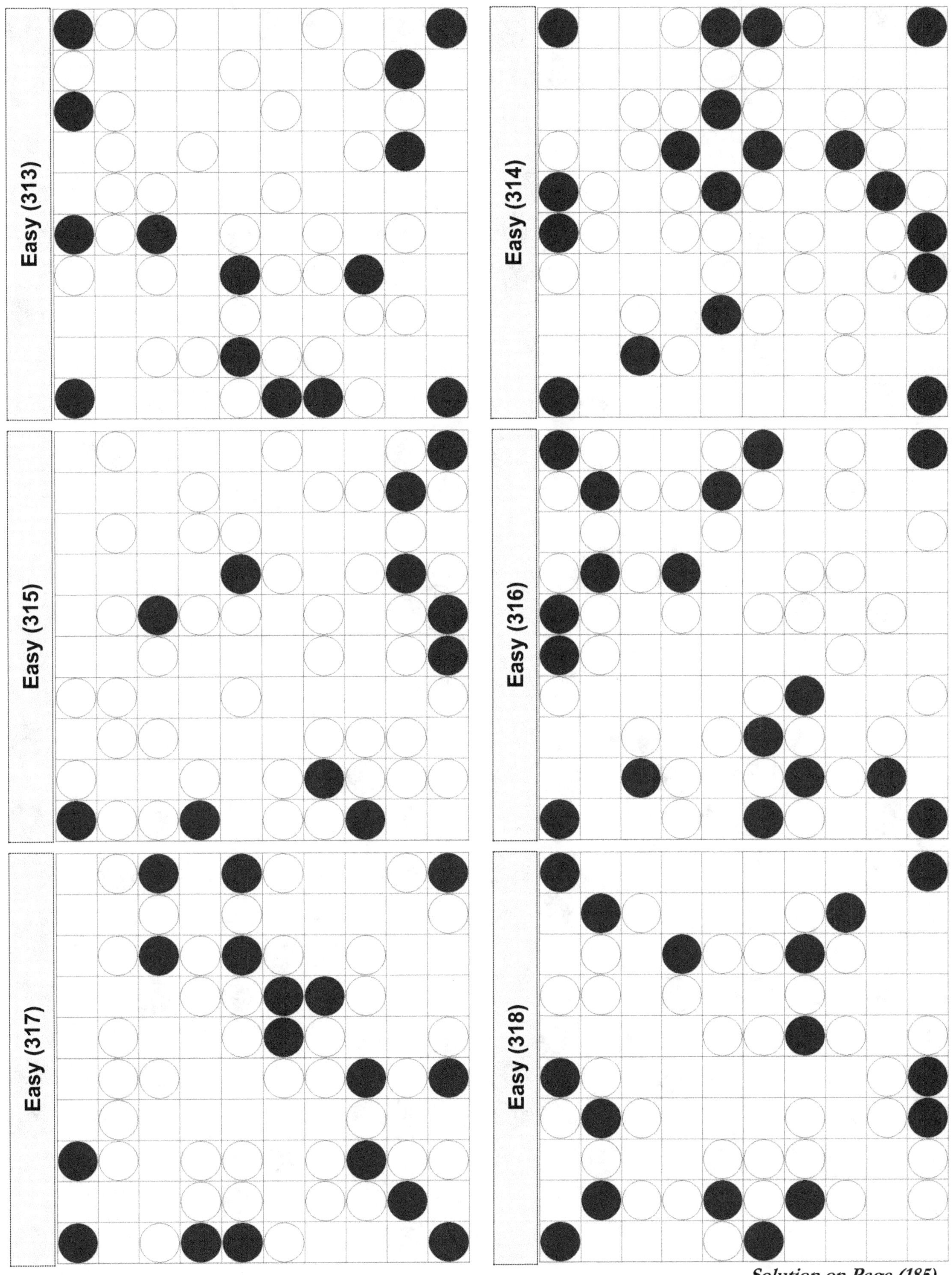

Solution on Page (185)

(55)

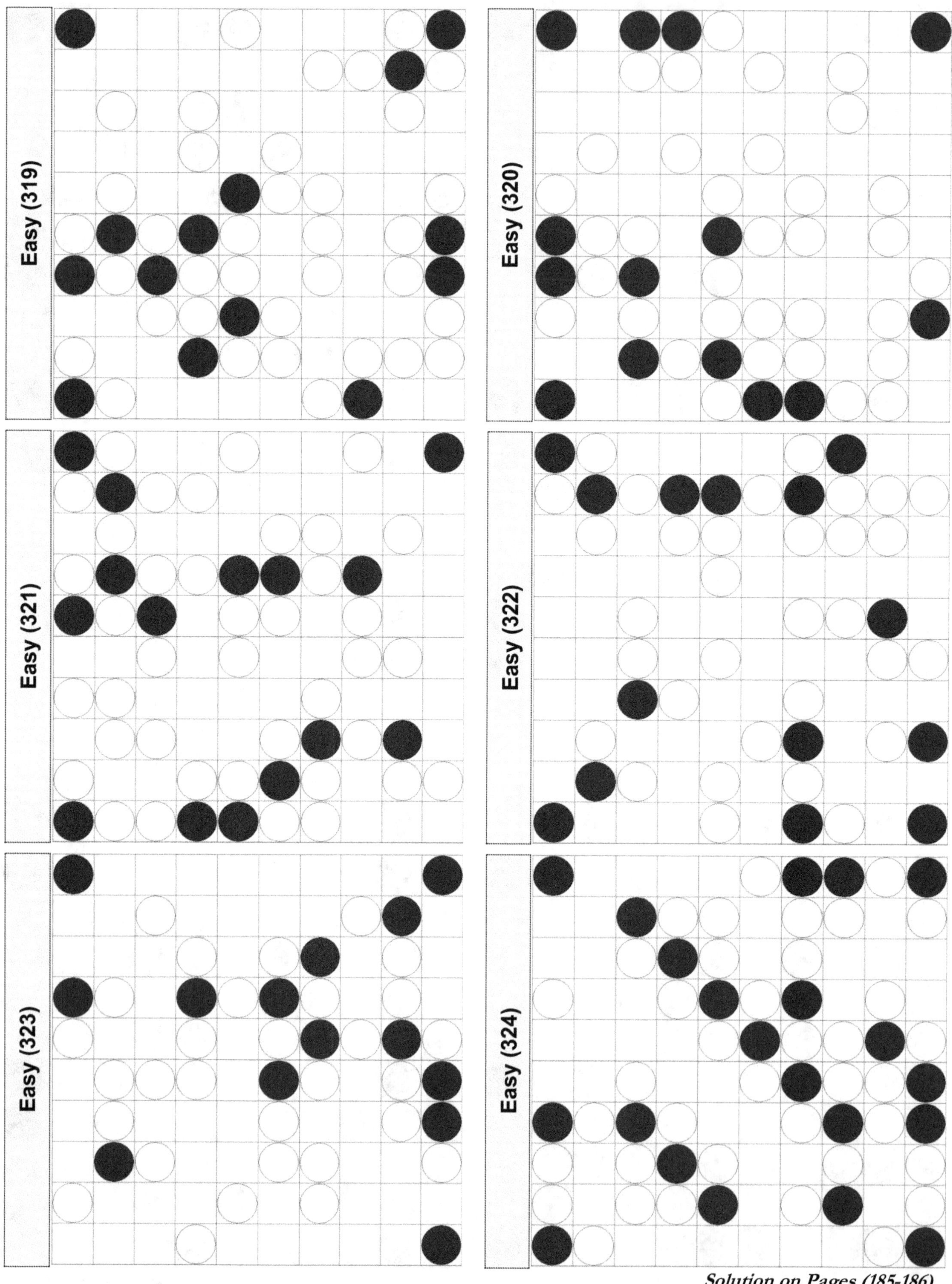

Solution on Pages (185-186)

(56)

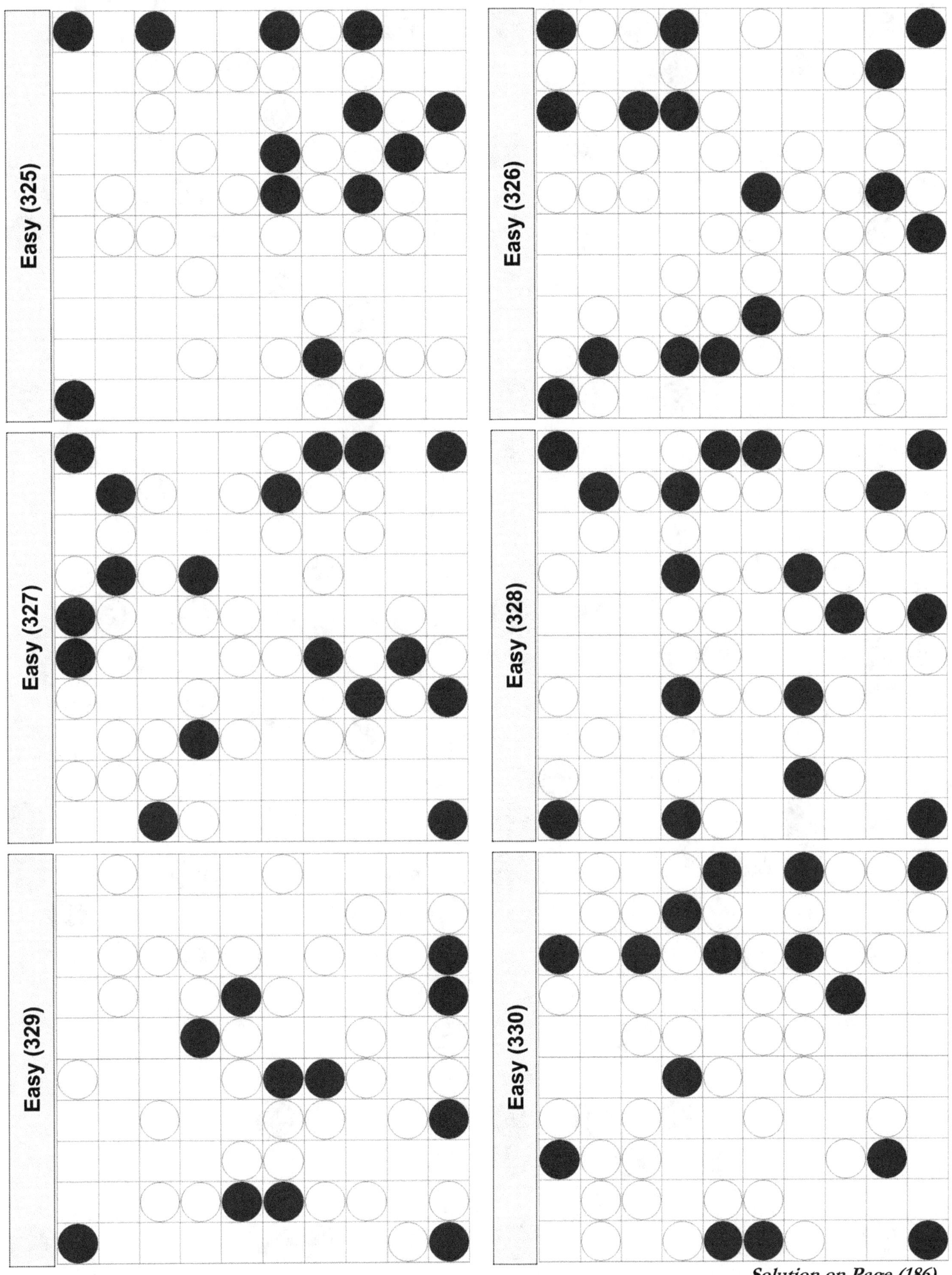

Solution on Page (186)

(57)

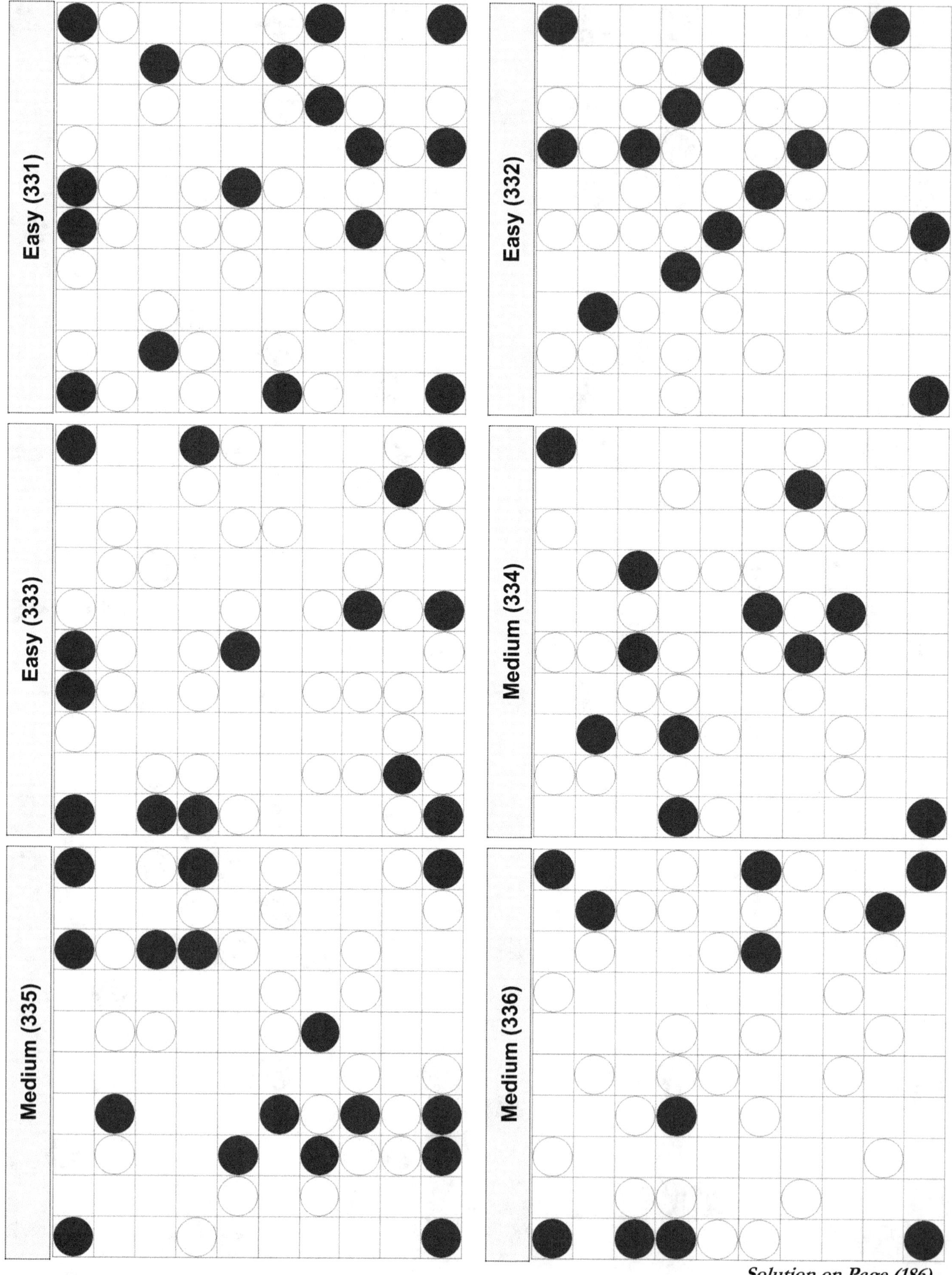

Solution on Page (186)

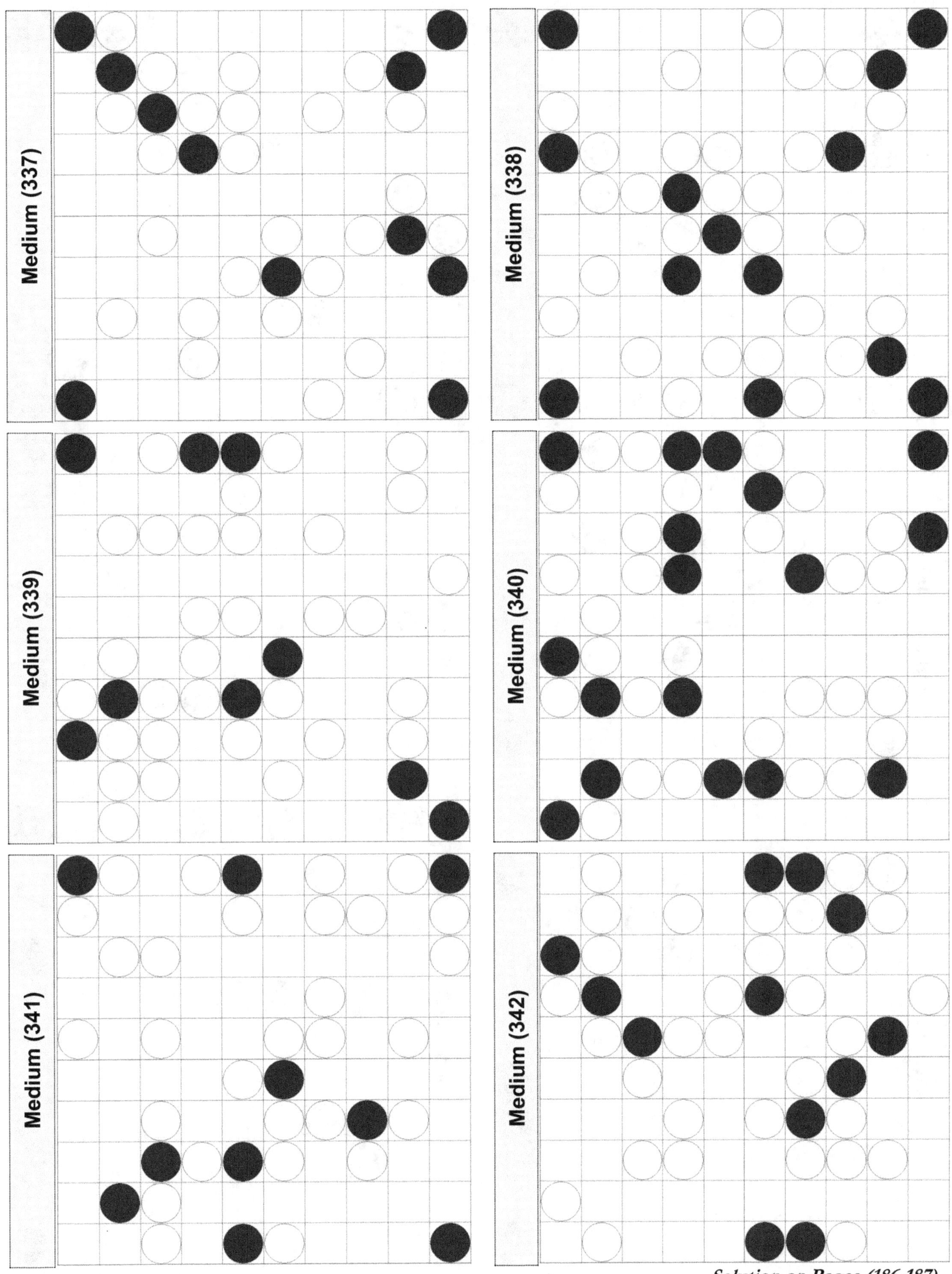

Medium (337)
Medium (338)
Medium (339)
Medium (340)
Medium (341)
Medium (342)
Solution on Pages (186-187)

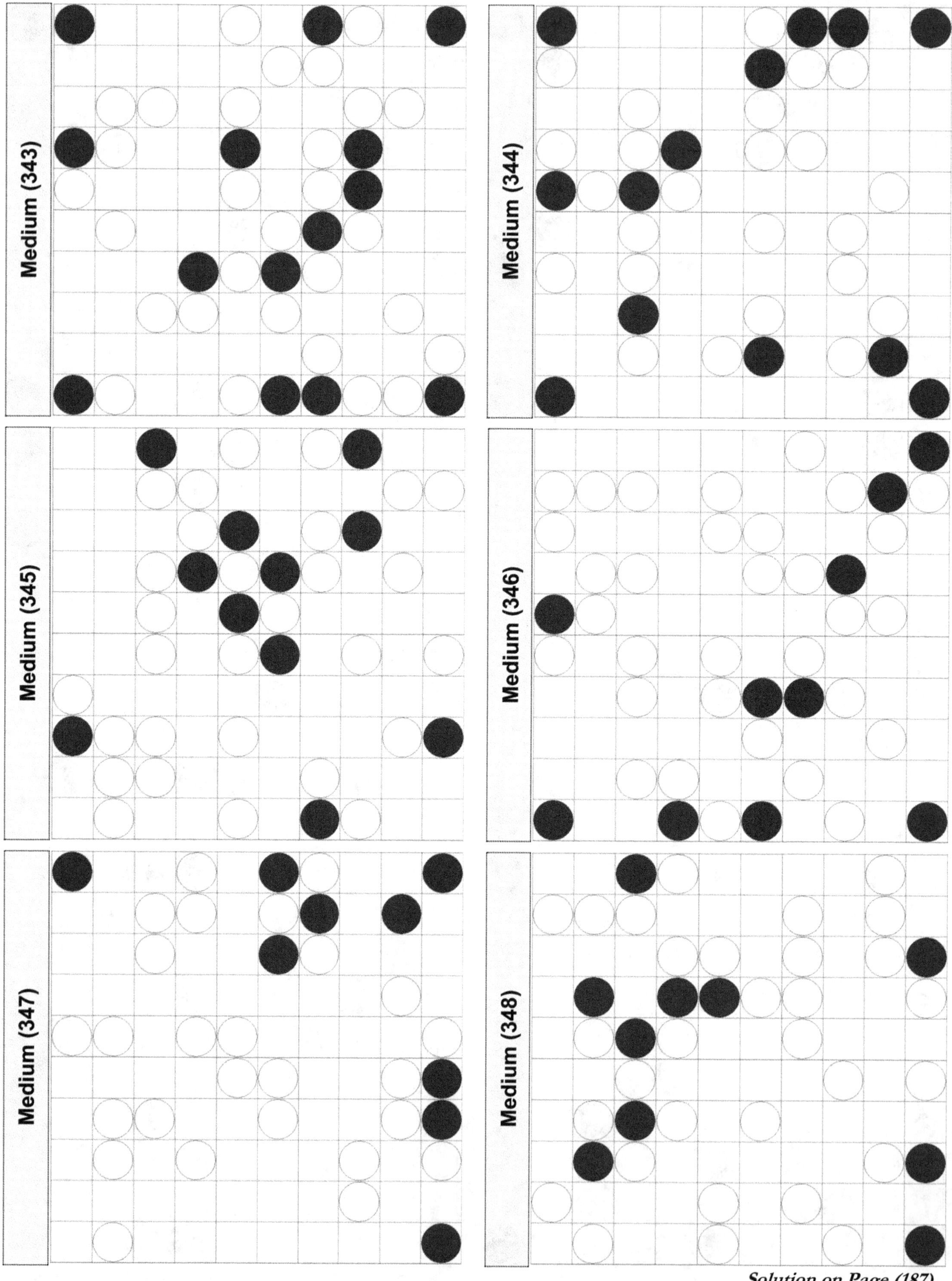

Solution on Page (187)

(60)

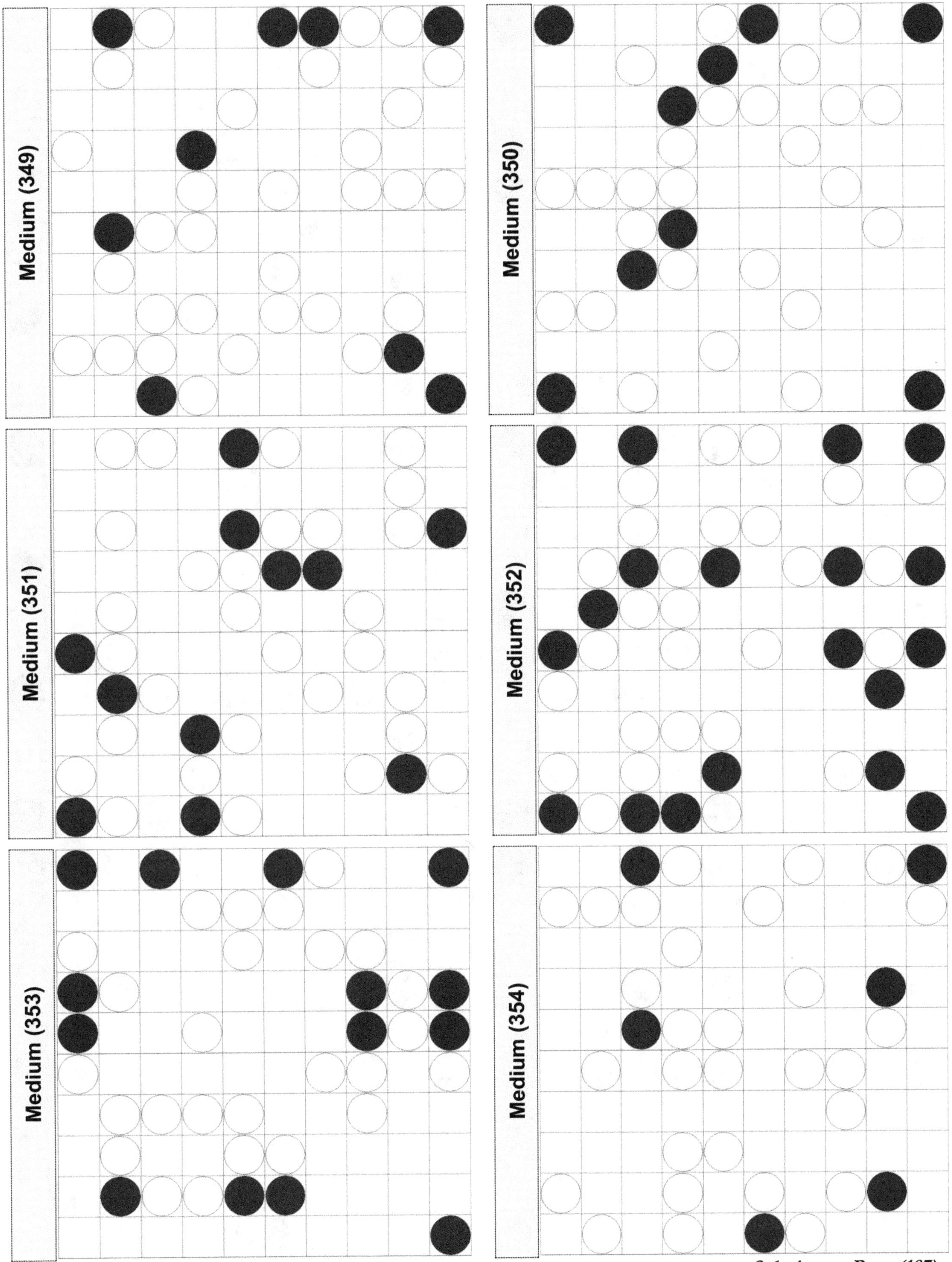

(61)

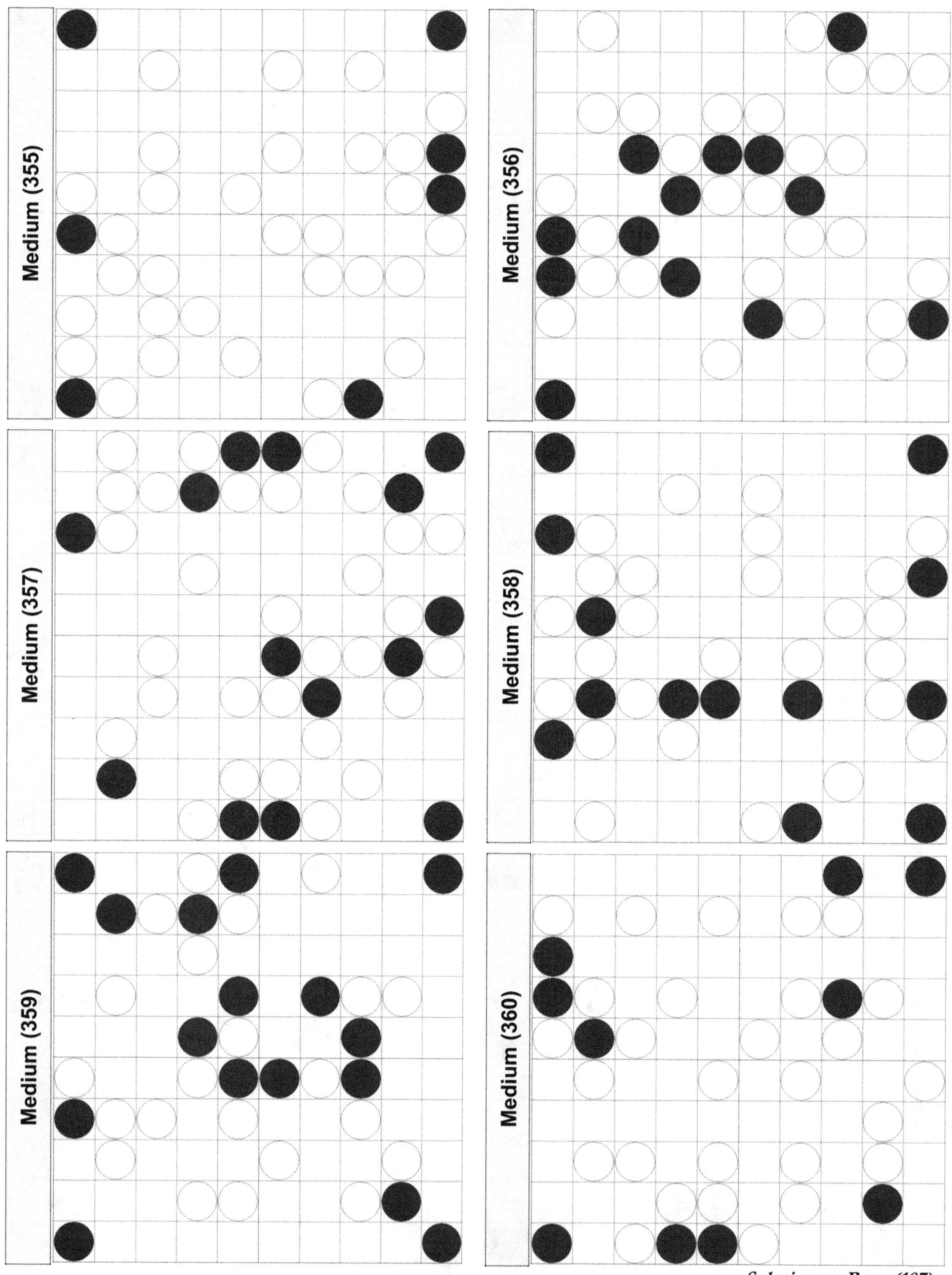

Medium (355)
Medium (356)
Medium (357)
Medium (358)
Medium (359)
Medium (360)
Solution on Page (187)

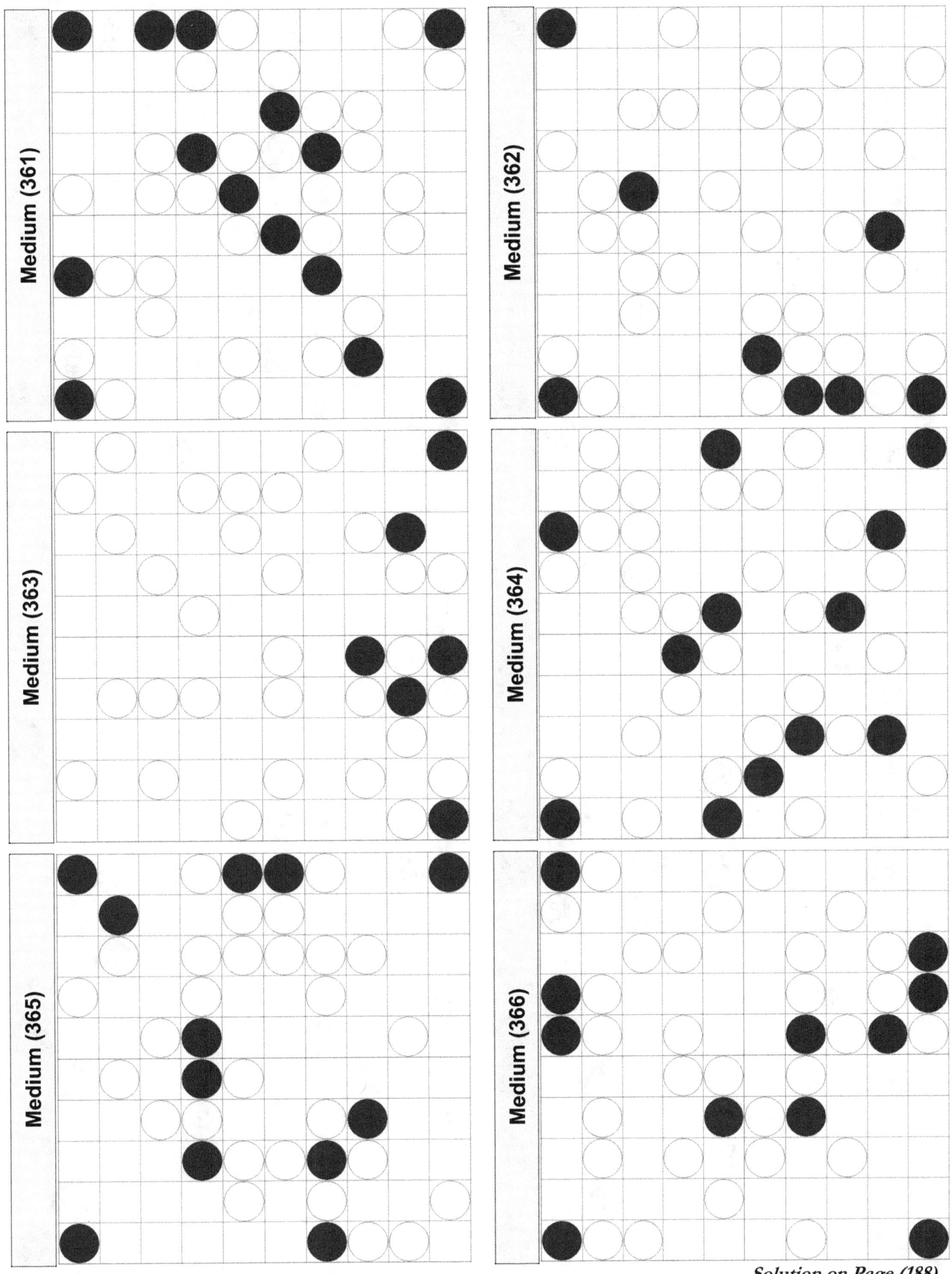

Medium (361)
Medium (362)
Medium (363)
Medium (364)
Medium (365)
Medium (366)
Solution on Page (188)

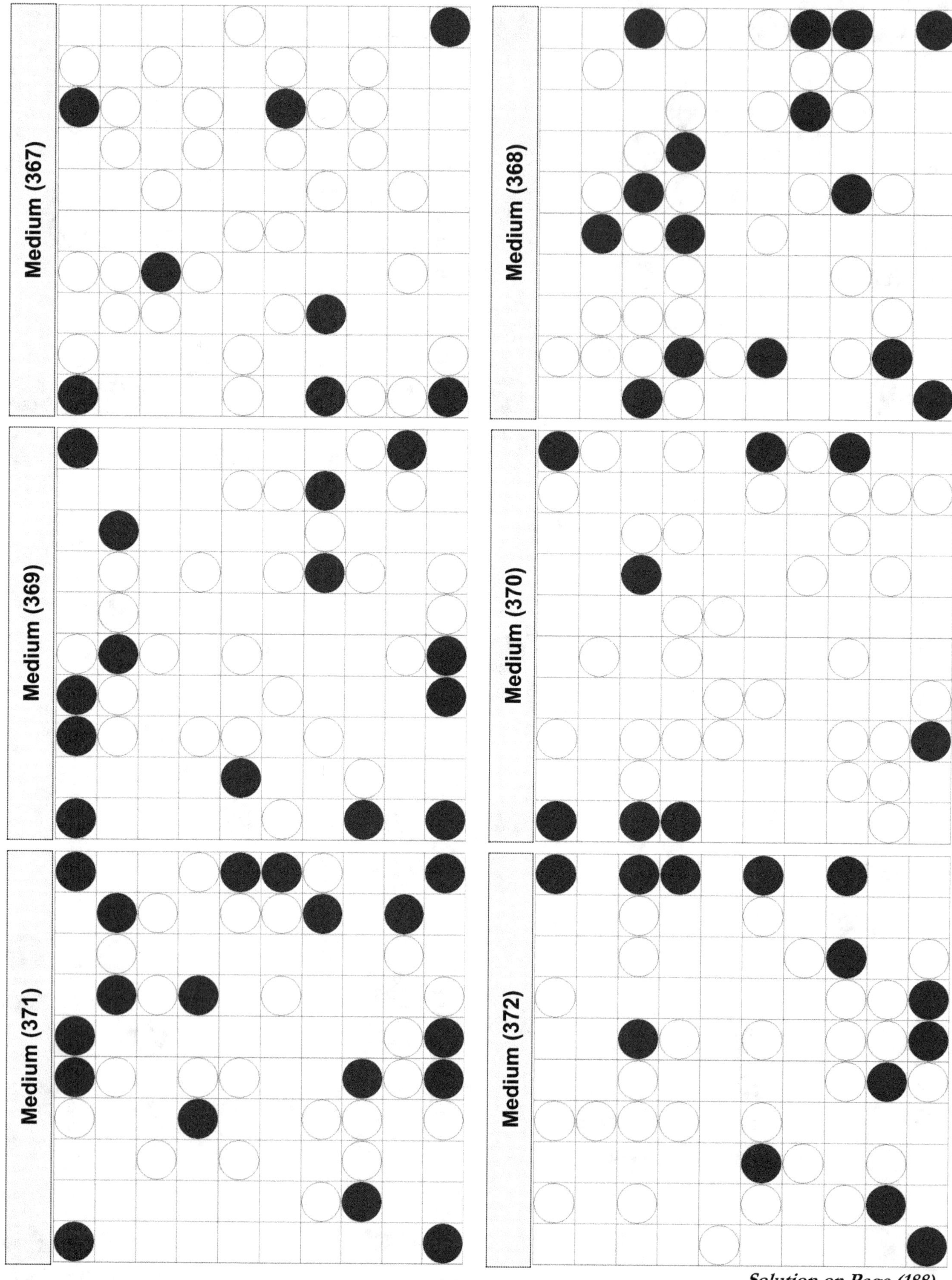

Solution on Page (188)

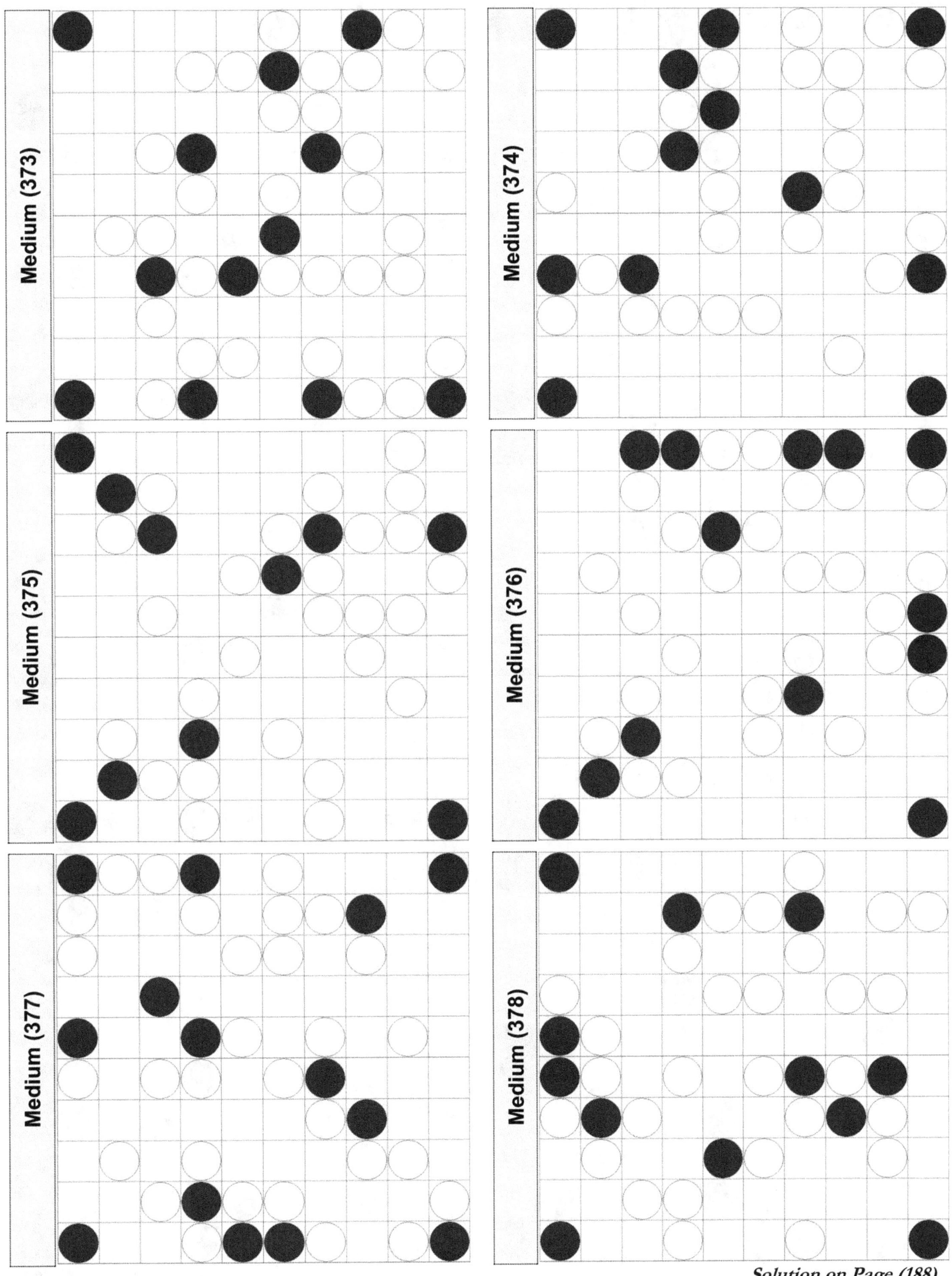

Solution on Page (188)

(65)

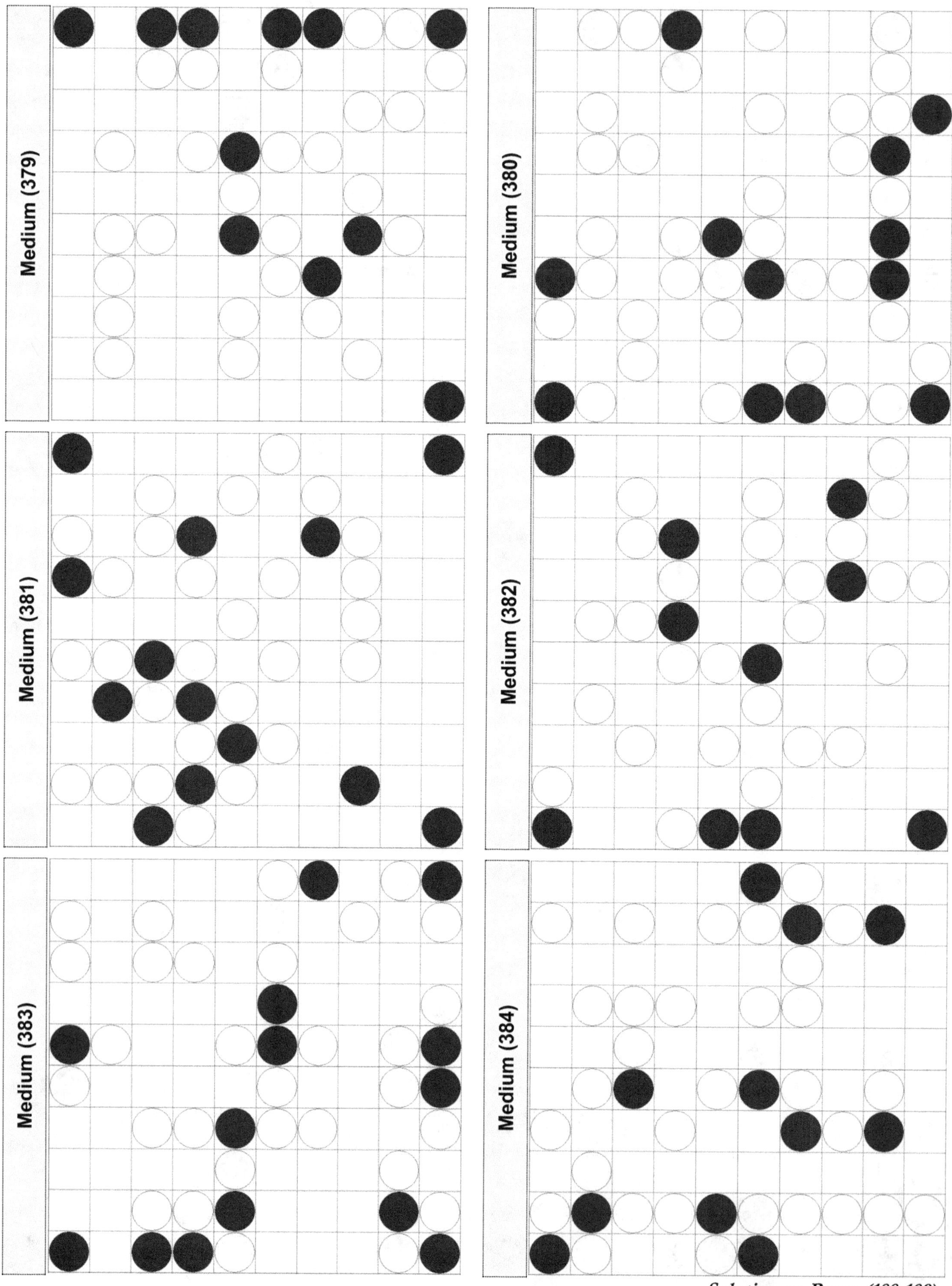

Medium (379)

Medium (380)

Medium (381)

Medium (382)

Medium (383)

Medium (384)

Solution on Pages (188-189)

(66)

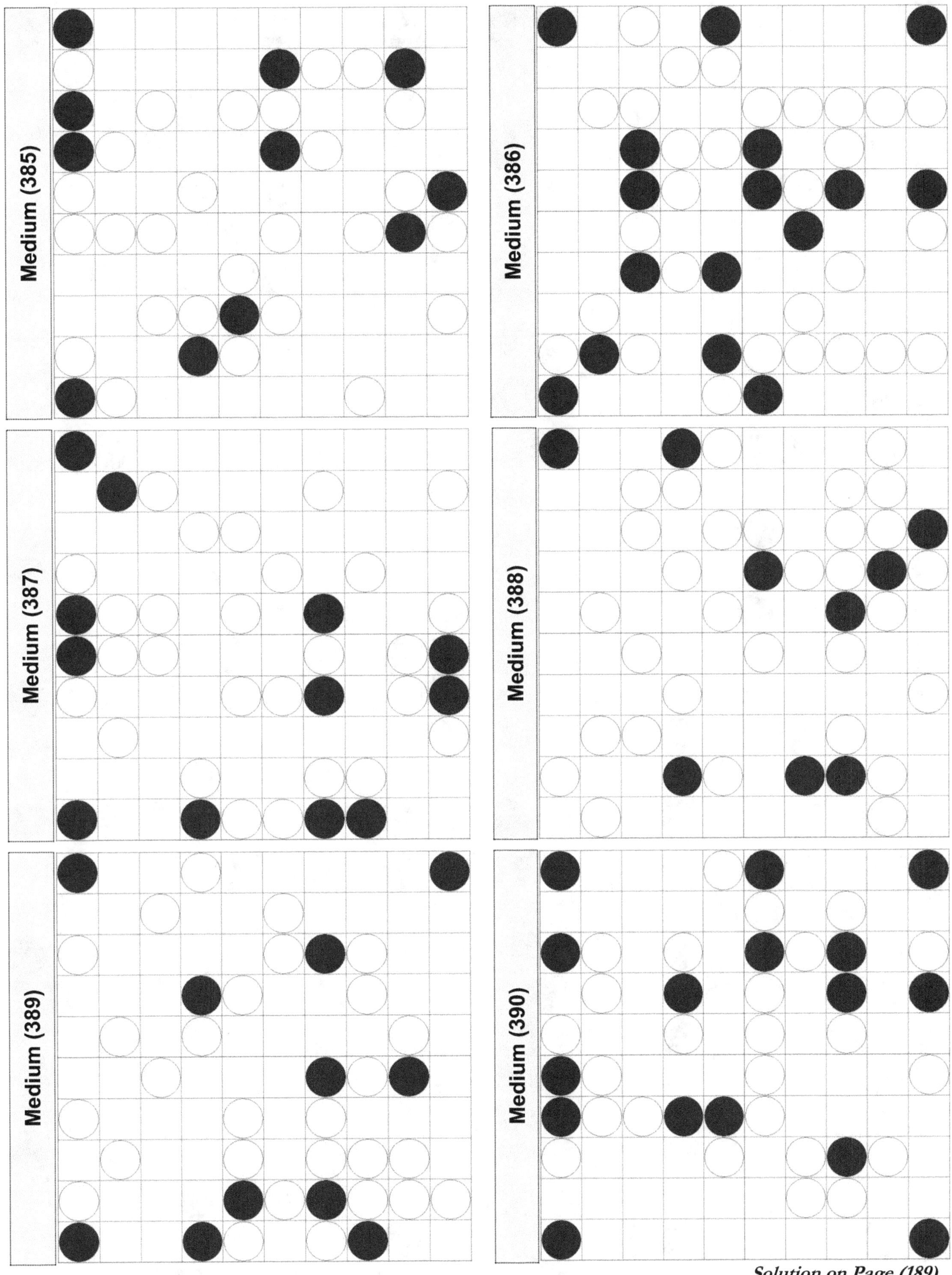

Solution on Page (189)

(67)

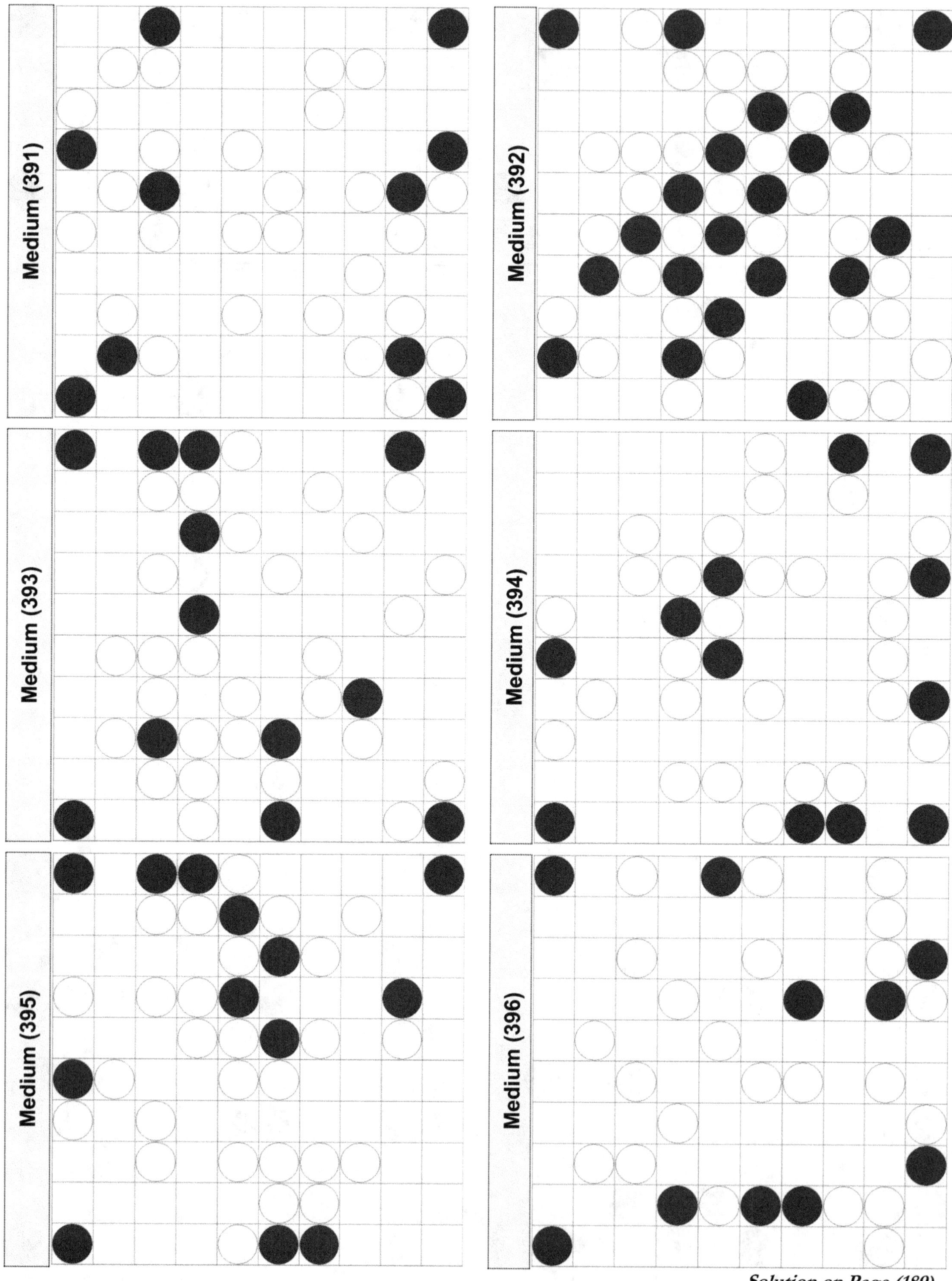

(68)

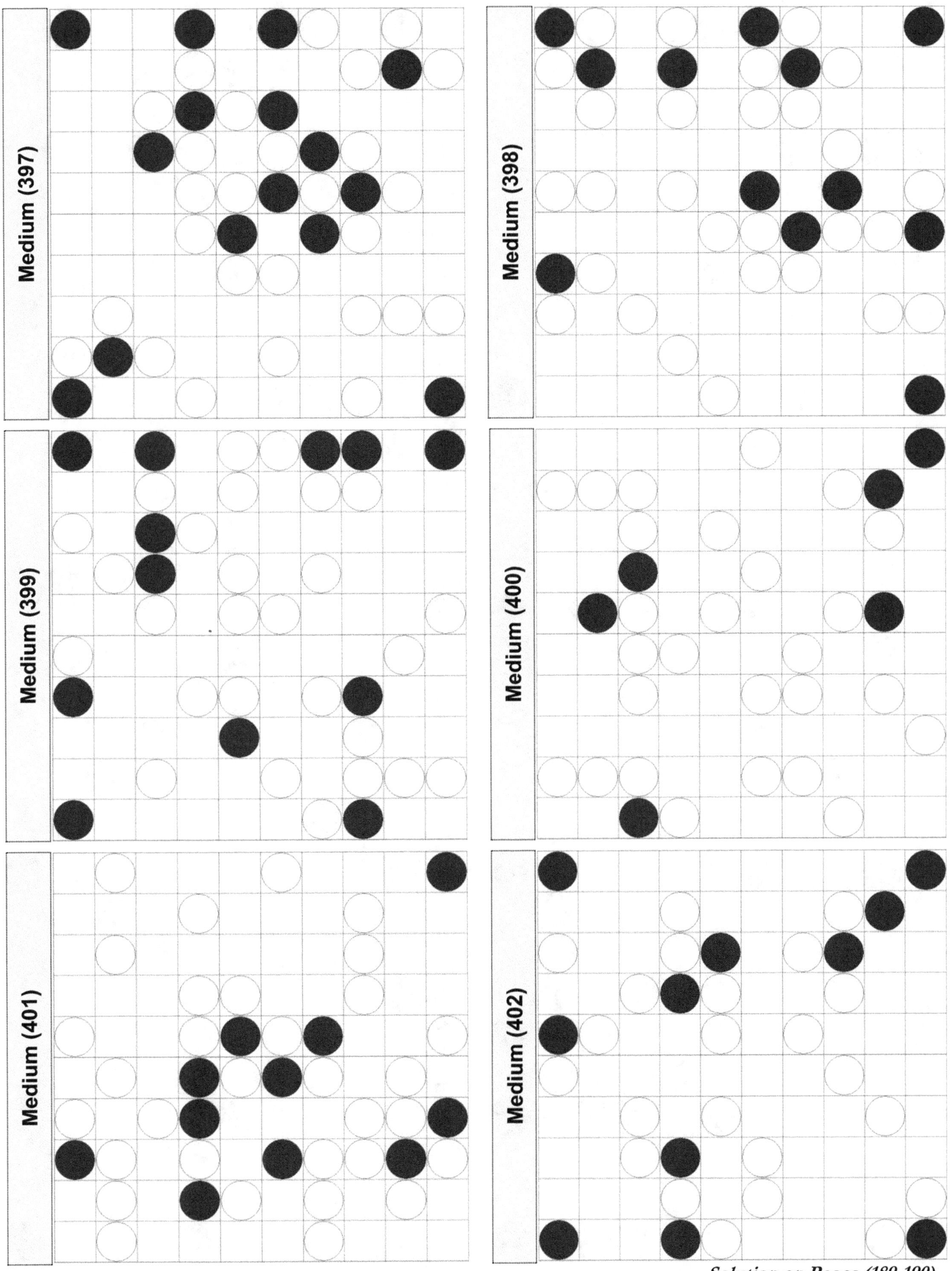

Solution on Pages (189-190)

(69)

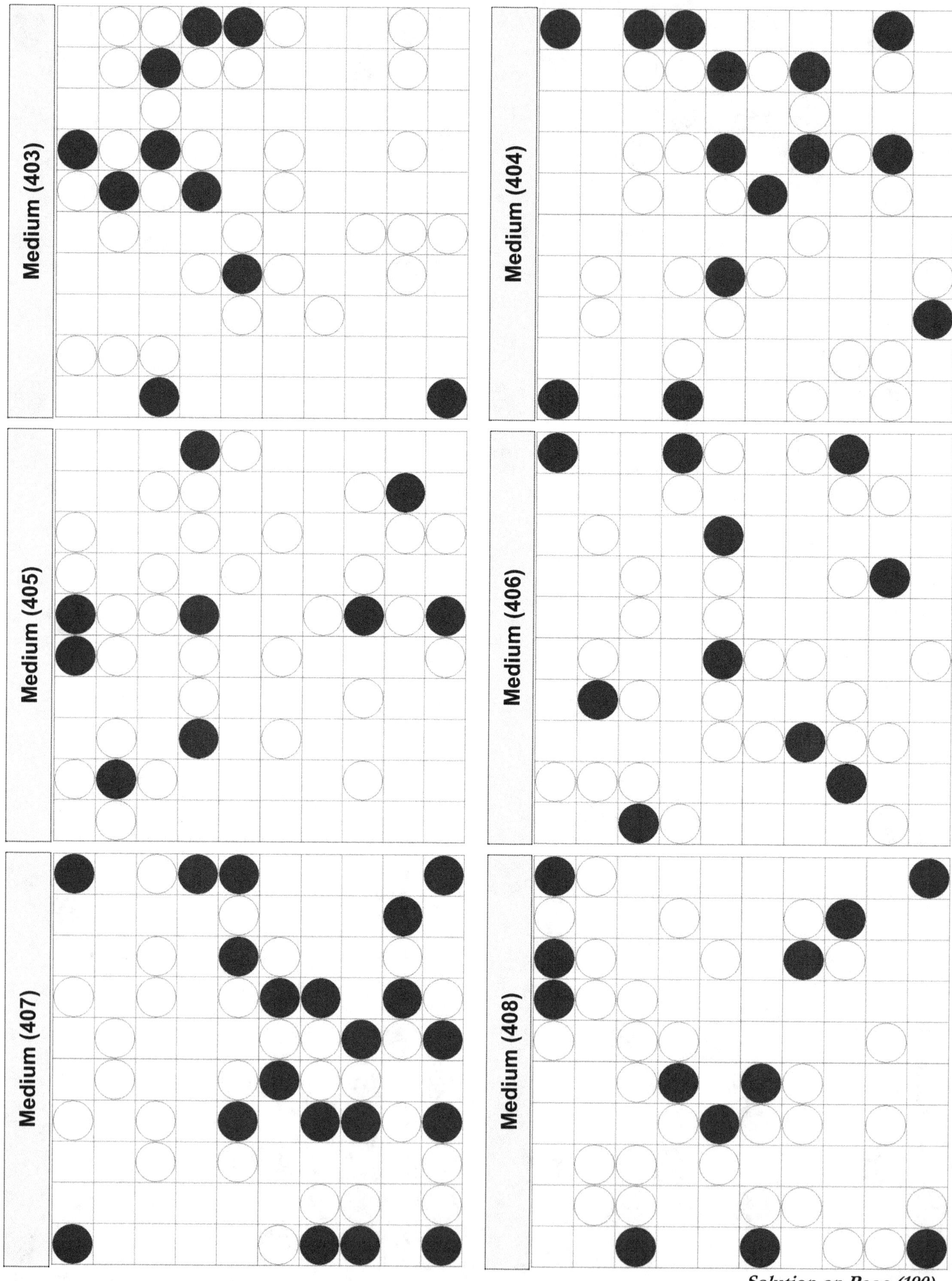

Solution on Page (190)

(70)

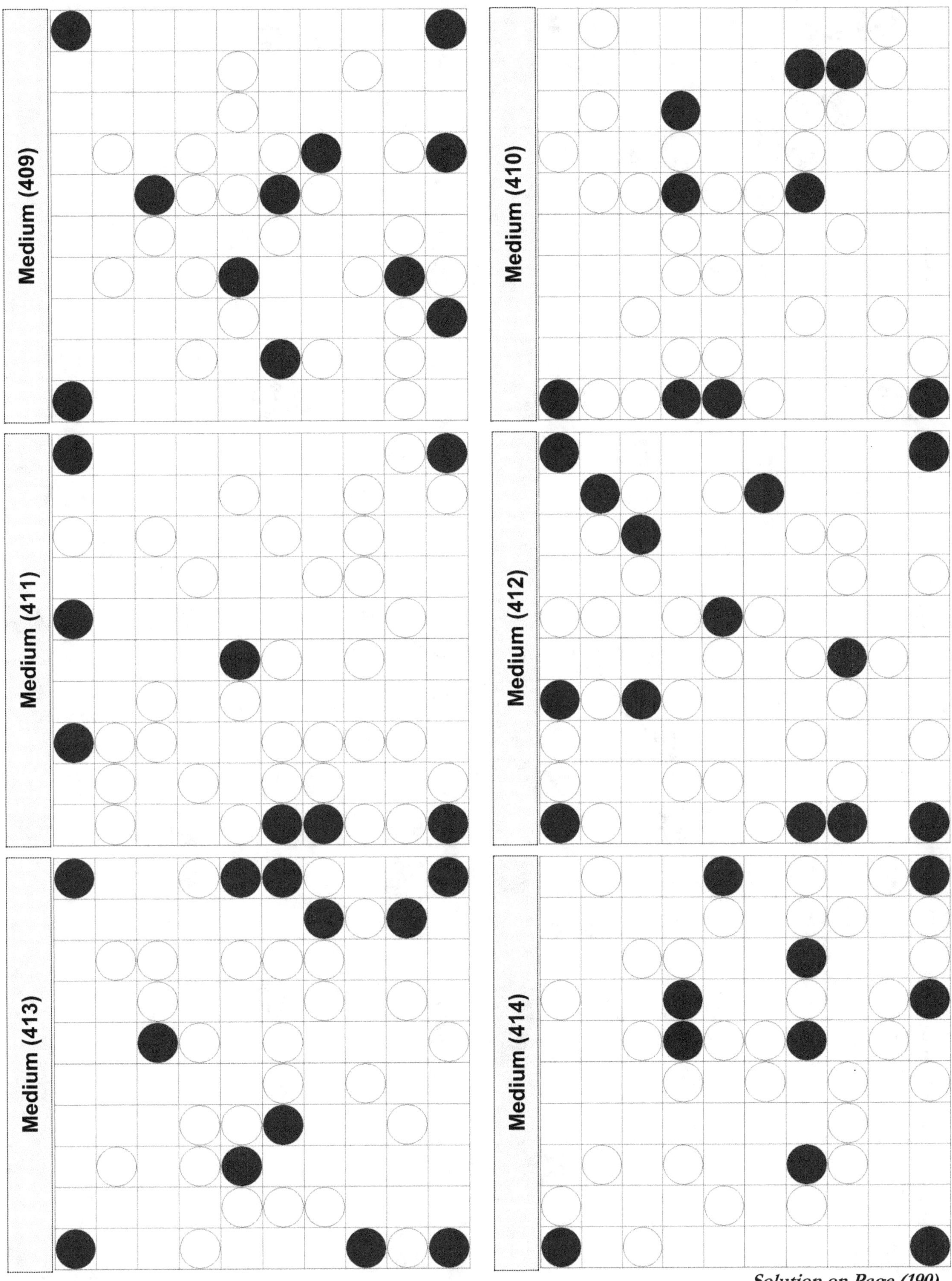

(71)

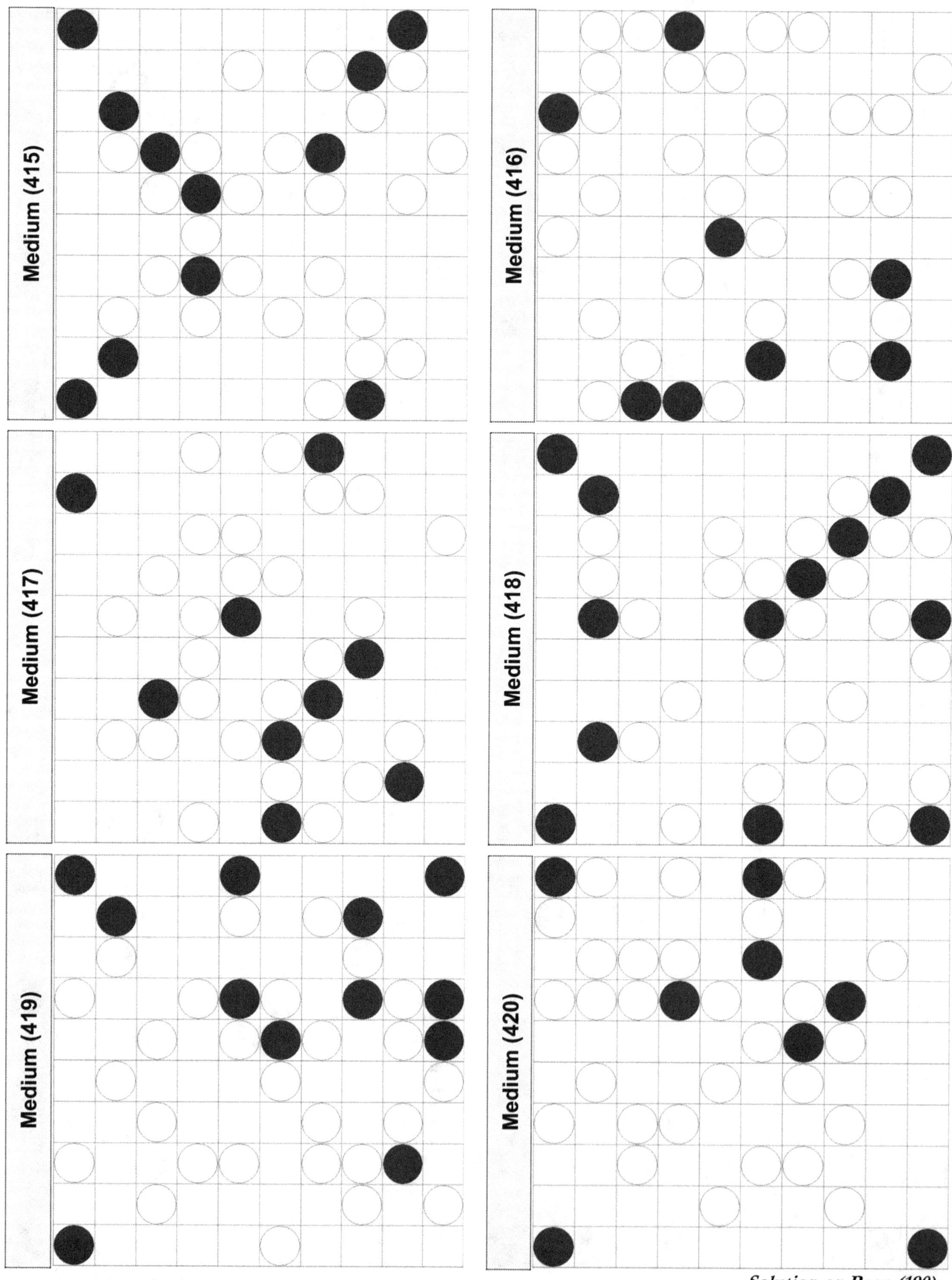

Medium (415)
Medium (416)
Medium (417)
Medium (418)
Medium (419)
Medium (420)
Solution on Page (190)

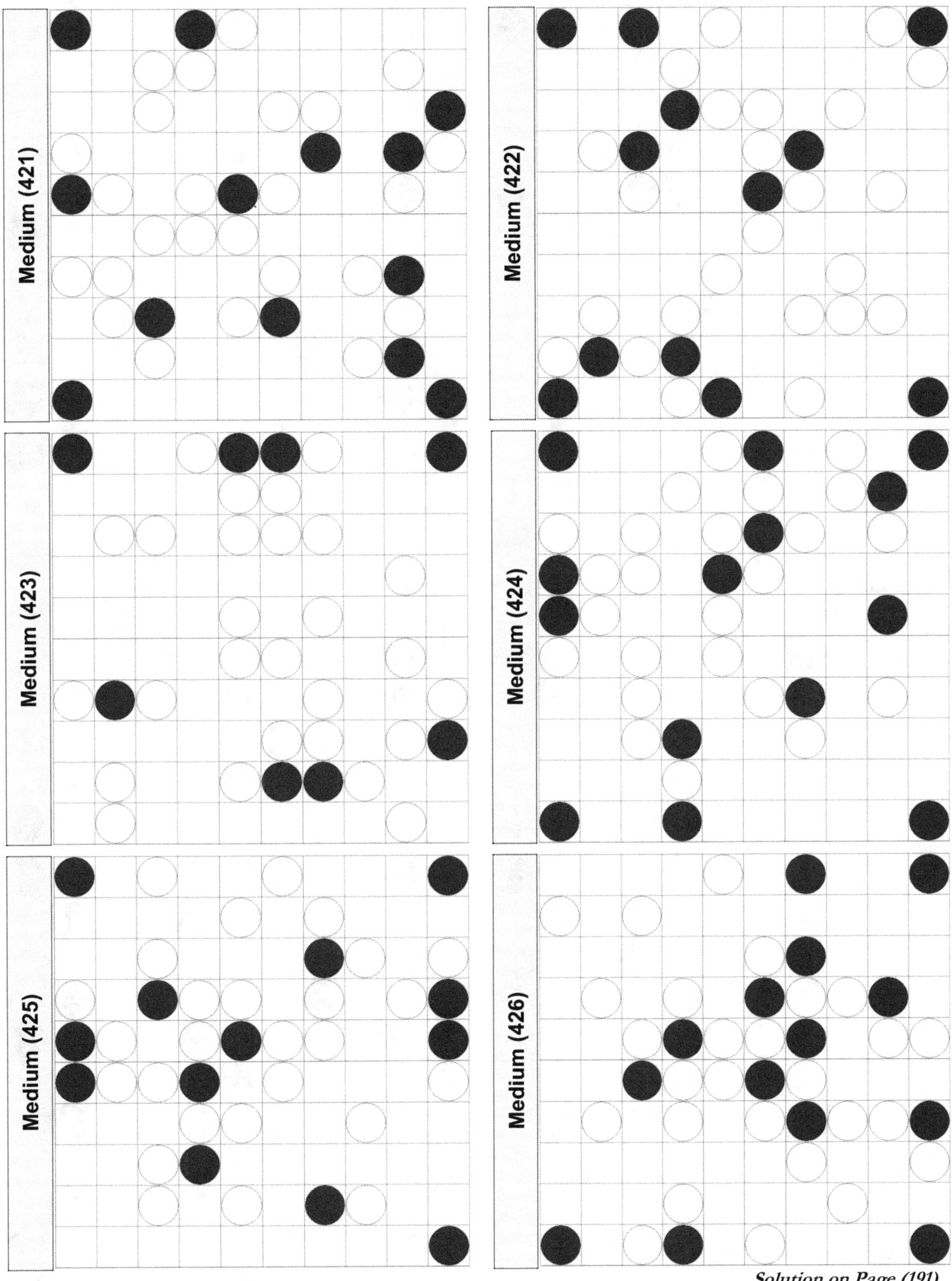

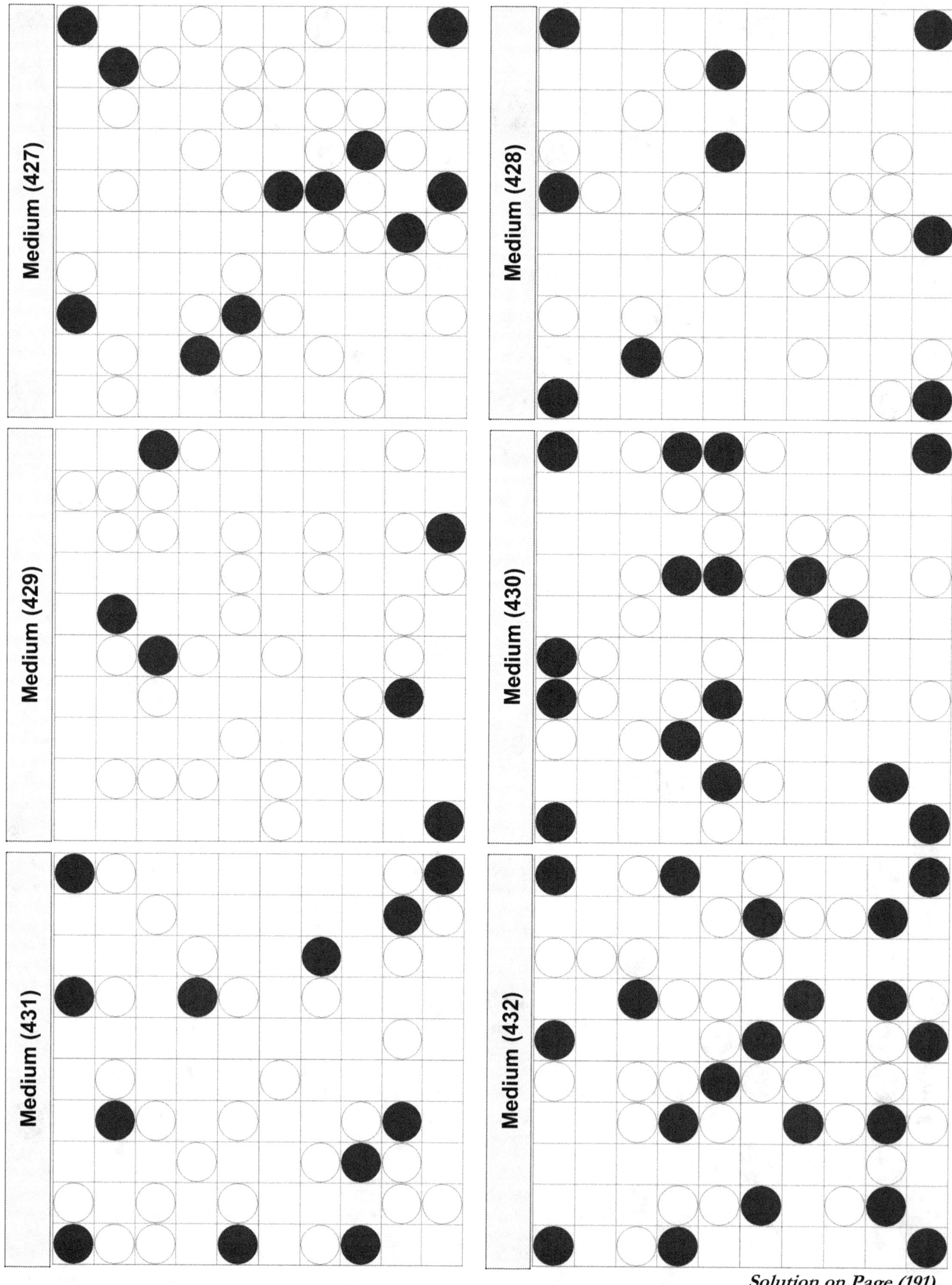

Solution on Page (191)

(74)

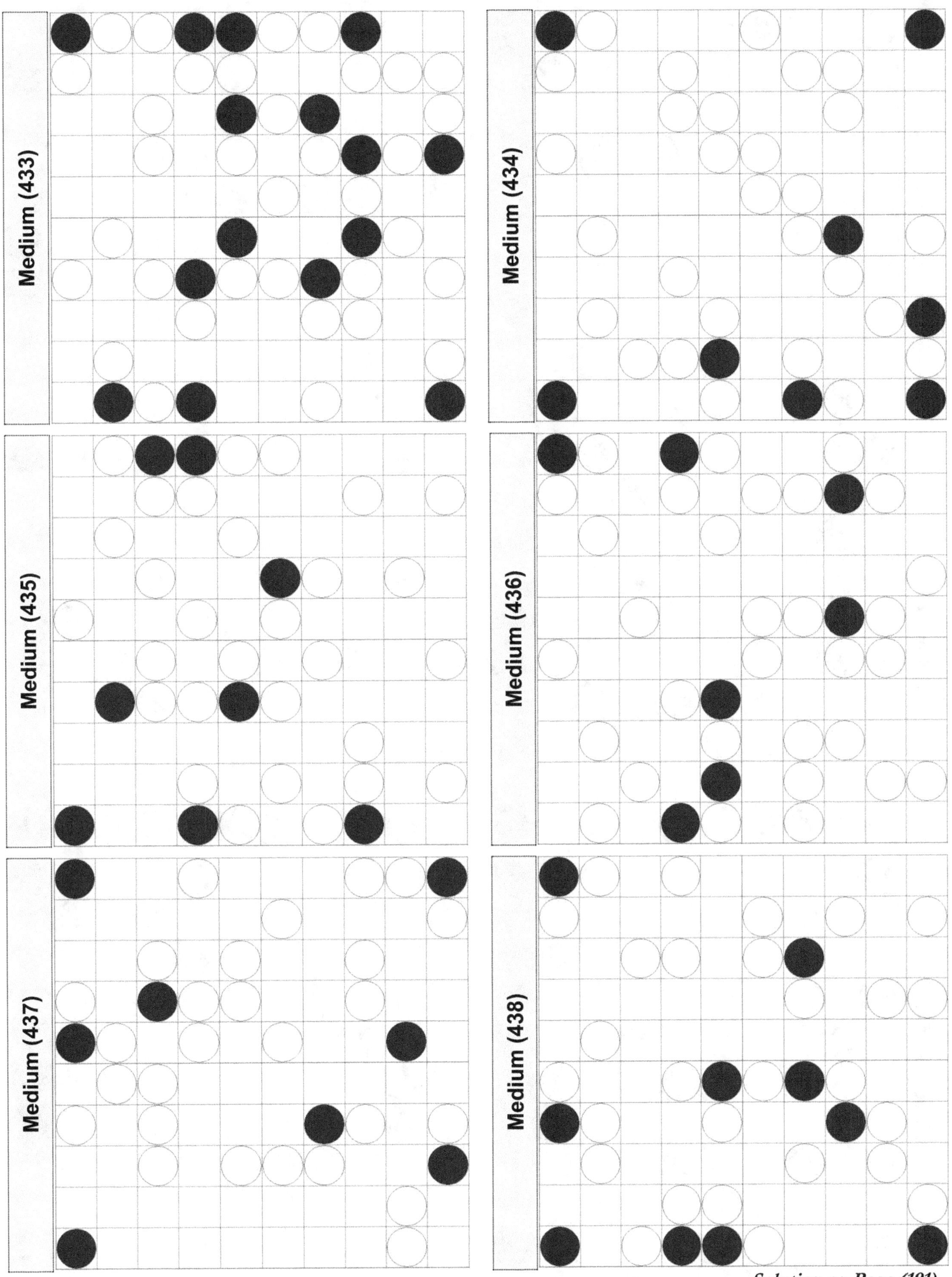

Medium (433)

Medium (434)

Medium (435)

Medium (436)

Medium (437)

Medium (438)

Solution on Page (191)

(75)

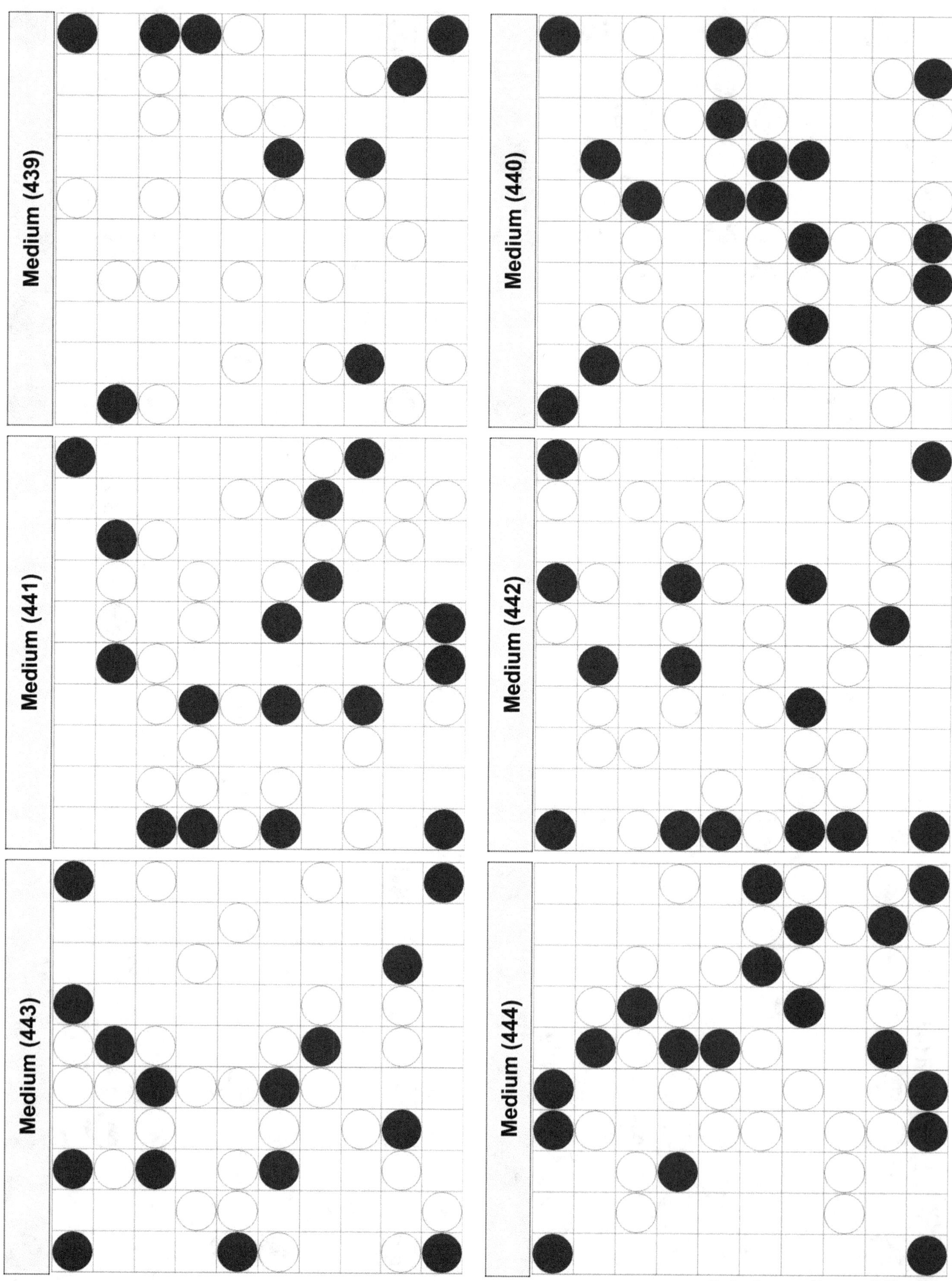

Solution on Pages (191-192)

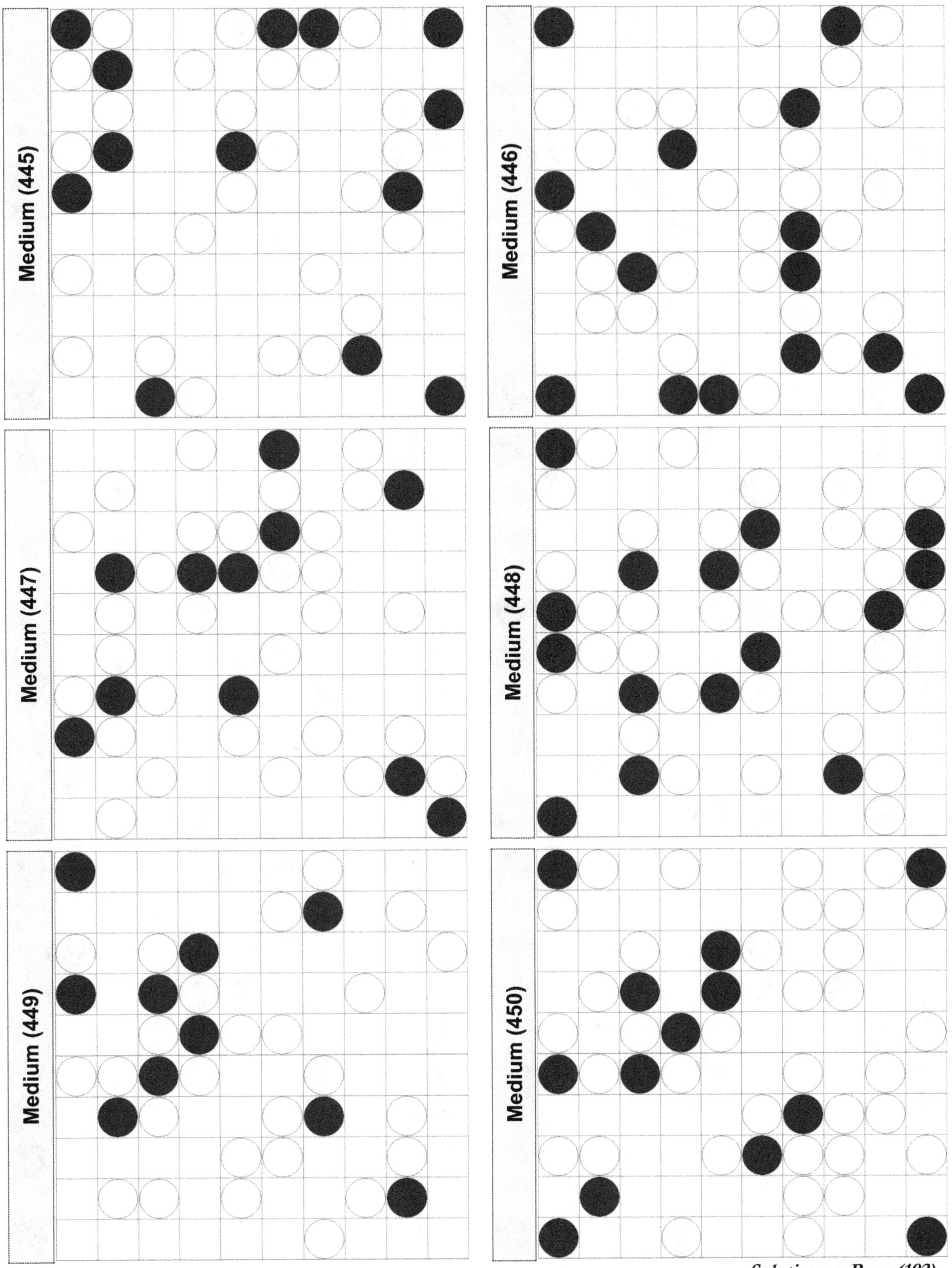

Solution on Page (192)

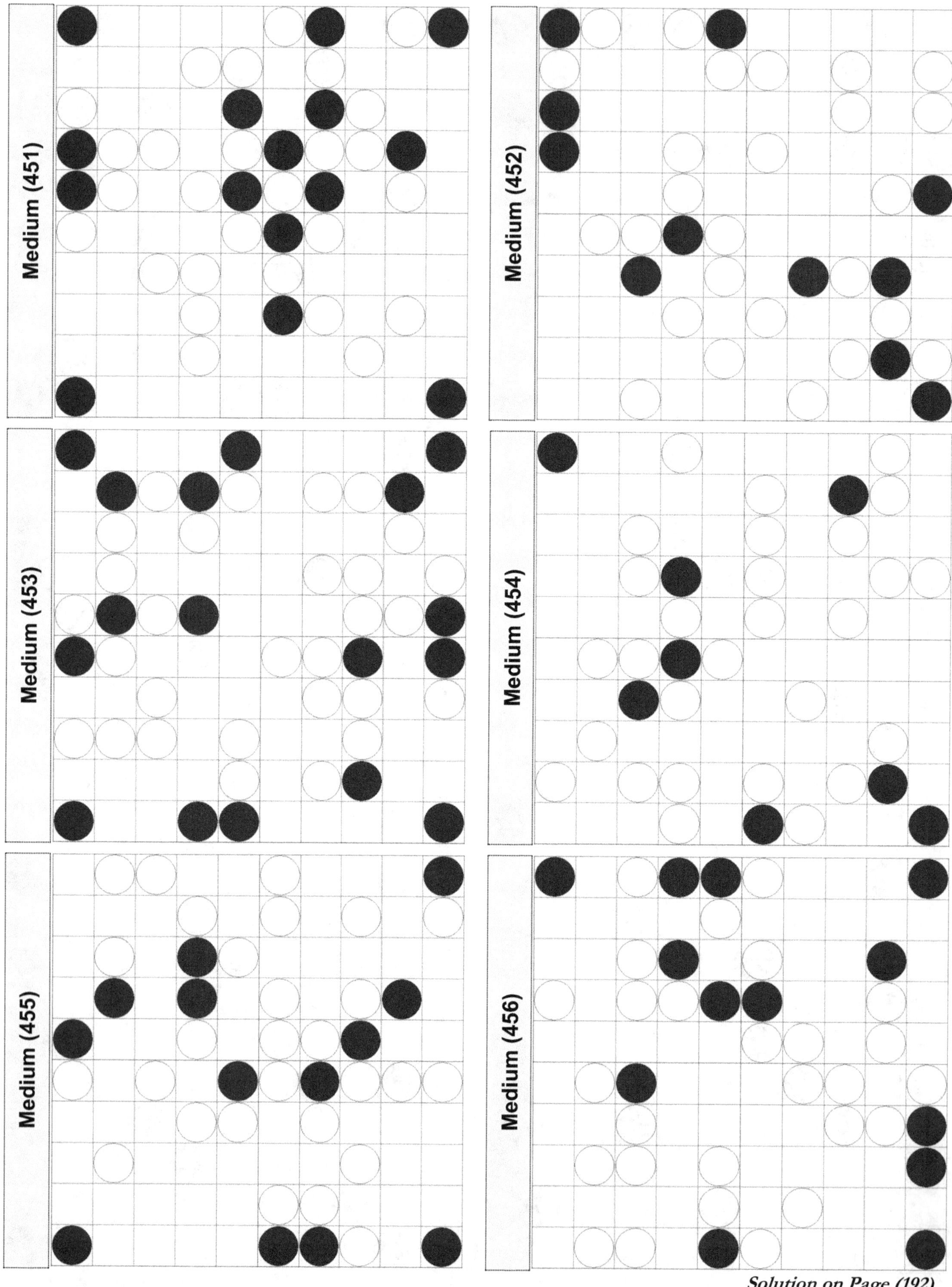

Medium (451)
Medium (452)
Medium (453)
Medium (454)
Medium (455)
Medium (456)
Solution on Page (192)

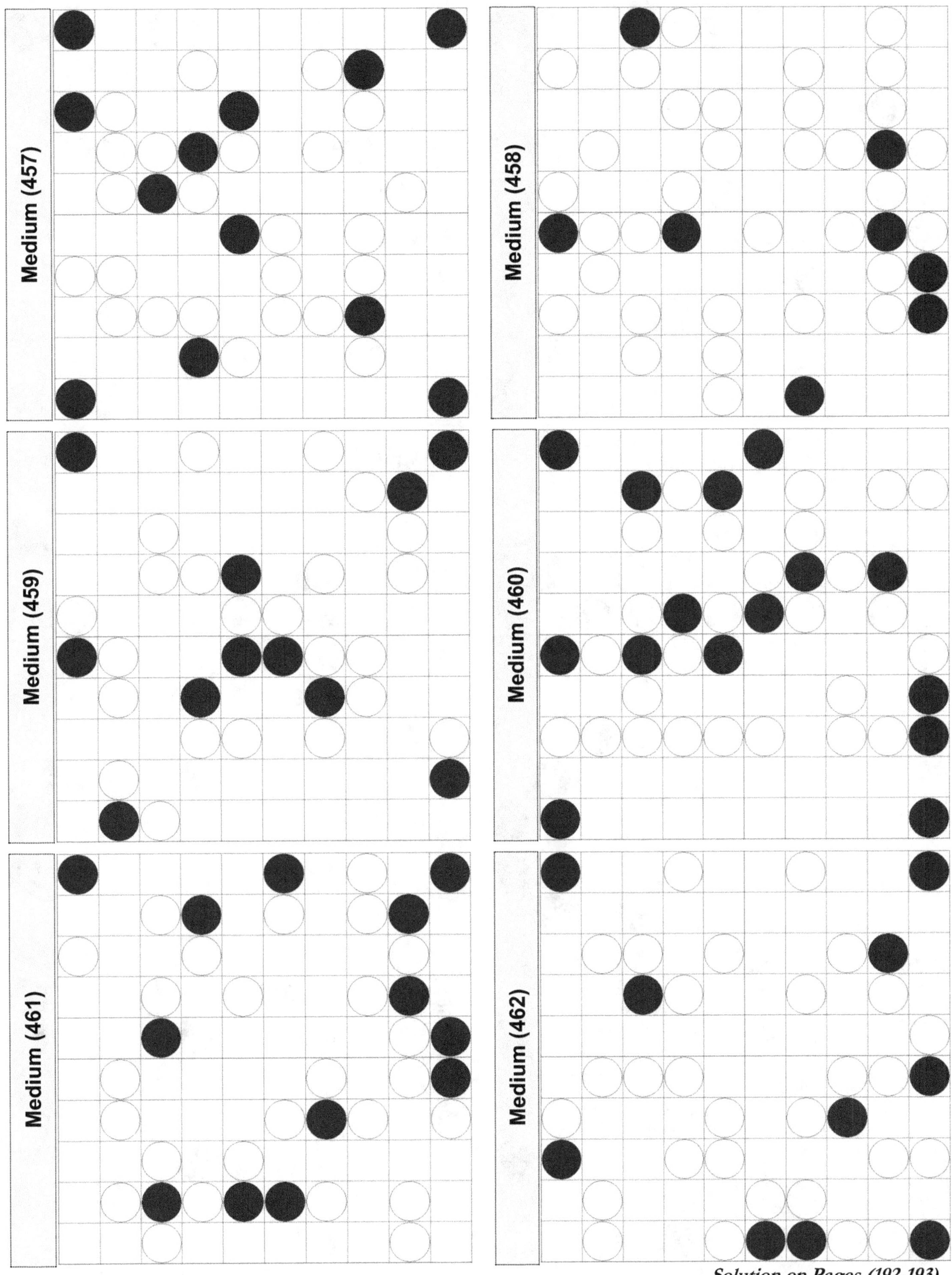

Solution on Pages (192-193)

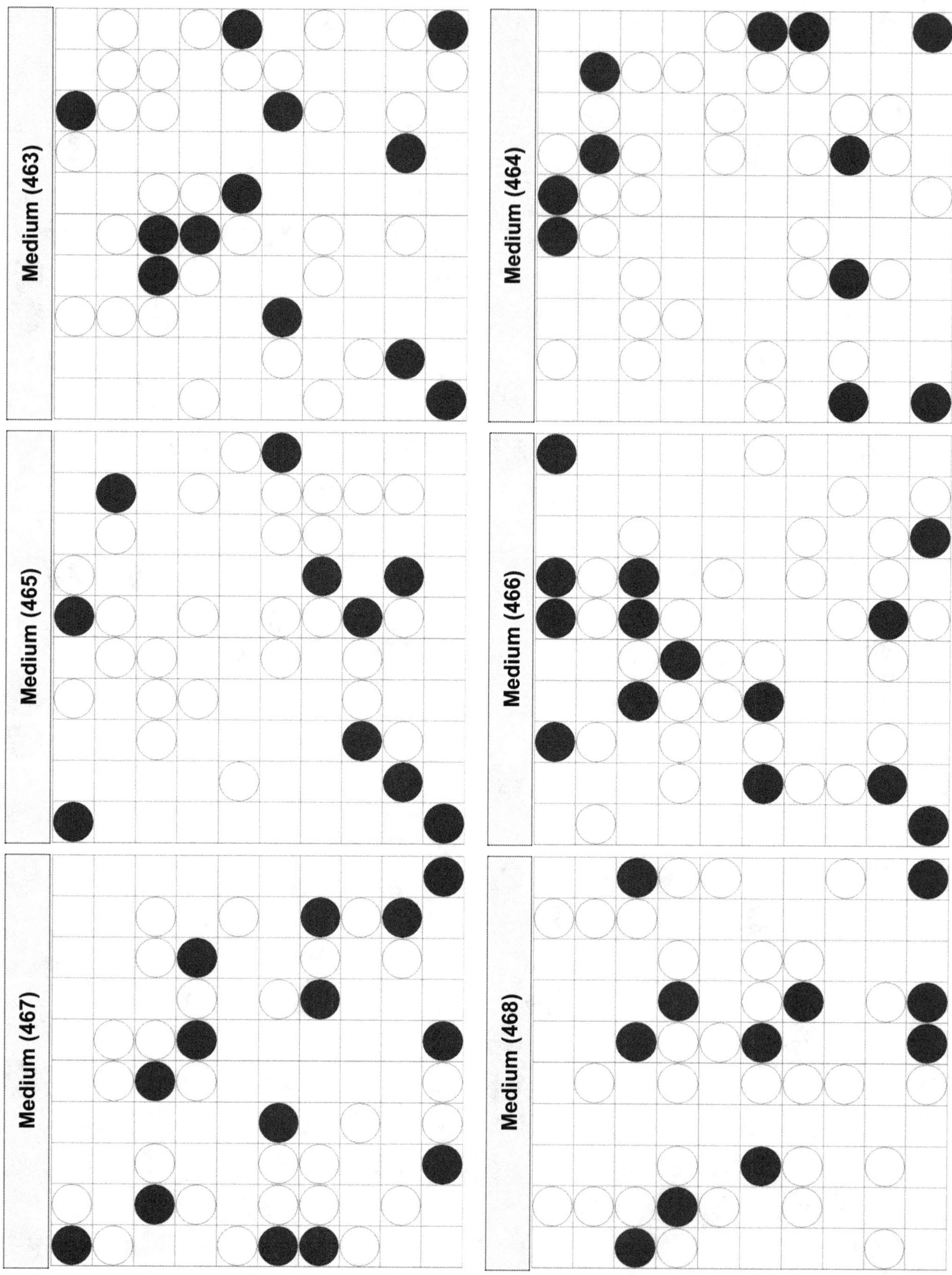

Medium (463)

Medium (464)

Medium (465)

Medium (466)

Medium (467)

Medium (468)

Solution on Page (193)

(80)

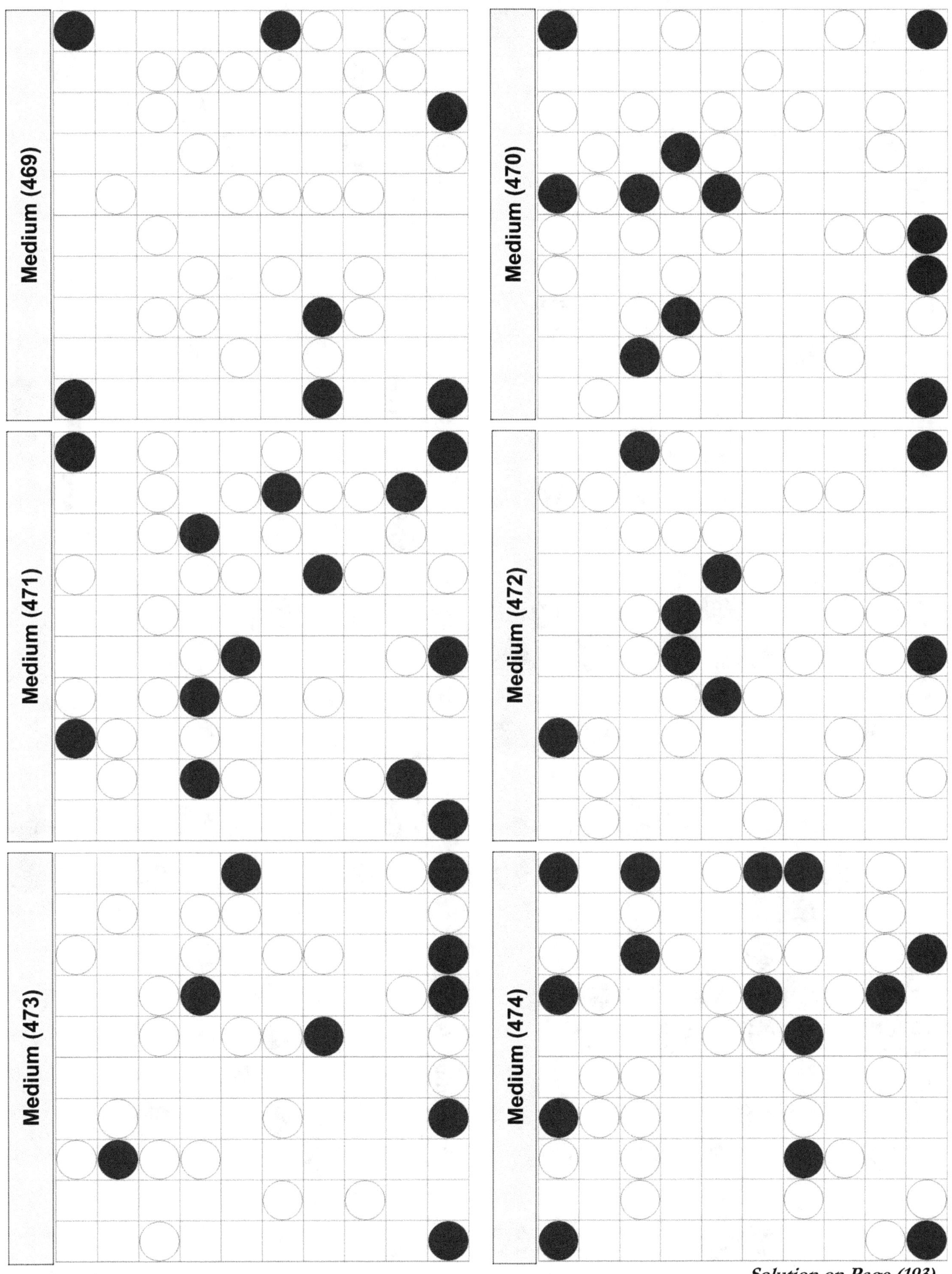

Solution on Page (193)

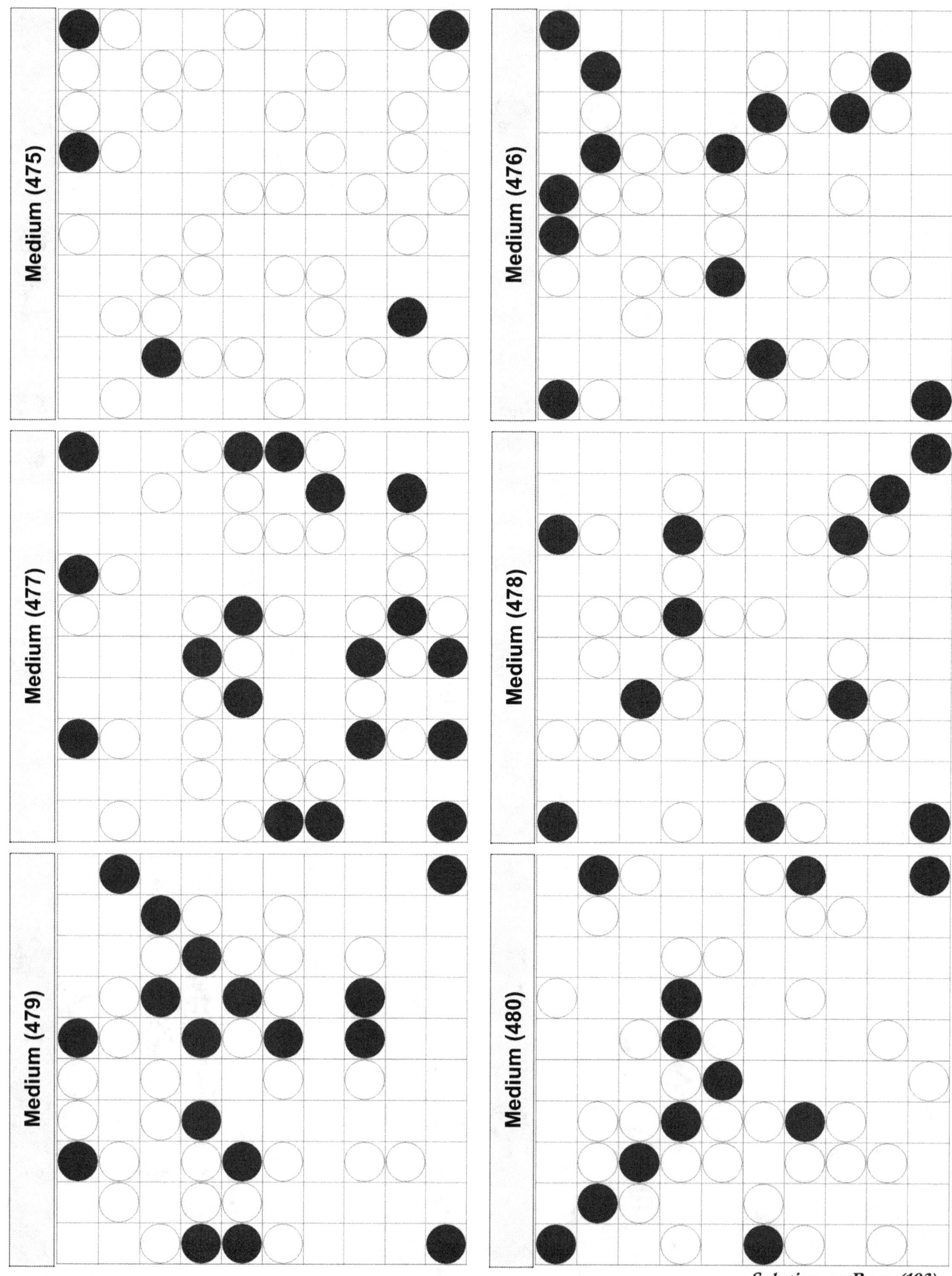

Solution on Page (193)

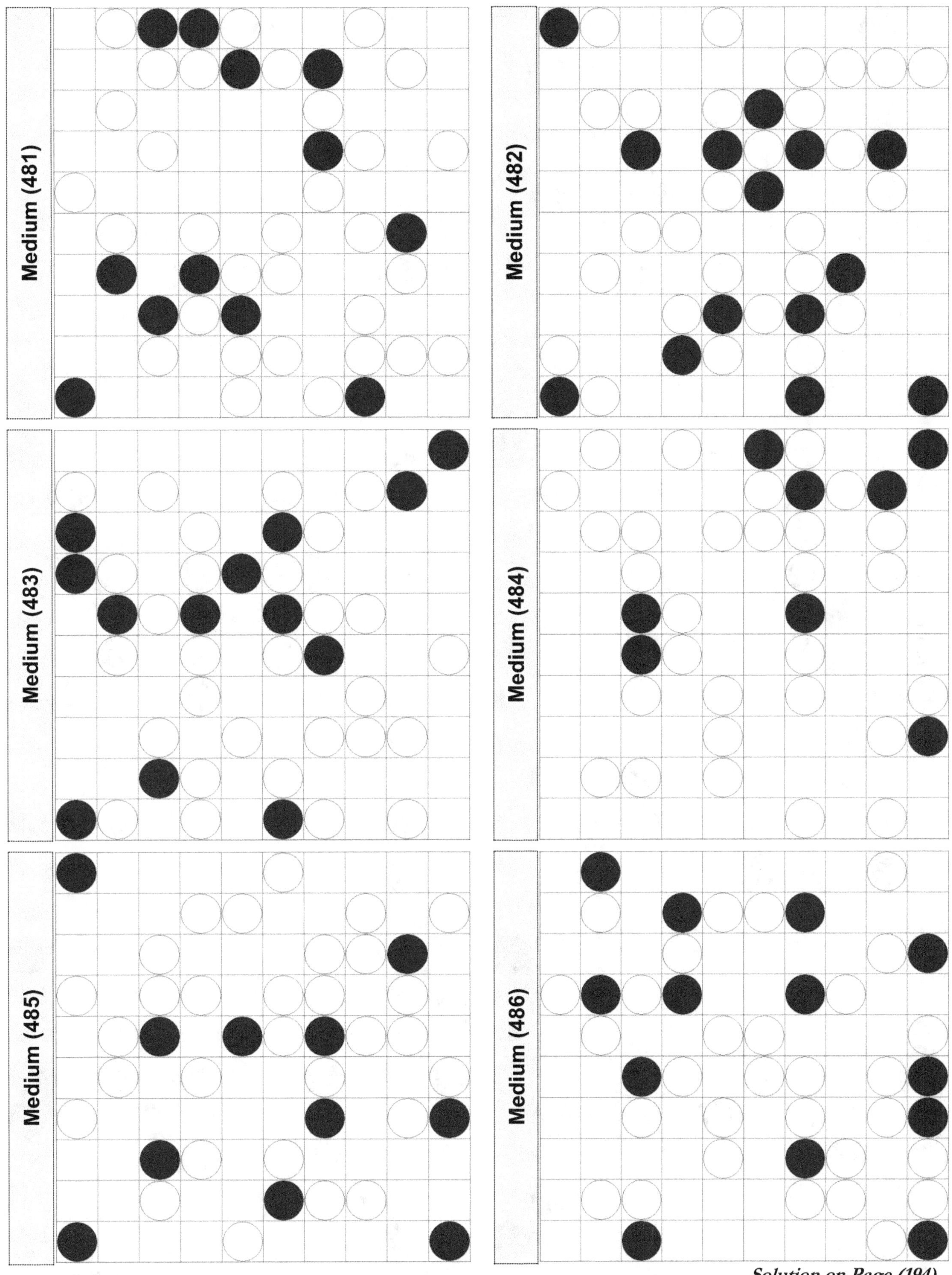

Medium (481)
Medium (482)
Medium (483)
Medium (484)
Medium (485)
Medium (486)
Solution on Page (194)

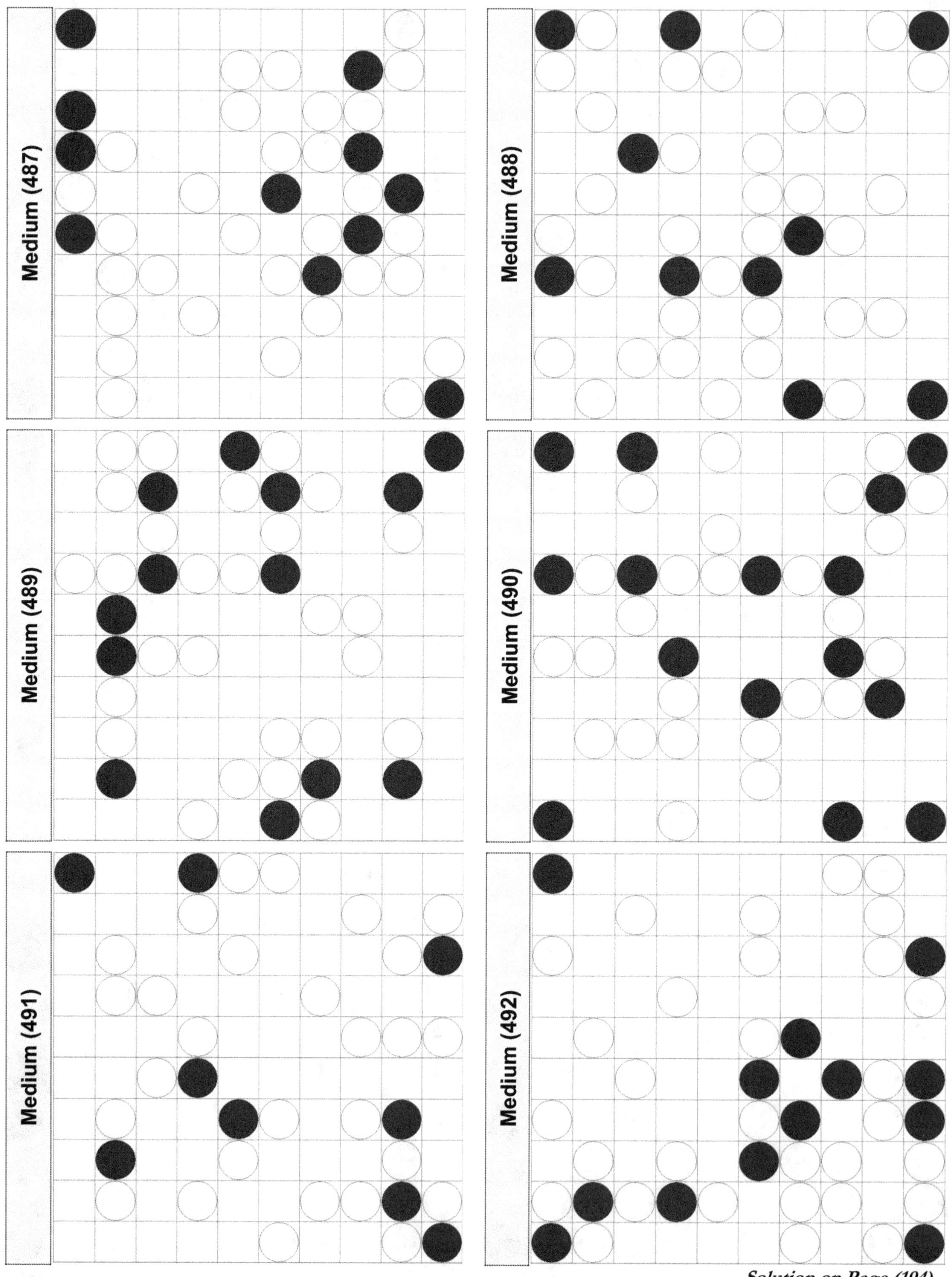

Solution on Page (194)

(84)

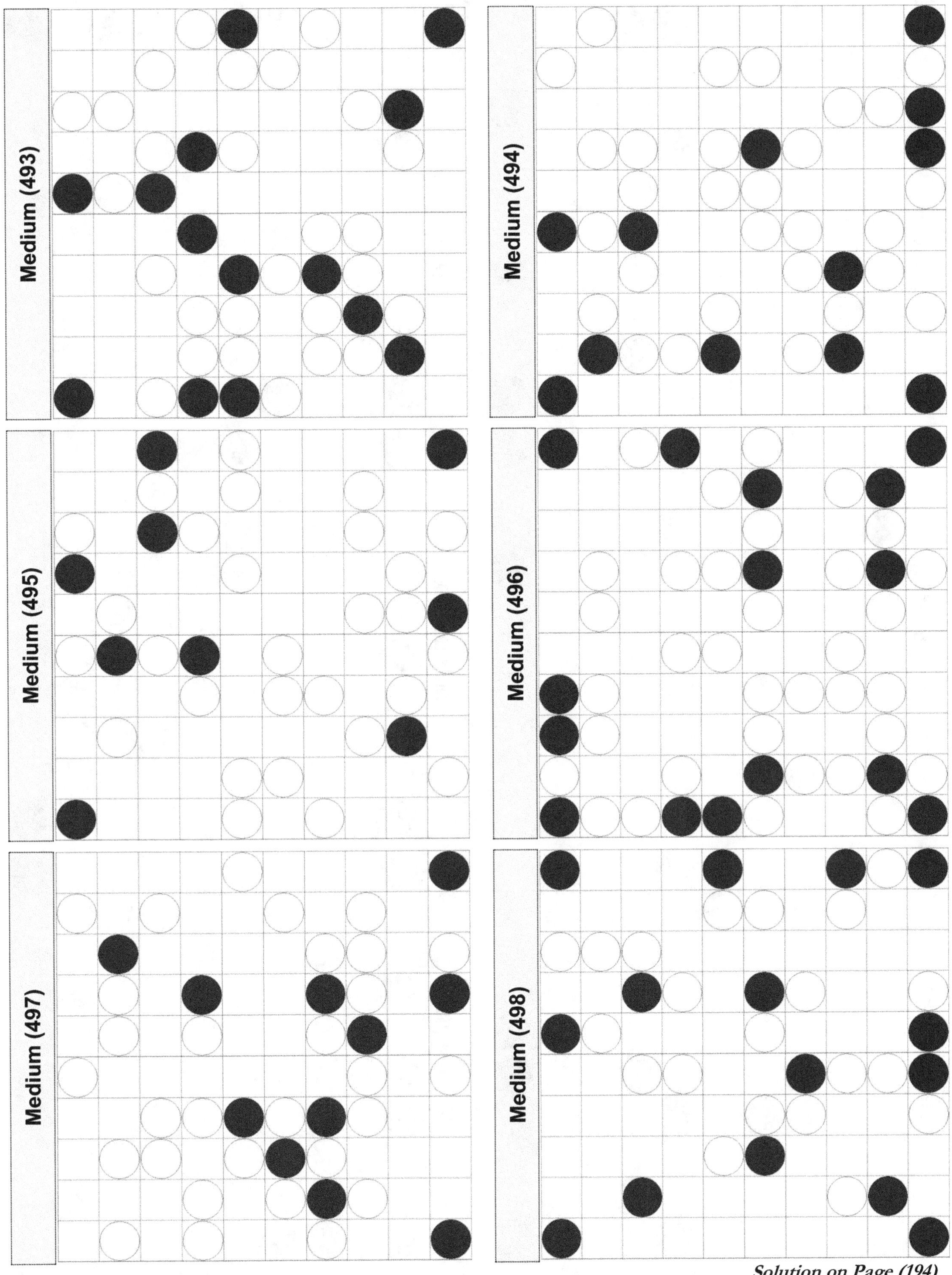

Solution on Page (194)

(85)

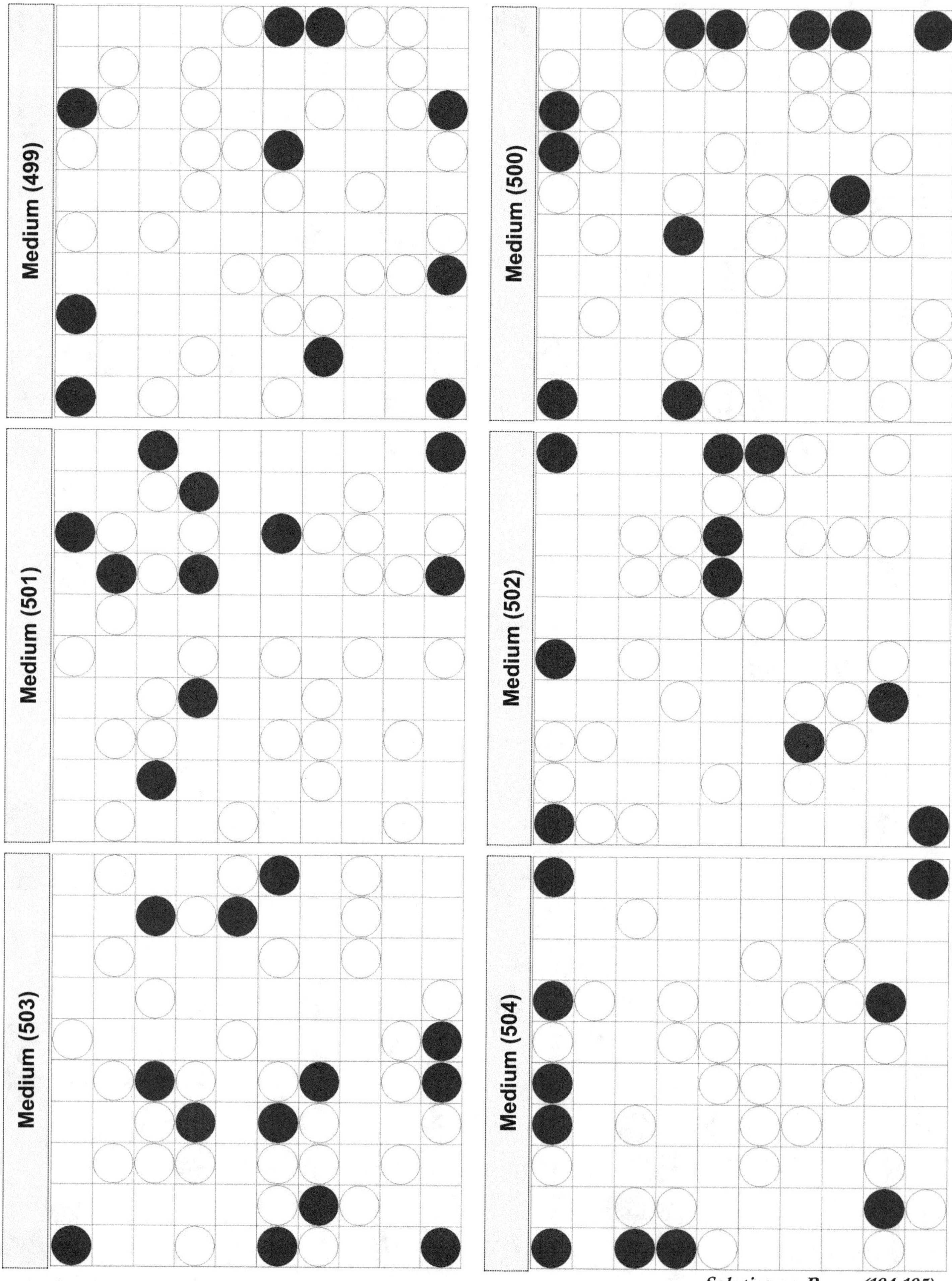

Solution on Pages (194-195)

(86)

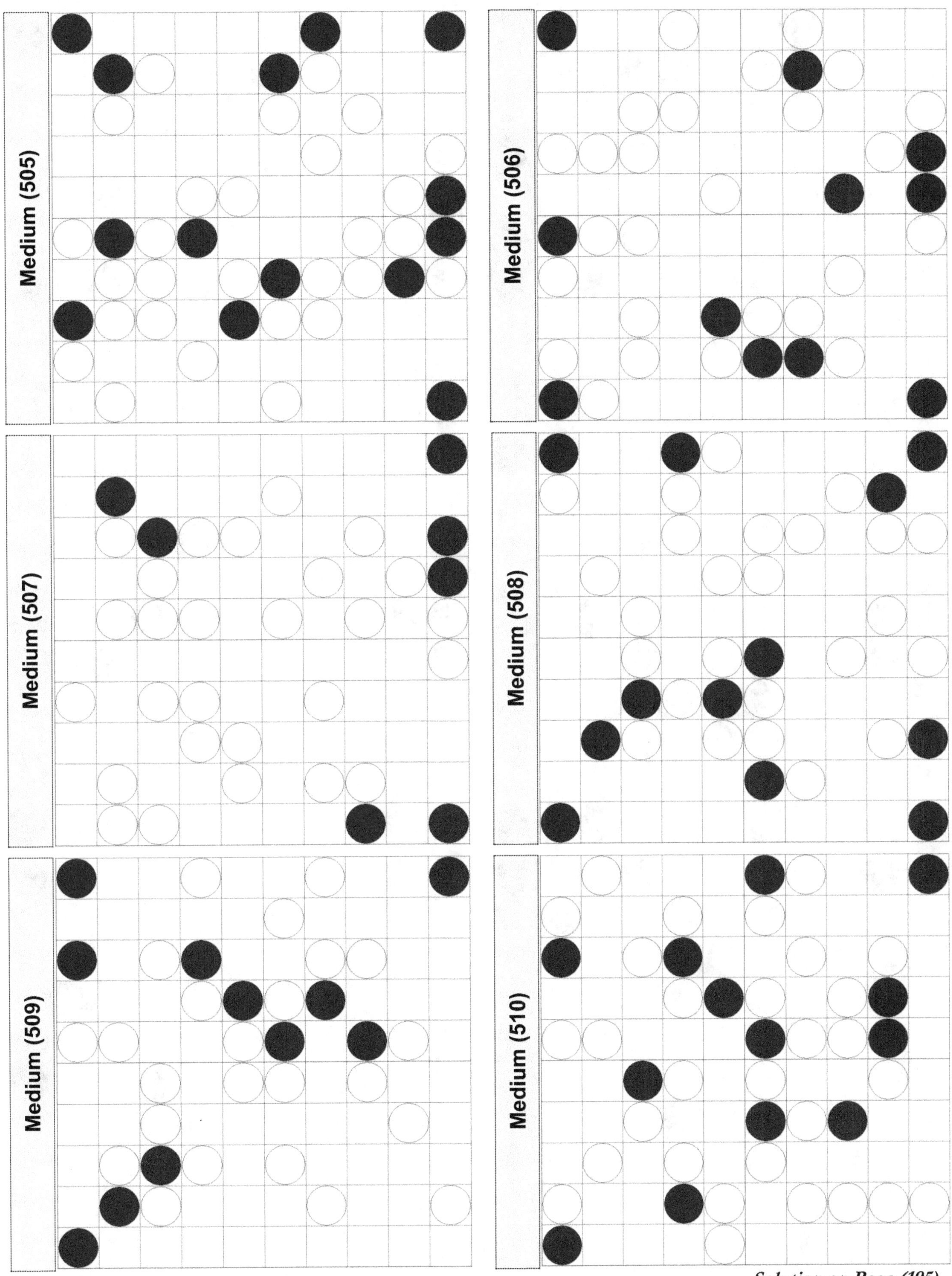

Solution on Page (195)

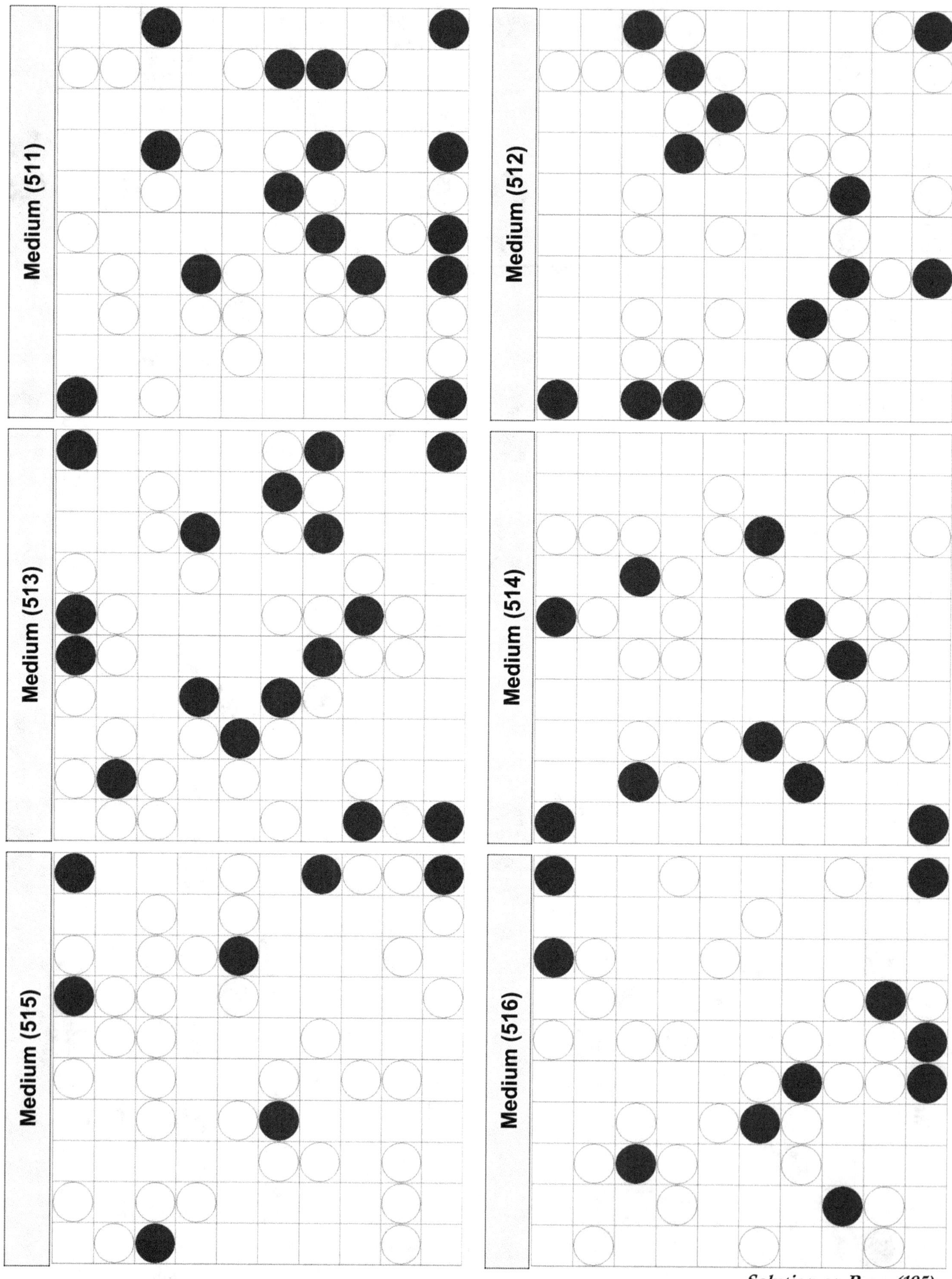

Solution on Page (195)

(88)

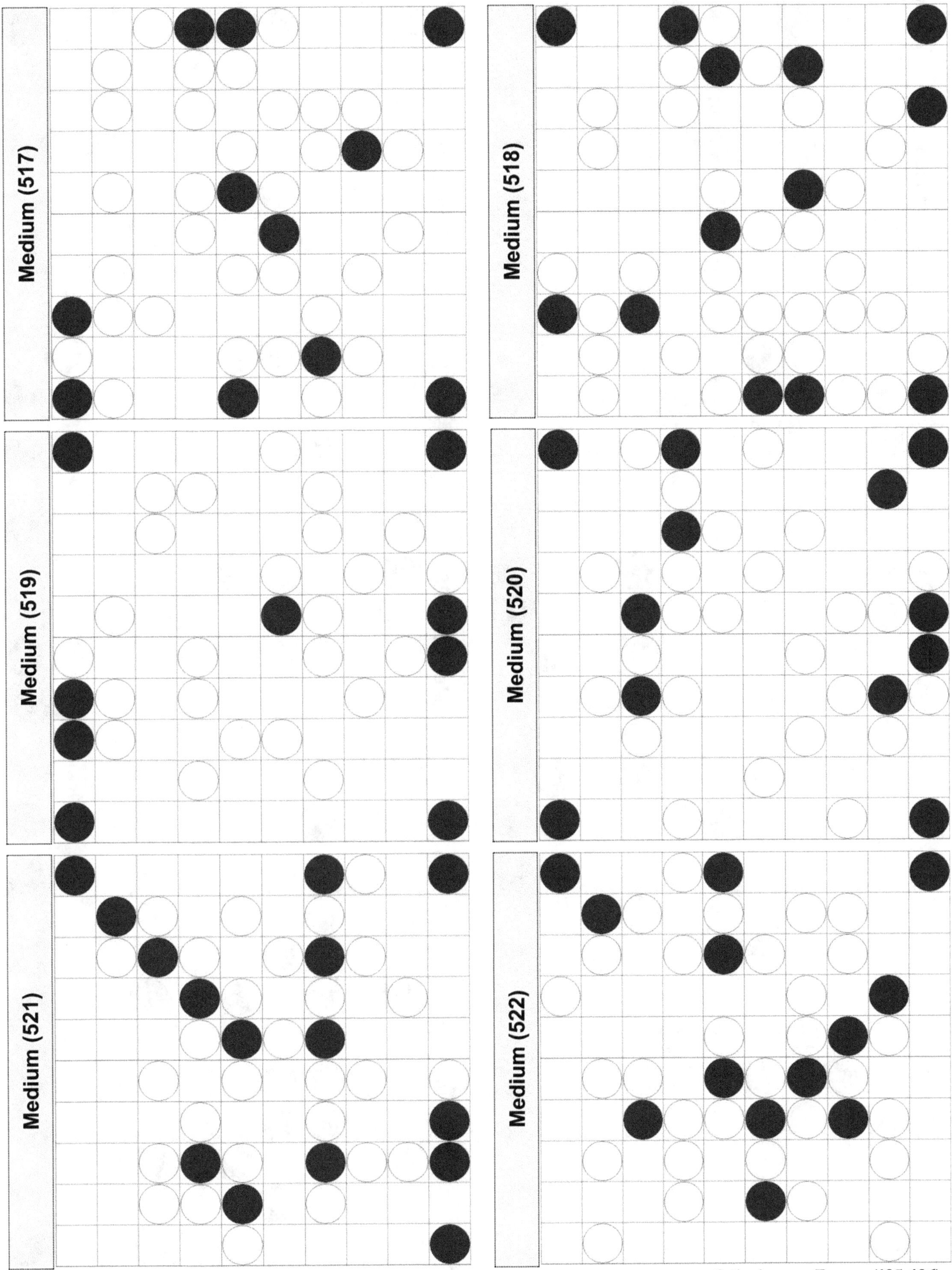

Solution on Pages (195-196)

(89)

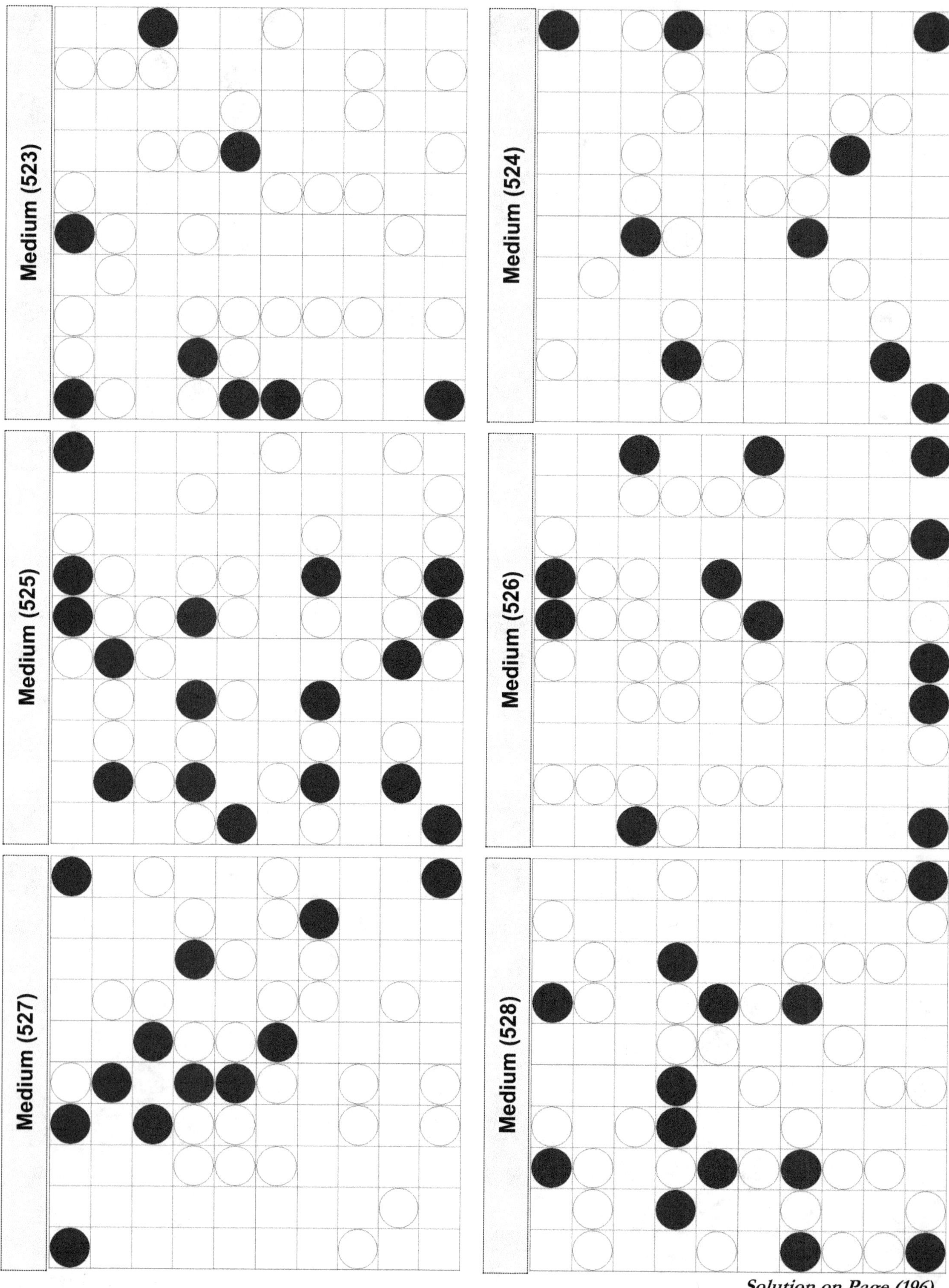

Medium (523)
Medium (524)
Medium (525)
Medium (526)
Medium (527)
Medium (528)
Solution on Page (196)

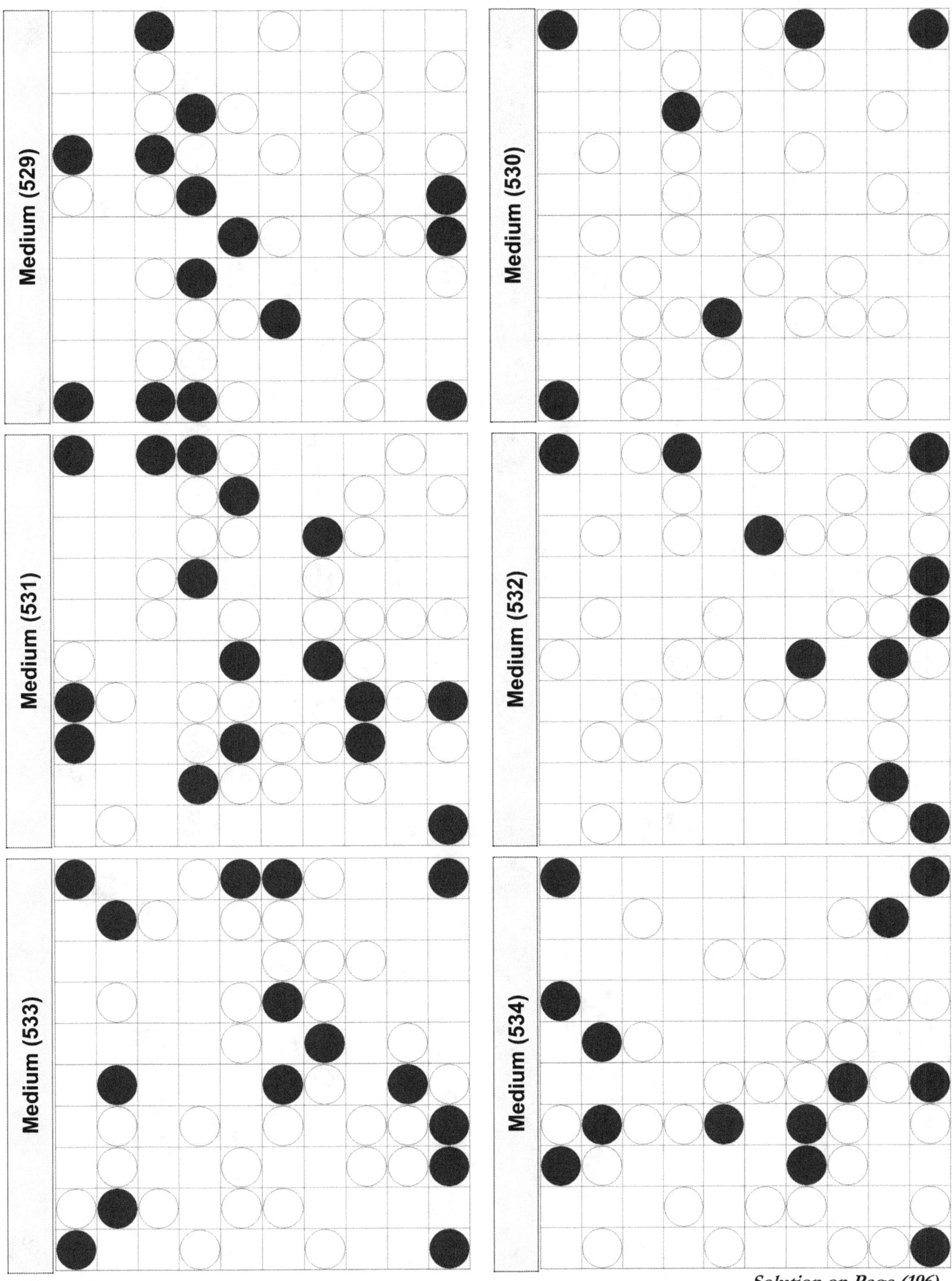

Solution on Page (196)

(91)

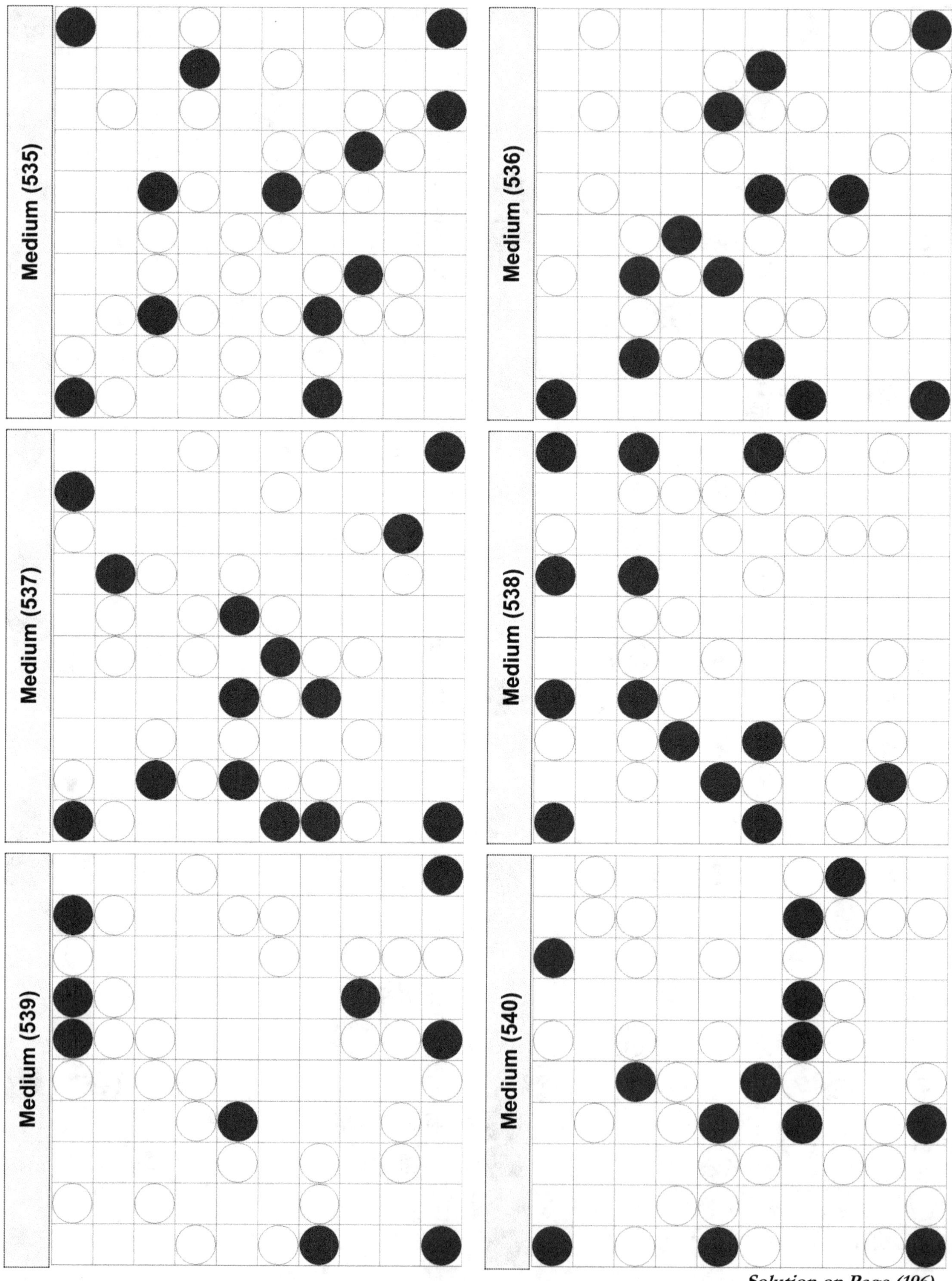

Solution on Page (196)

(92)

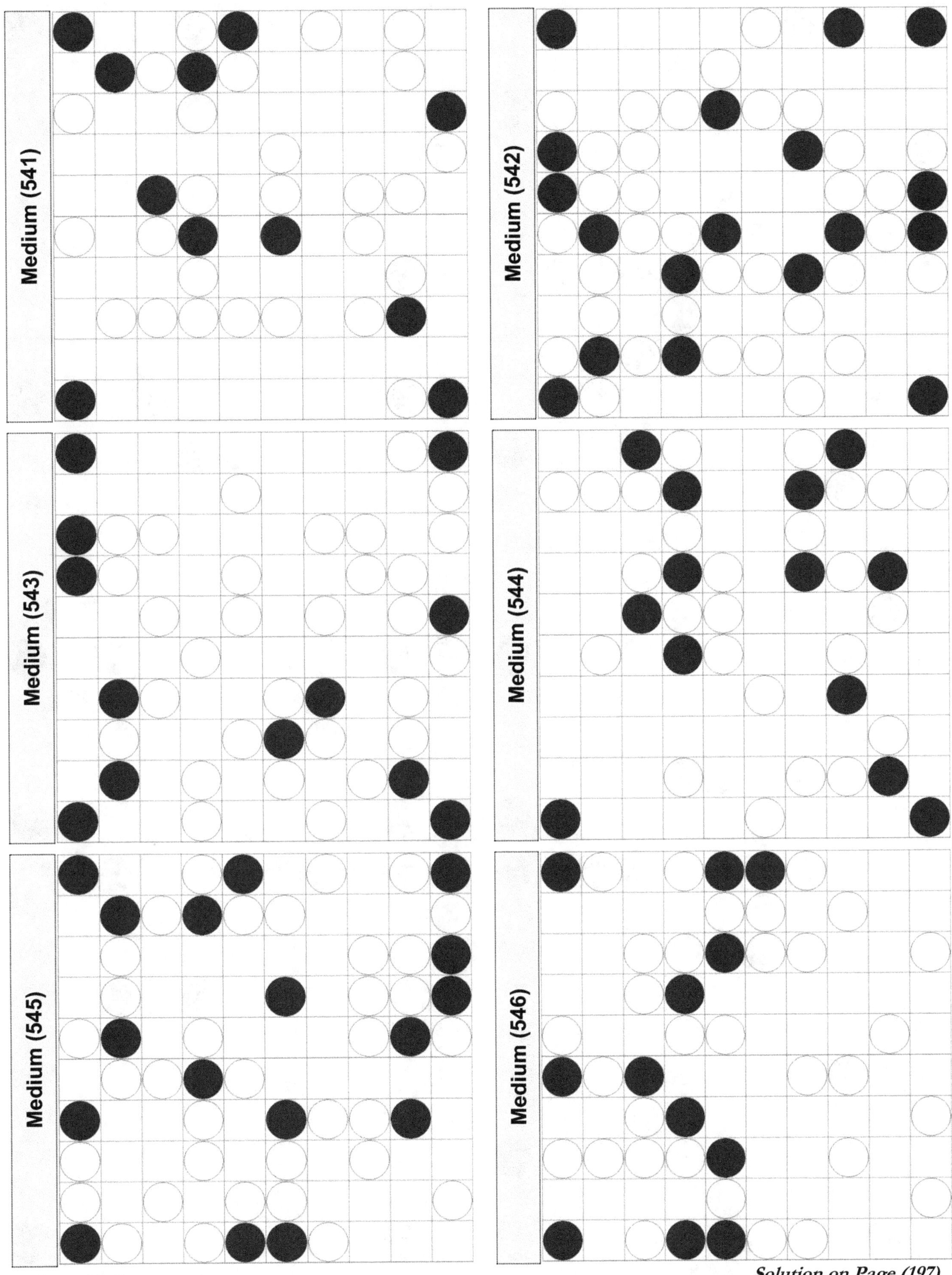

Solution on Page (197)

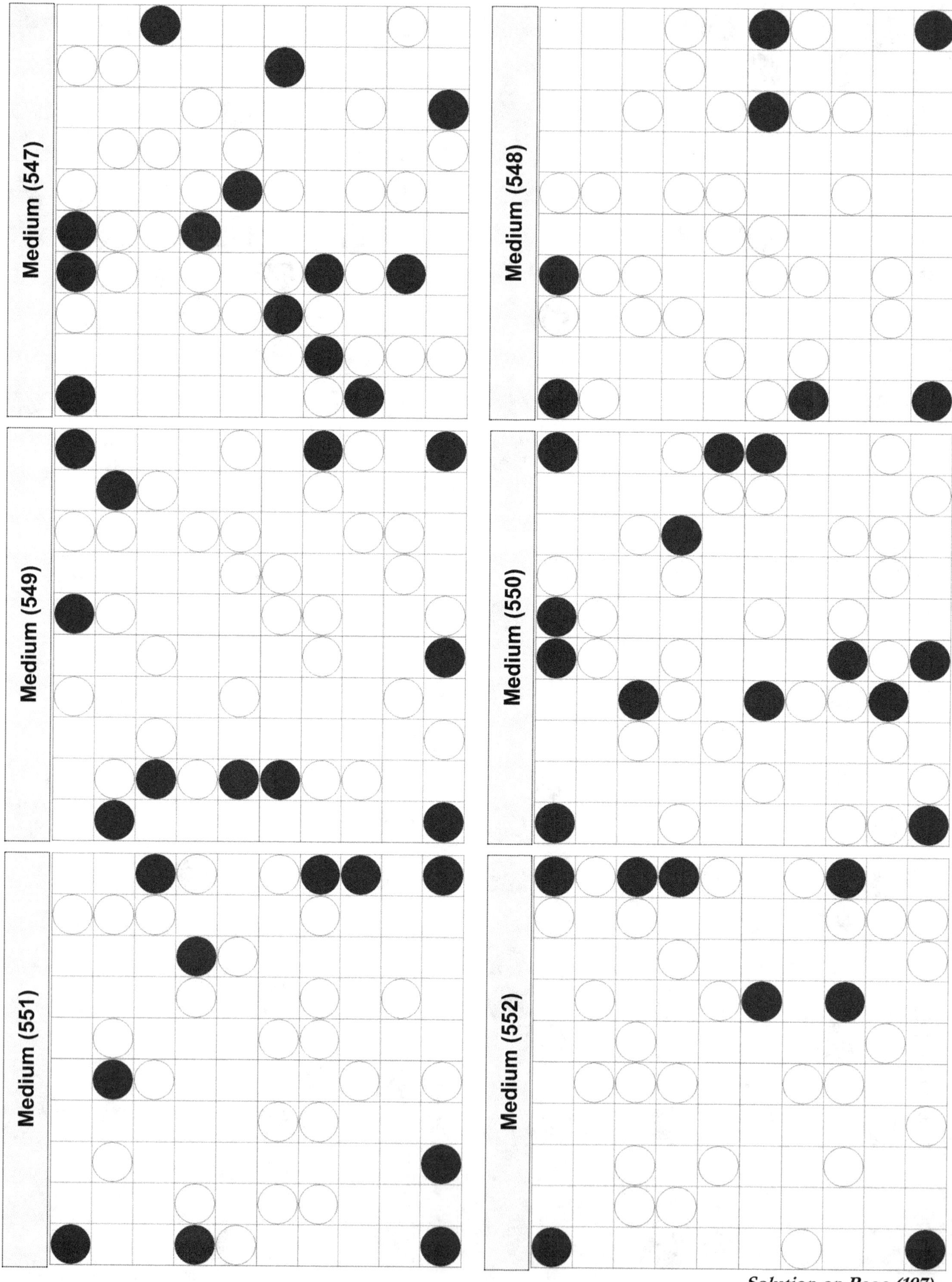

Solution on Page (197)

(94)

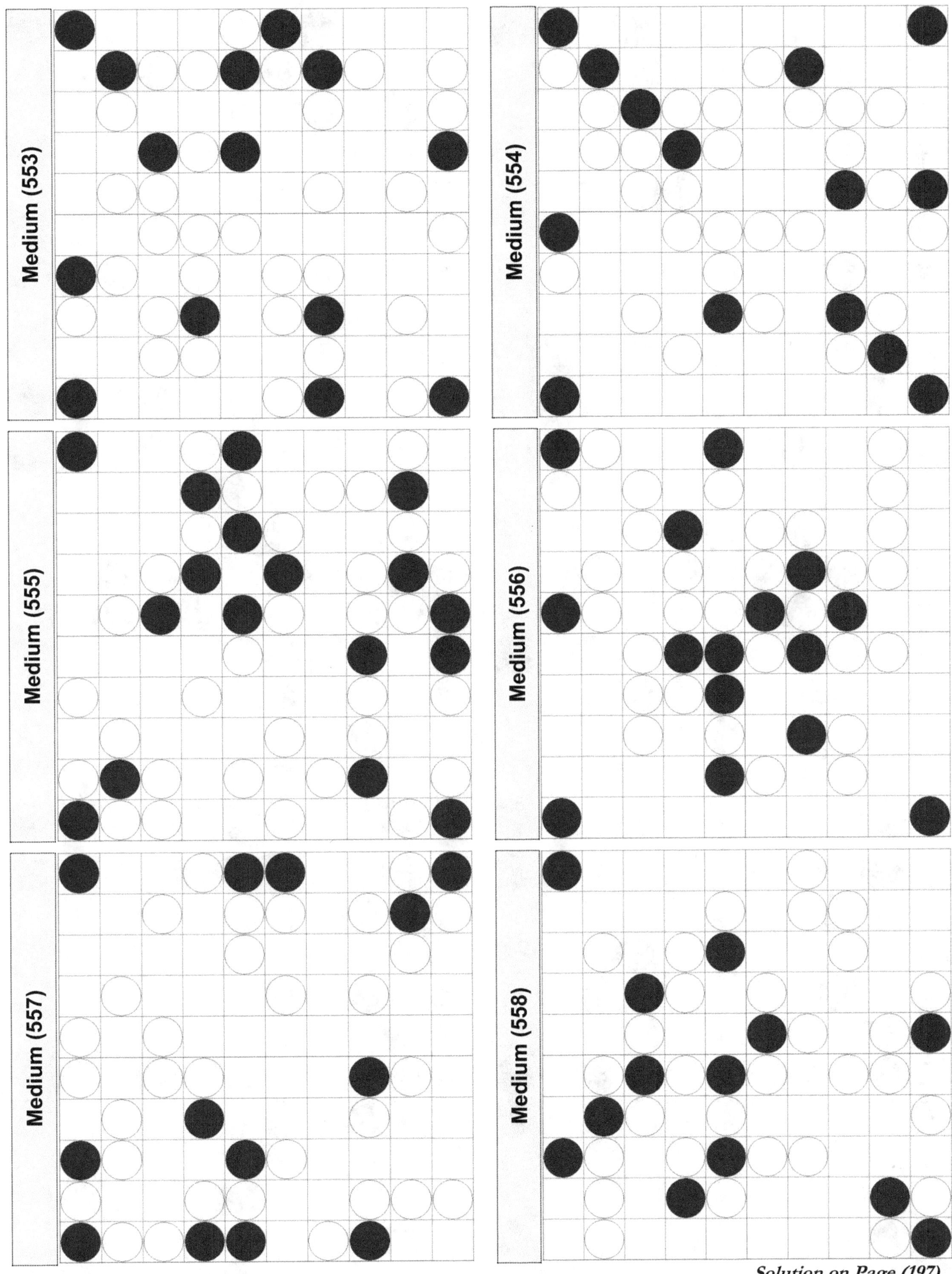

Solution on Page (197)

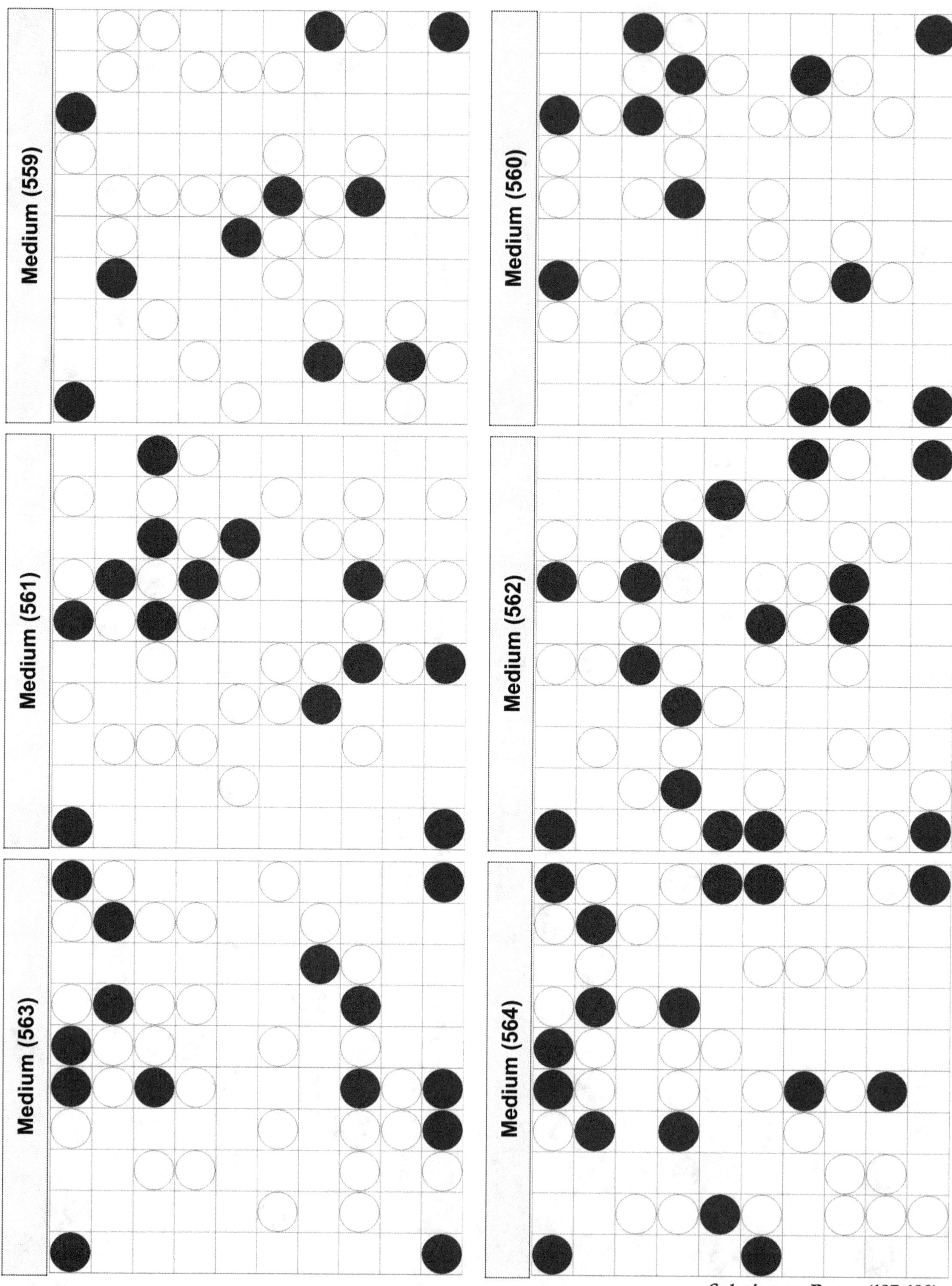

Solution on Pages (197-198)

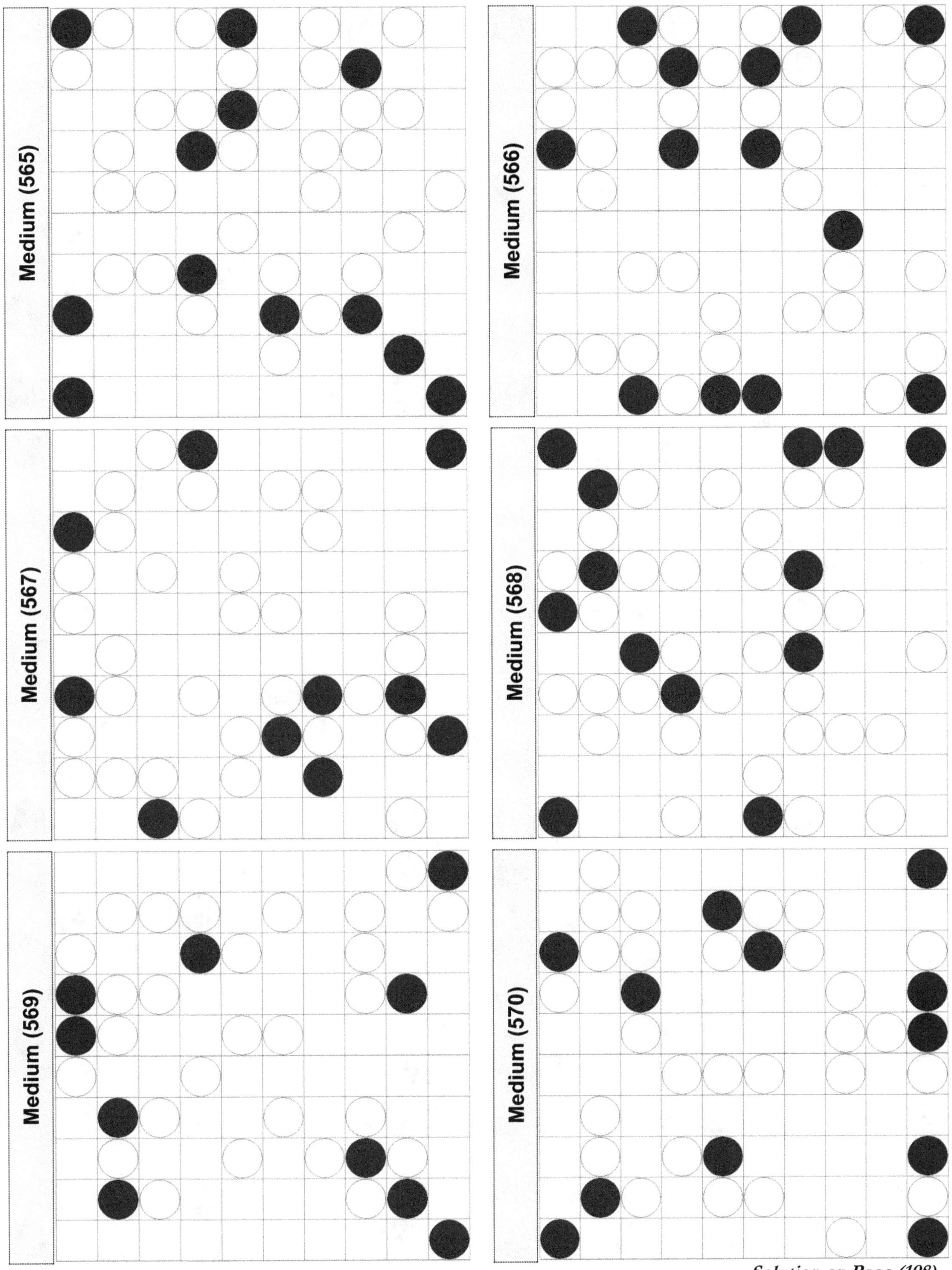

Solution on Page (198)

(97)

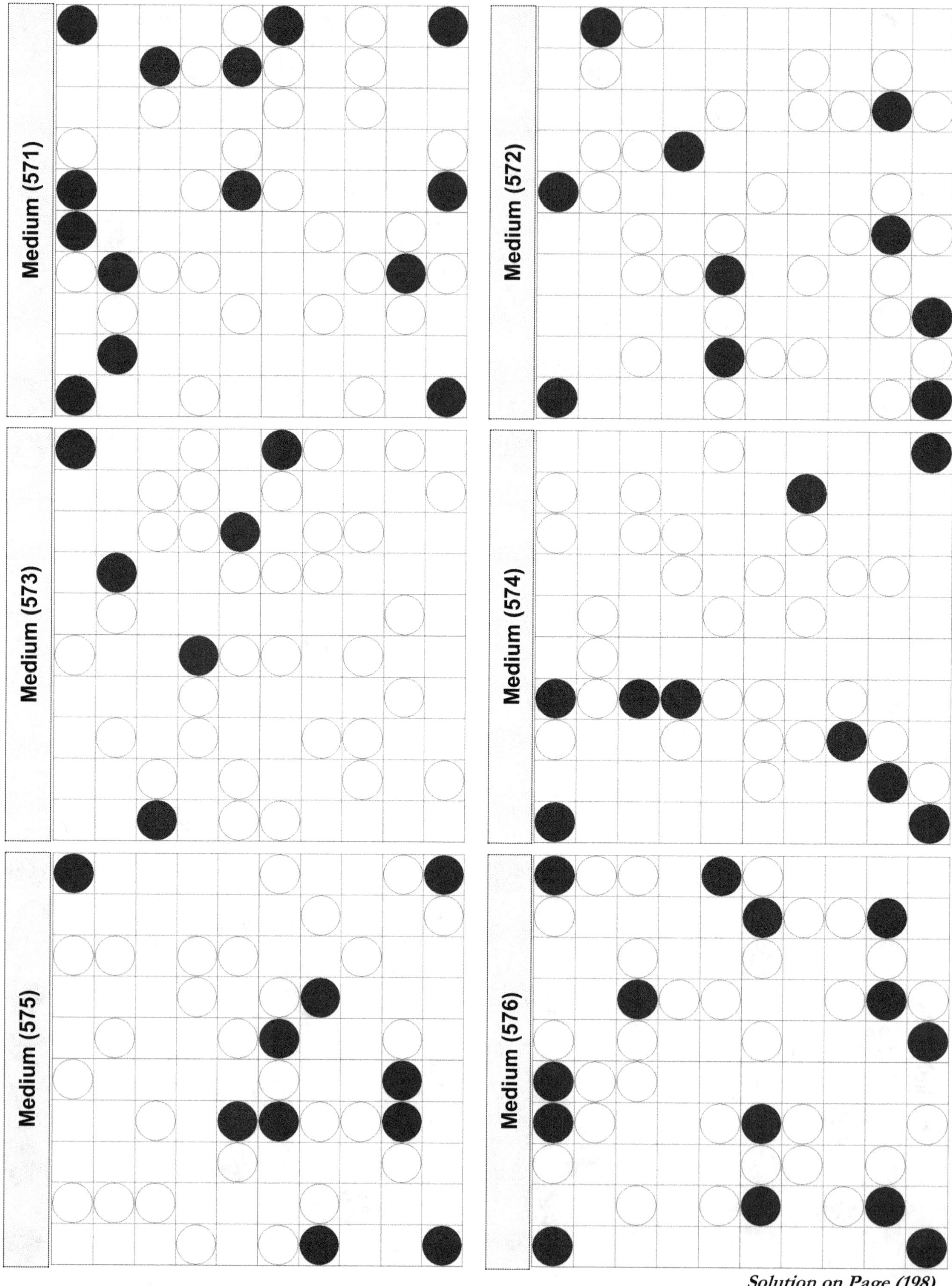

Solution on Page (198)

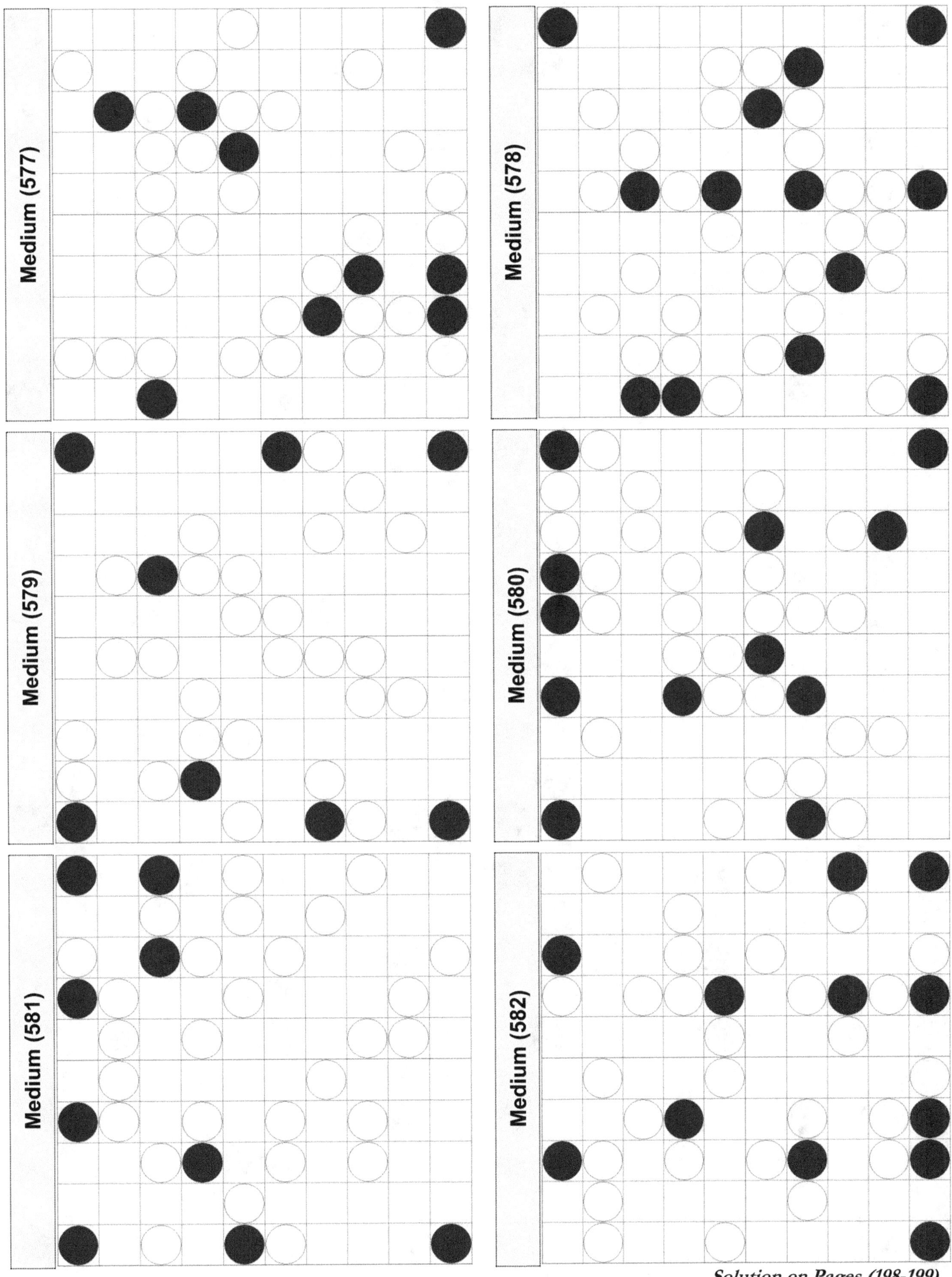

Solution on Pages (198-199)

(99)

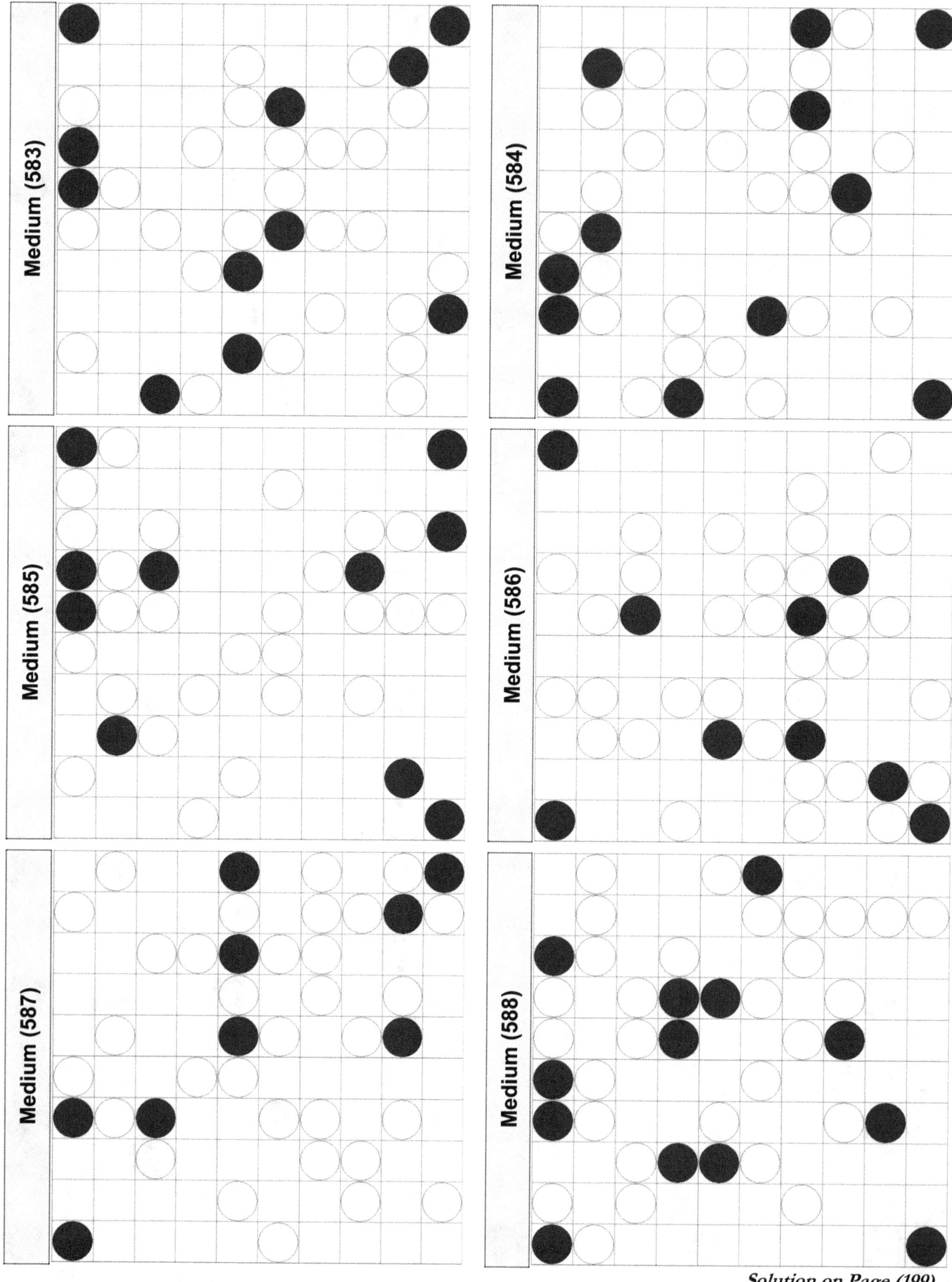

Solution on Page (199)

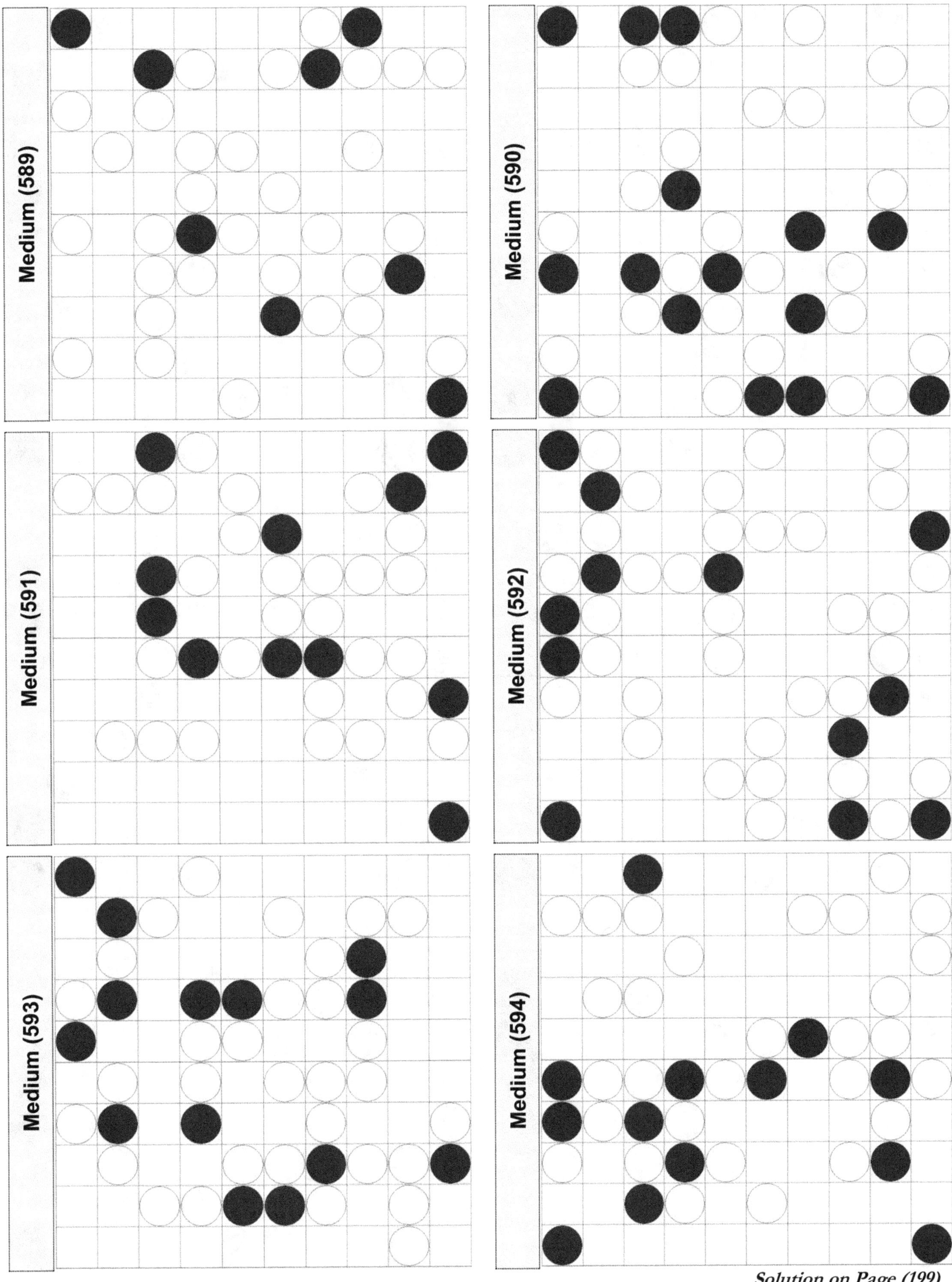

Solution on Page (199)

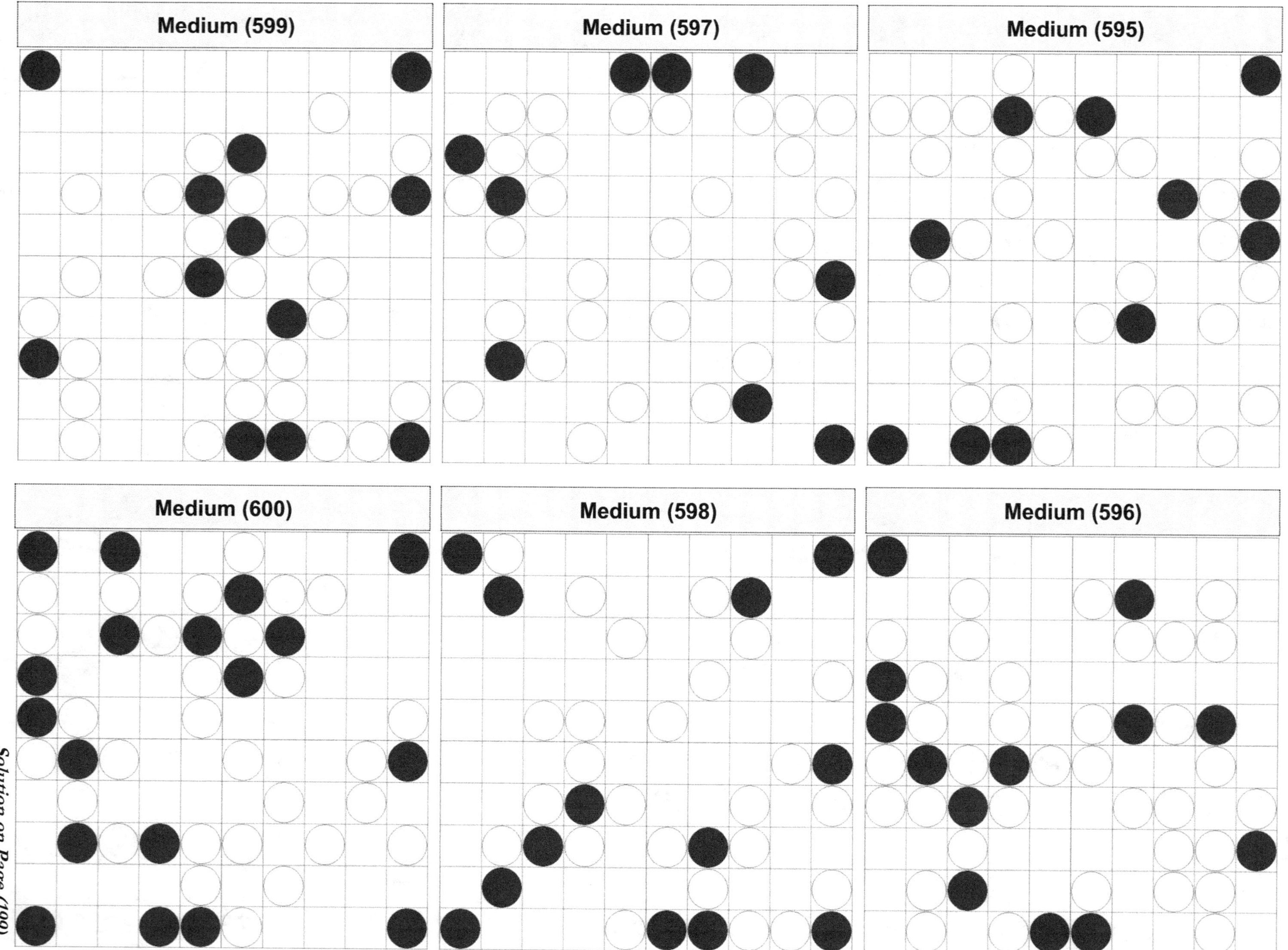

Solution on Page (199)

(102)

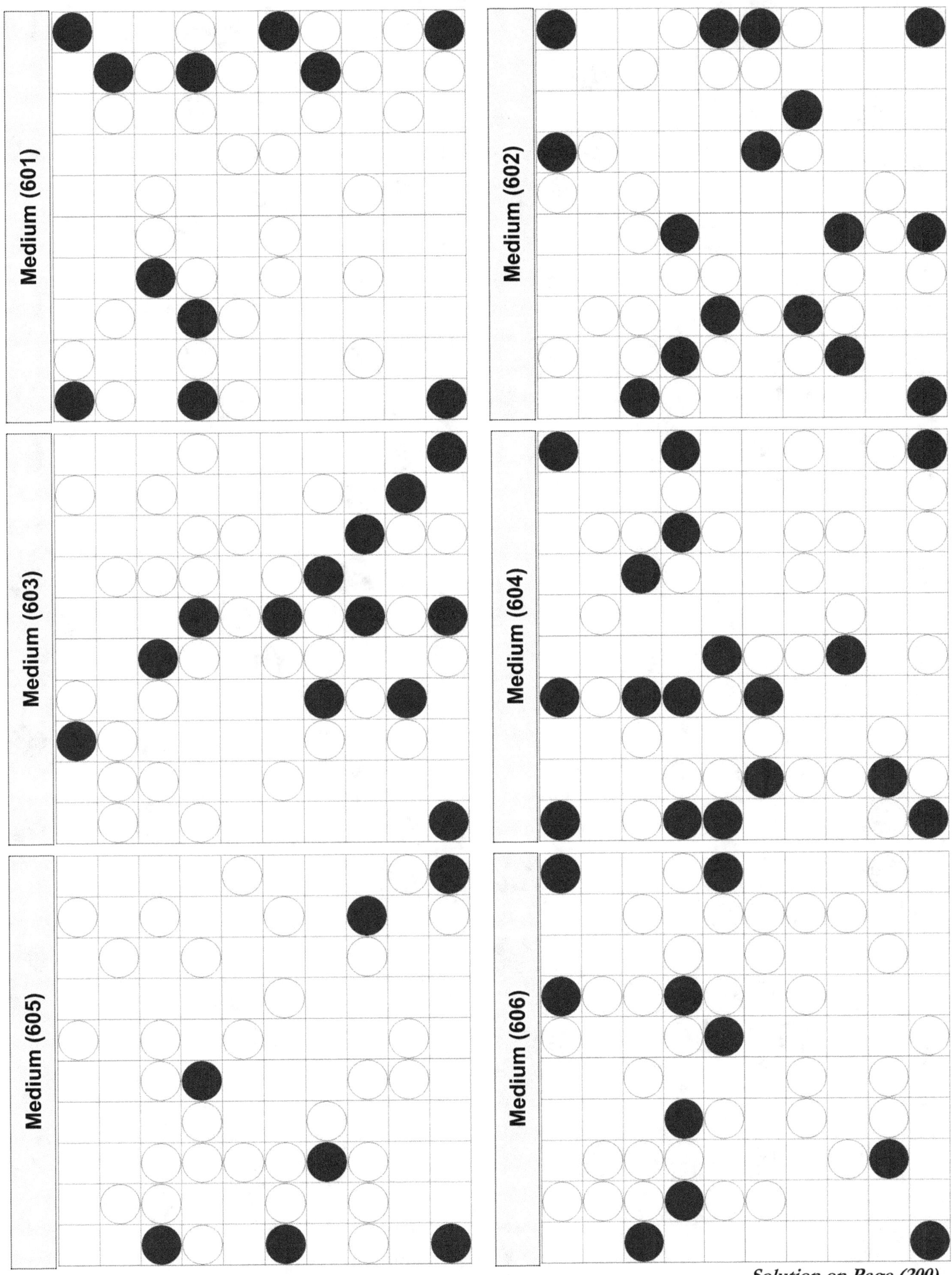

Medium (601)
Medium (602)
Medium (603)
Medium (604)
Medium (605)
Medium (606)
Solution on Page (200)

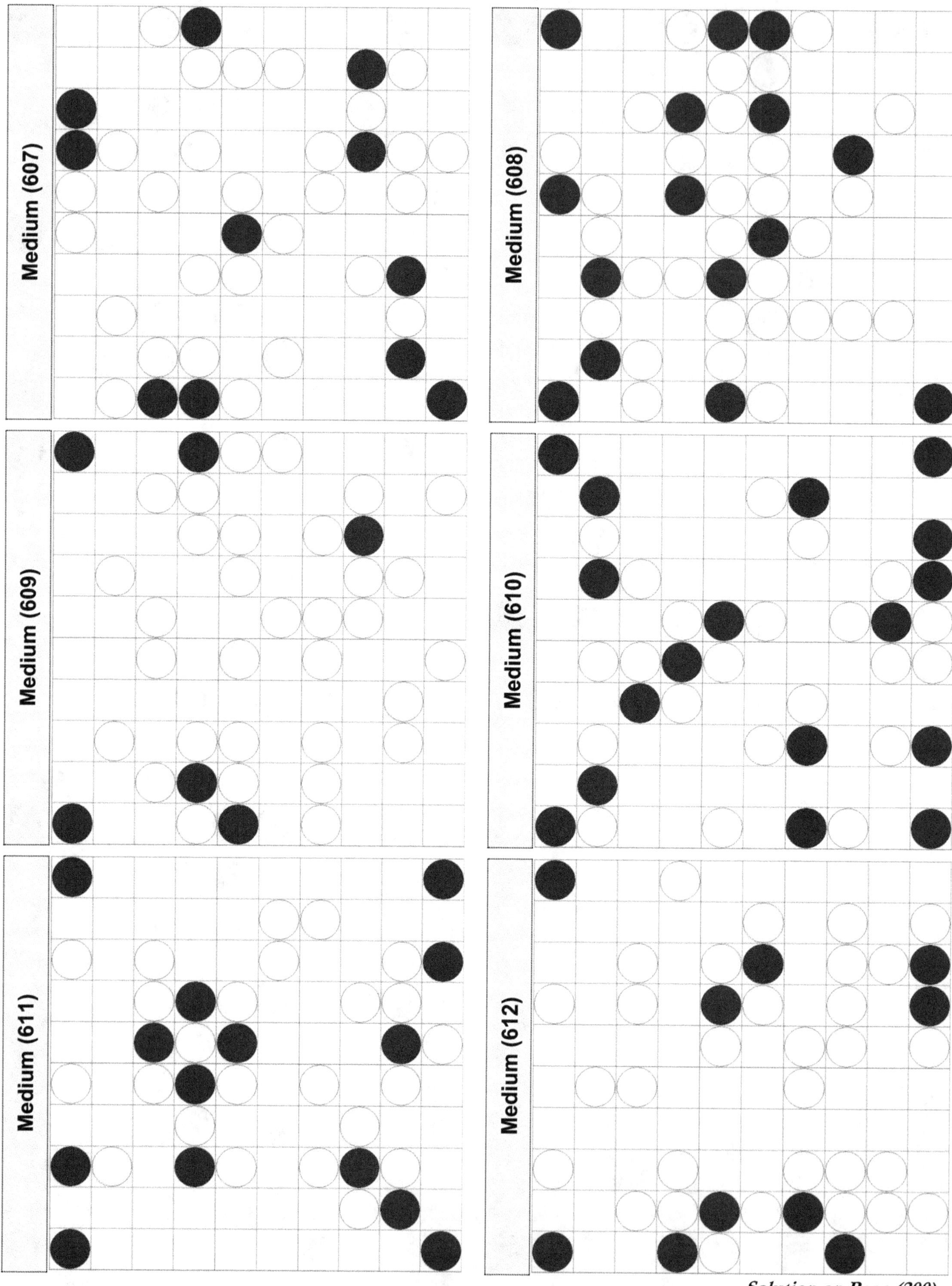

Medium (607)
Medium (608)
Medium (609)
Medium (610)
Medium (611)
Medium (612)
Solution on Page (200)

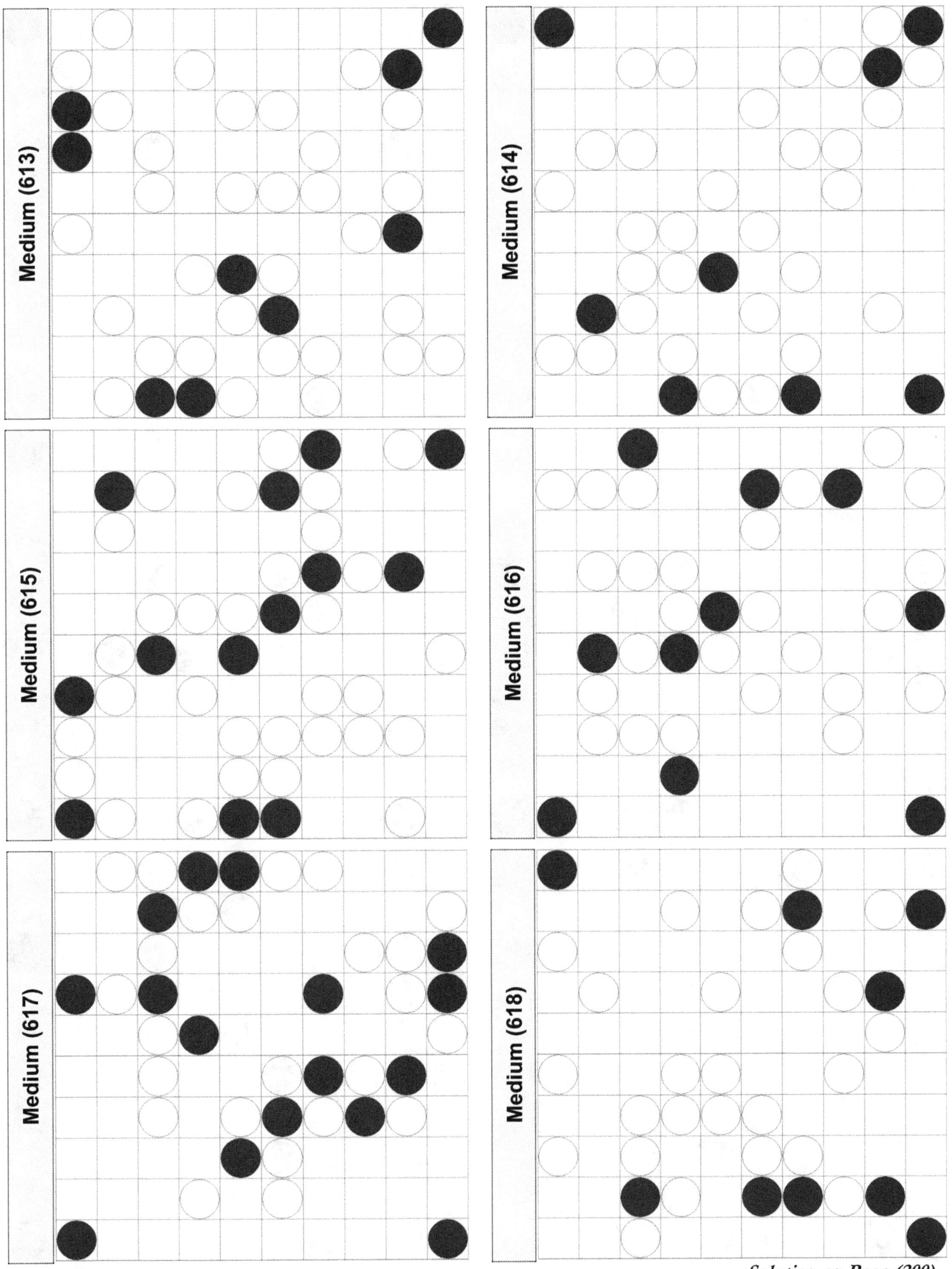

Solution on Page (200)

(105)

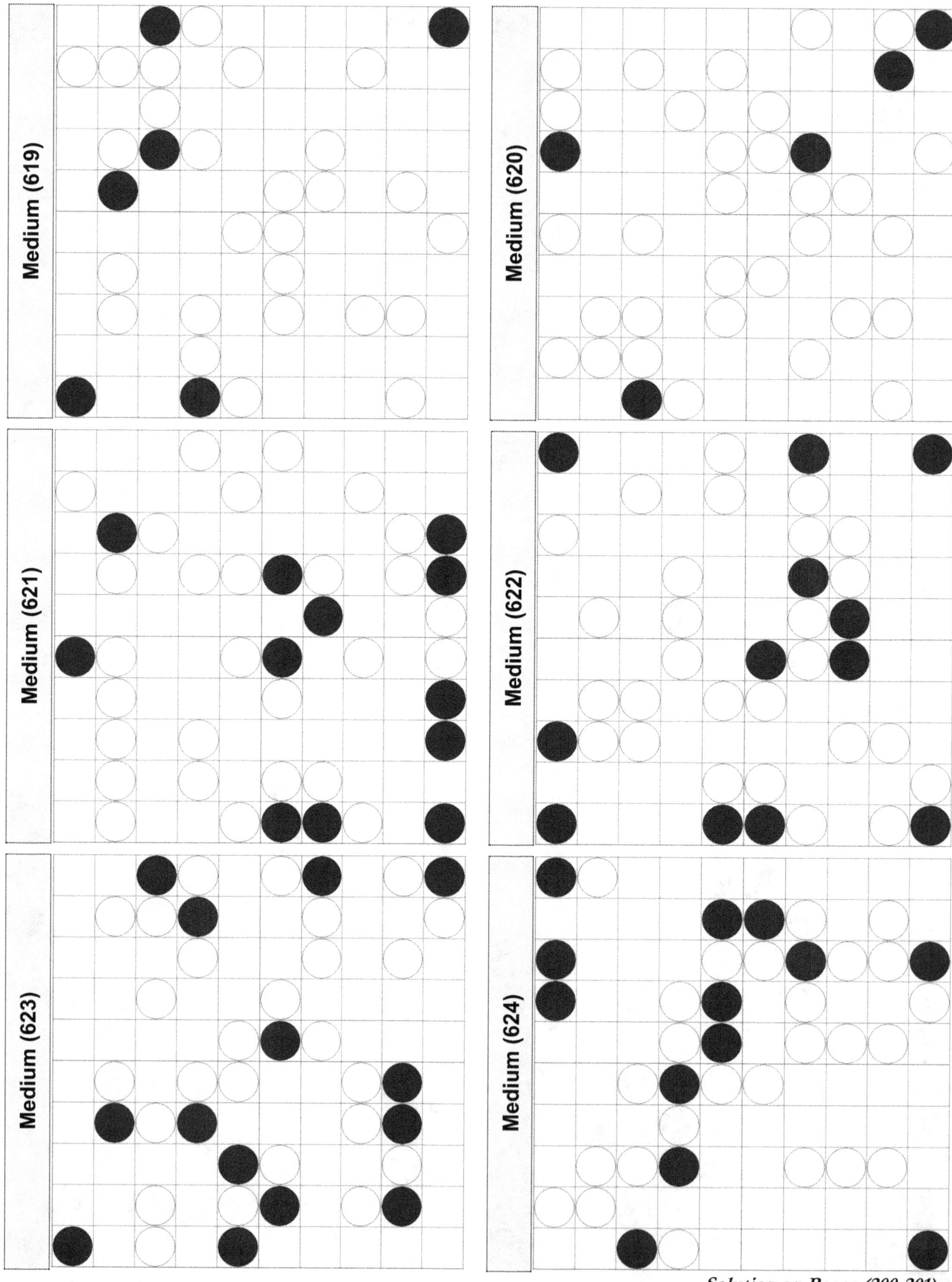

Solution on Pages (200-201)

(106)

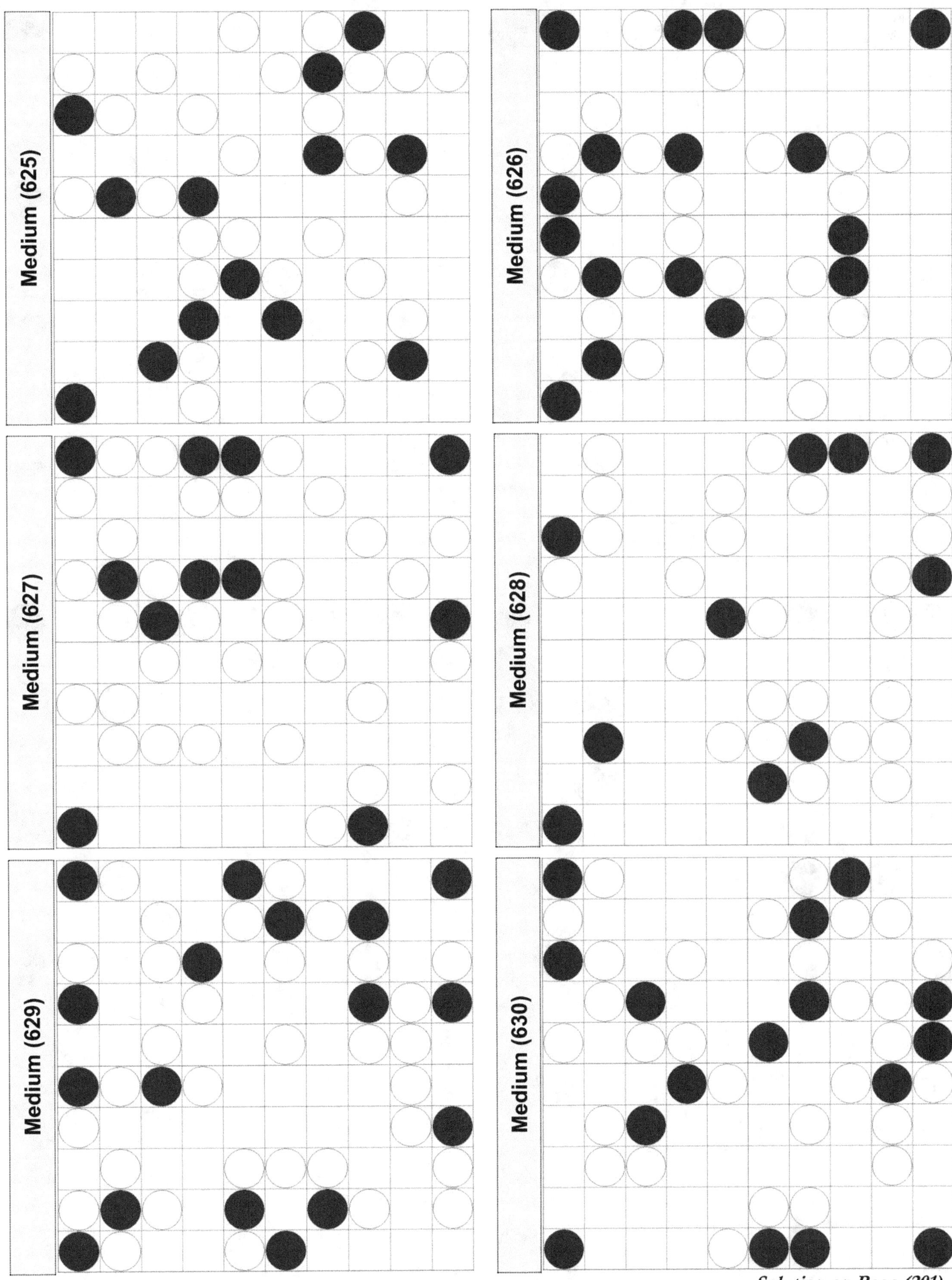

Solution on Page (201)

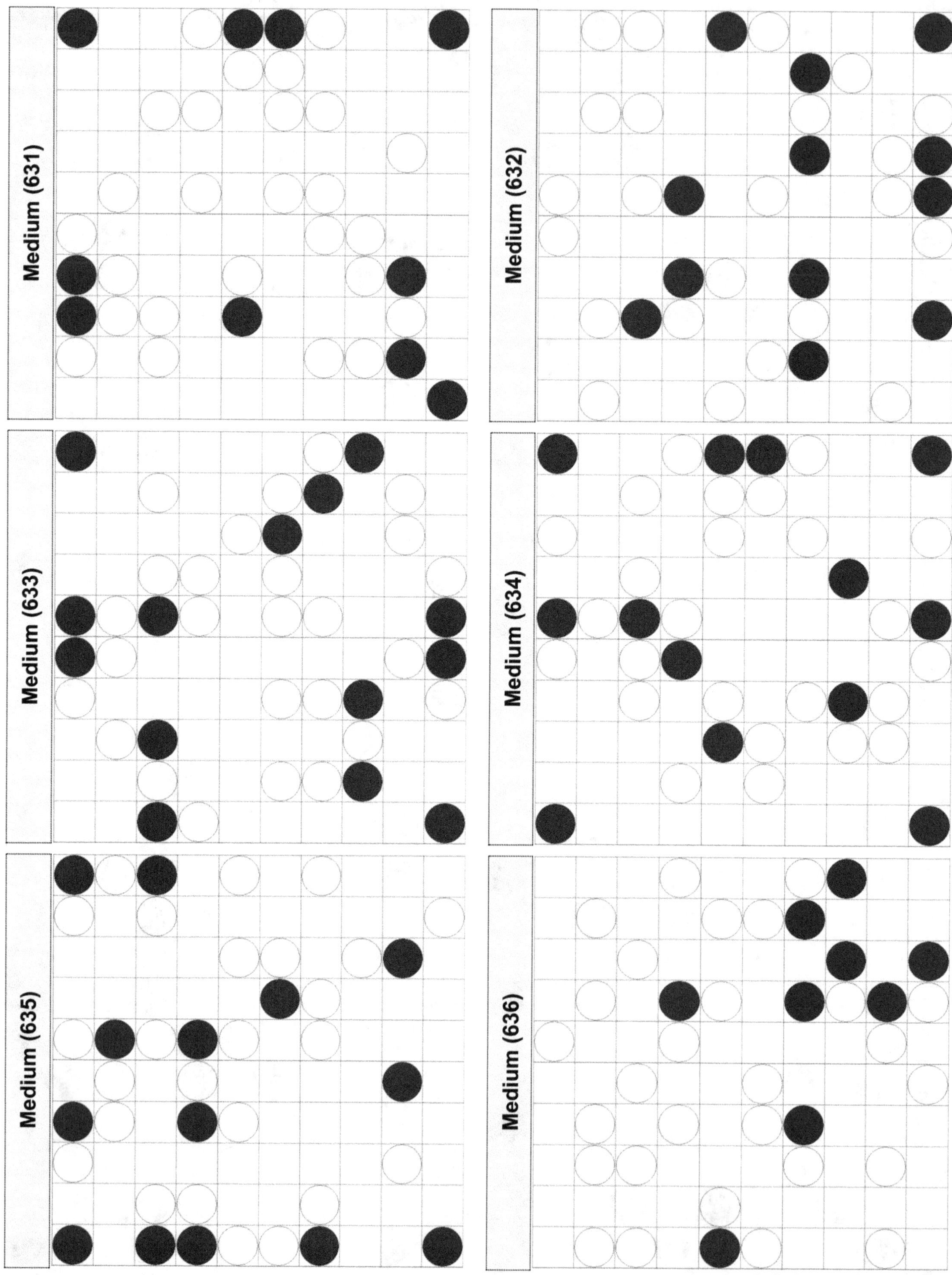

(108)

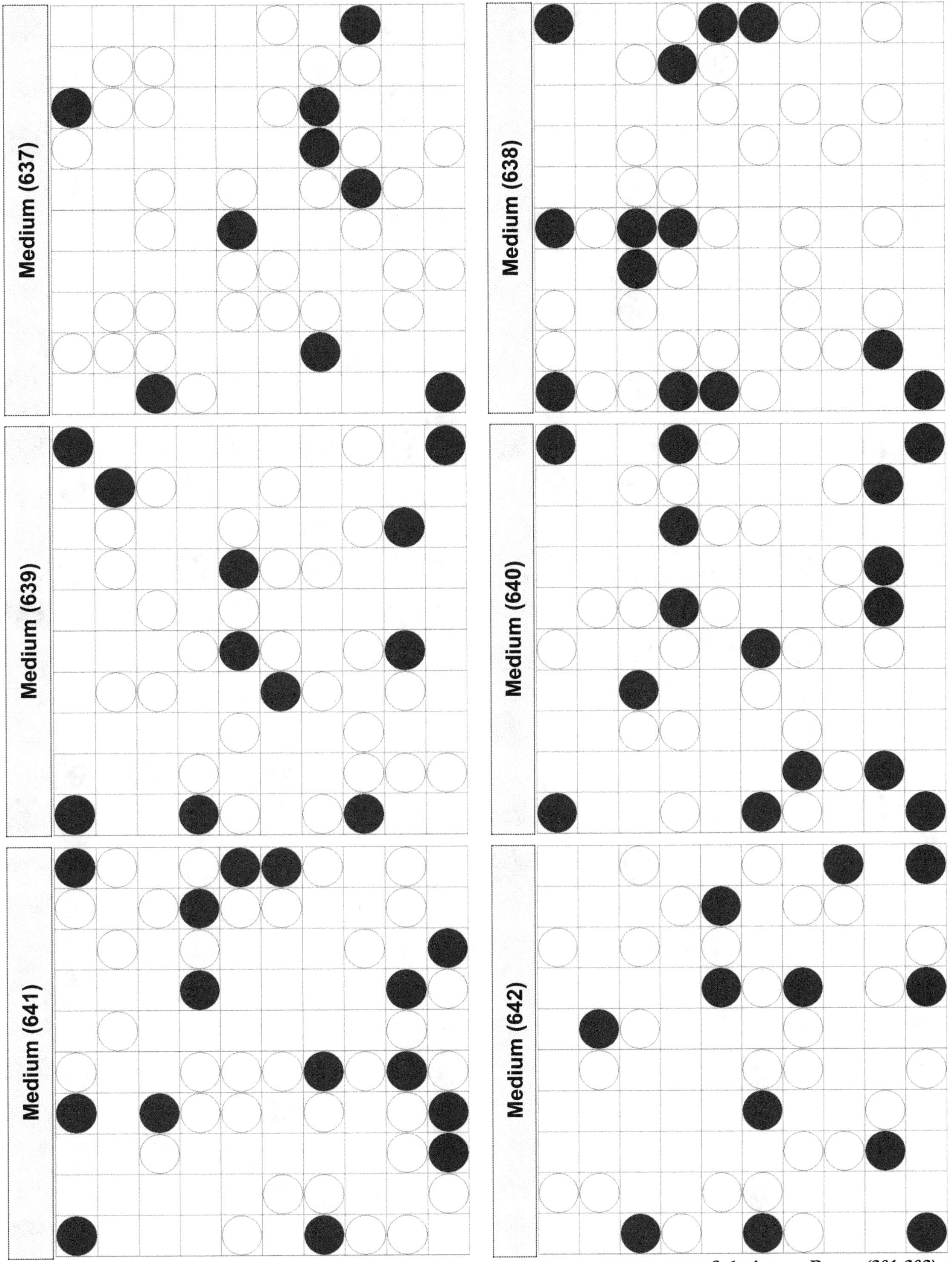

Solution on Pages (201-202)

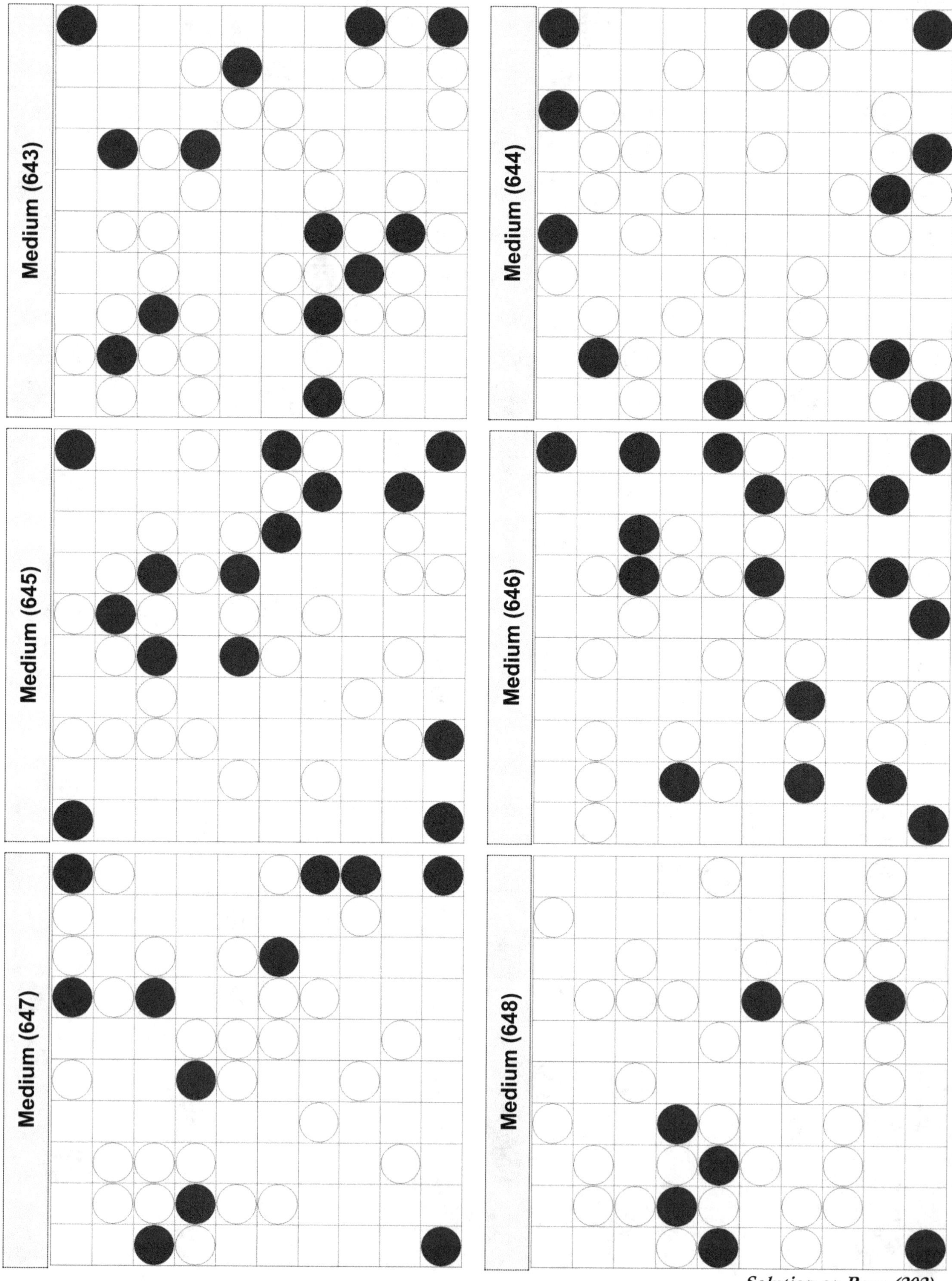

Solution on Page (202)

(110)

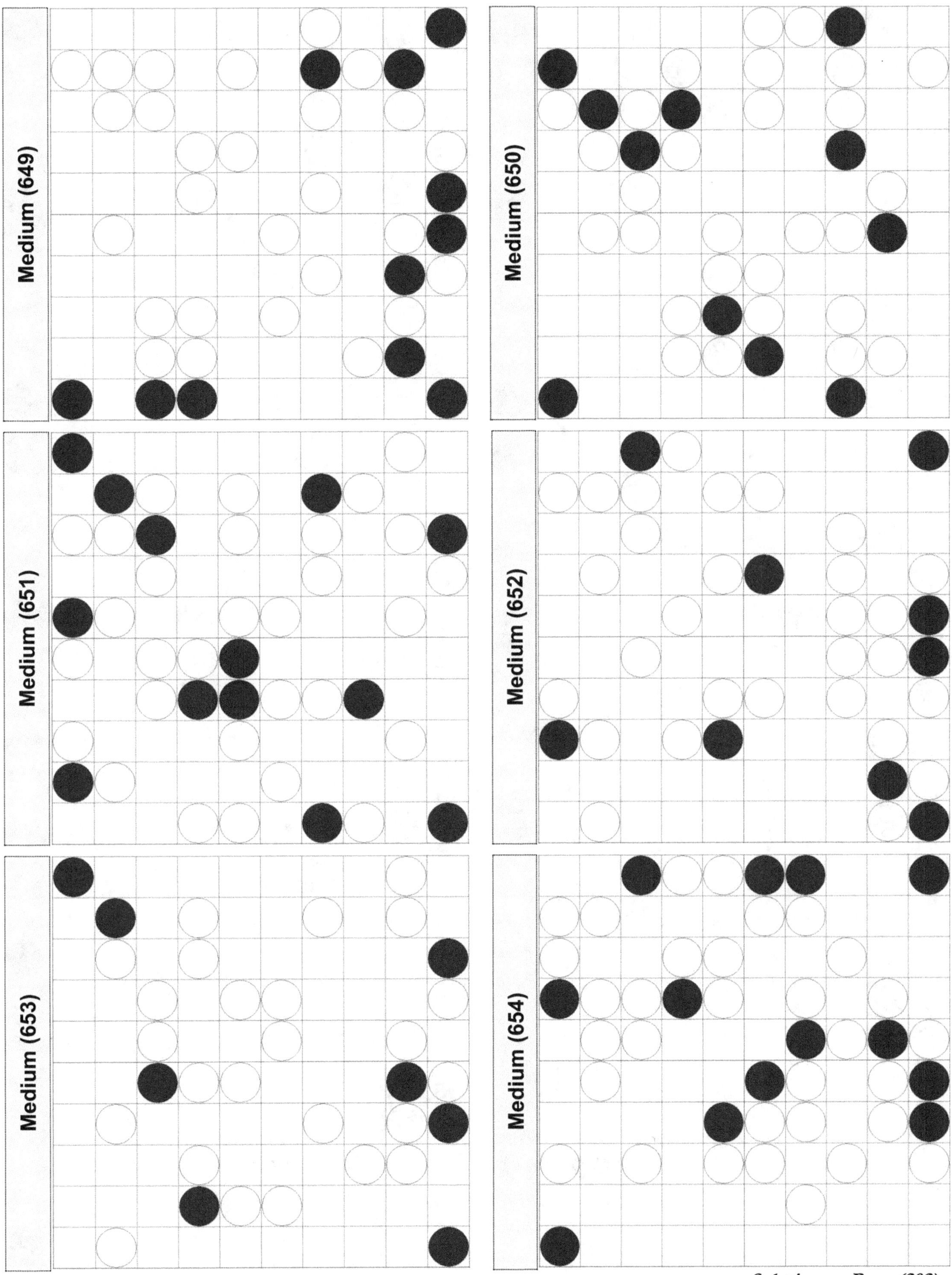

Solution on Page (202)

(111)

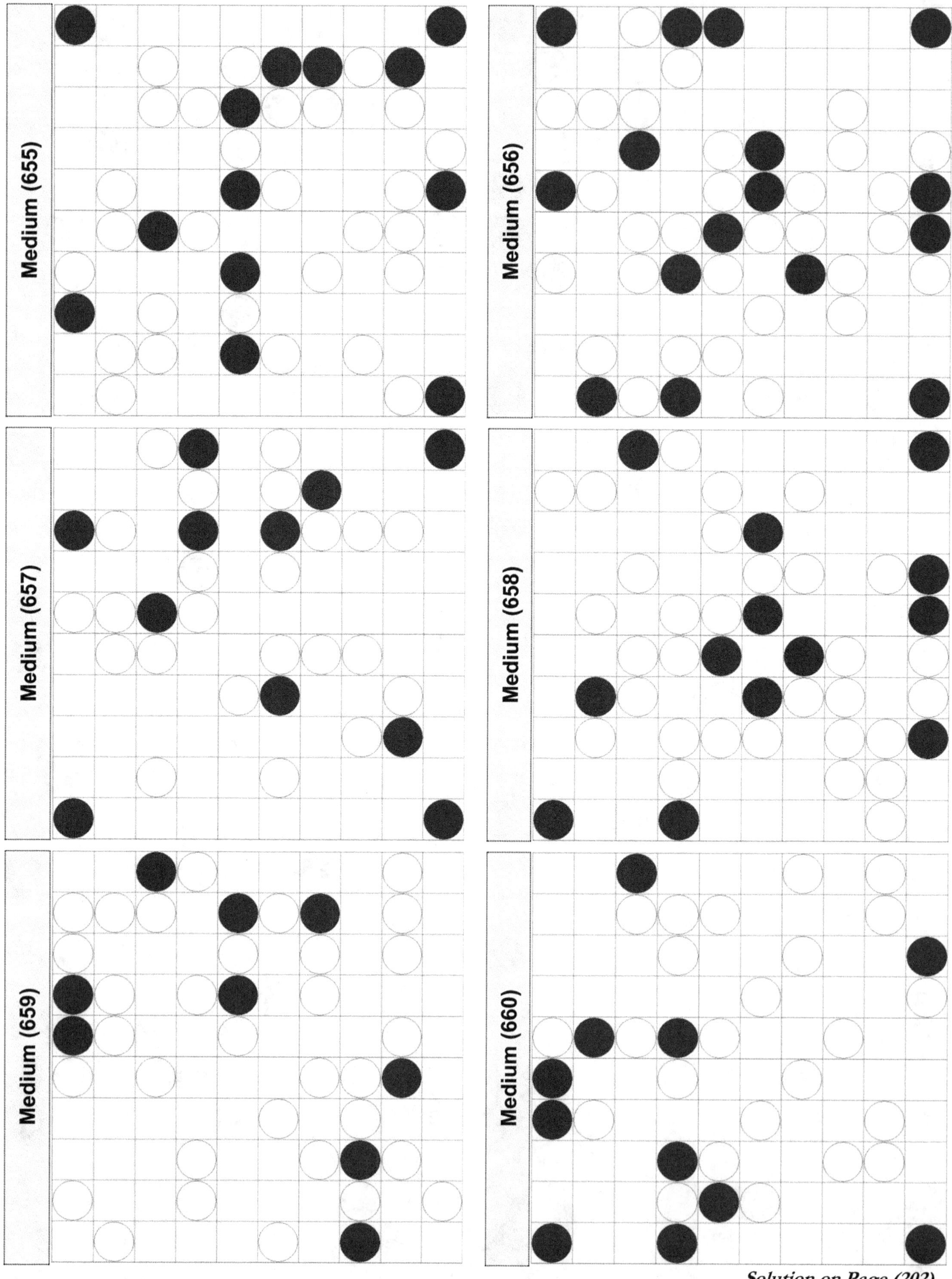

Medium (655)

Medium (656)

Medium (657)

Medium (658)

Medium (659)

Medium (660)

Solution on Page (202)

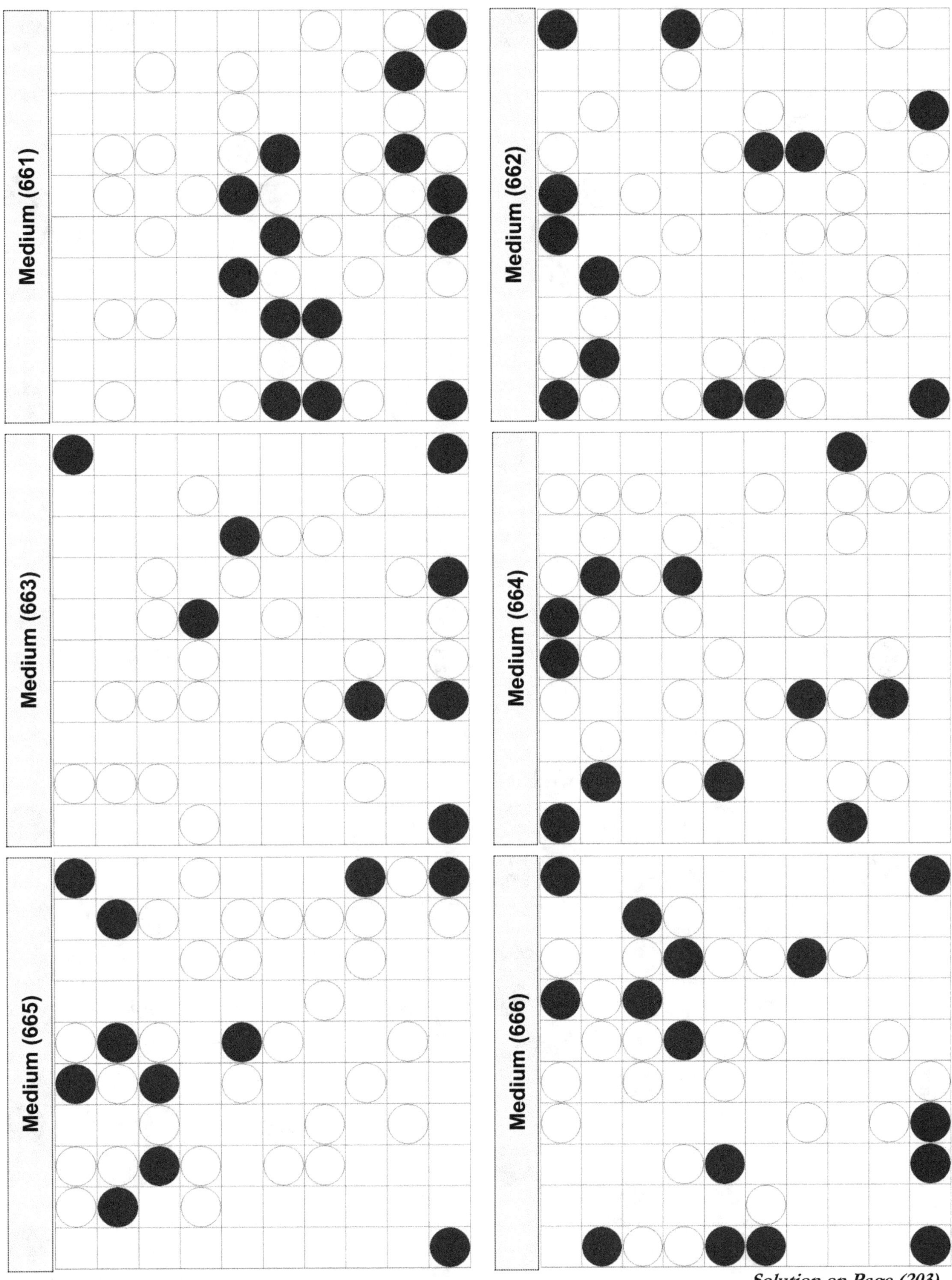

(113)

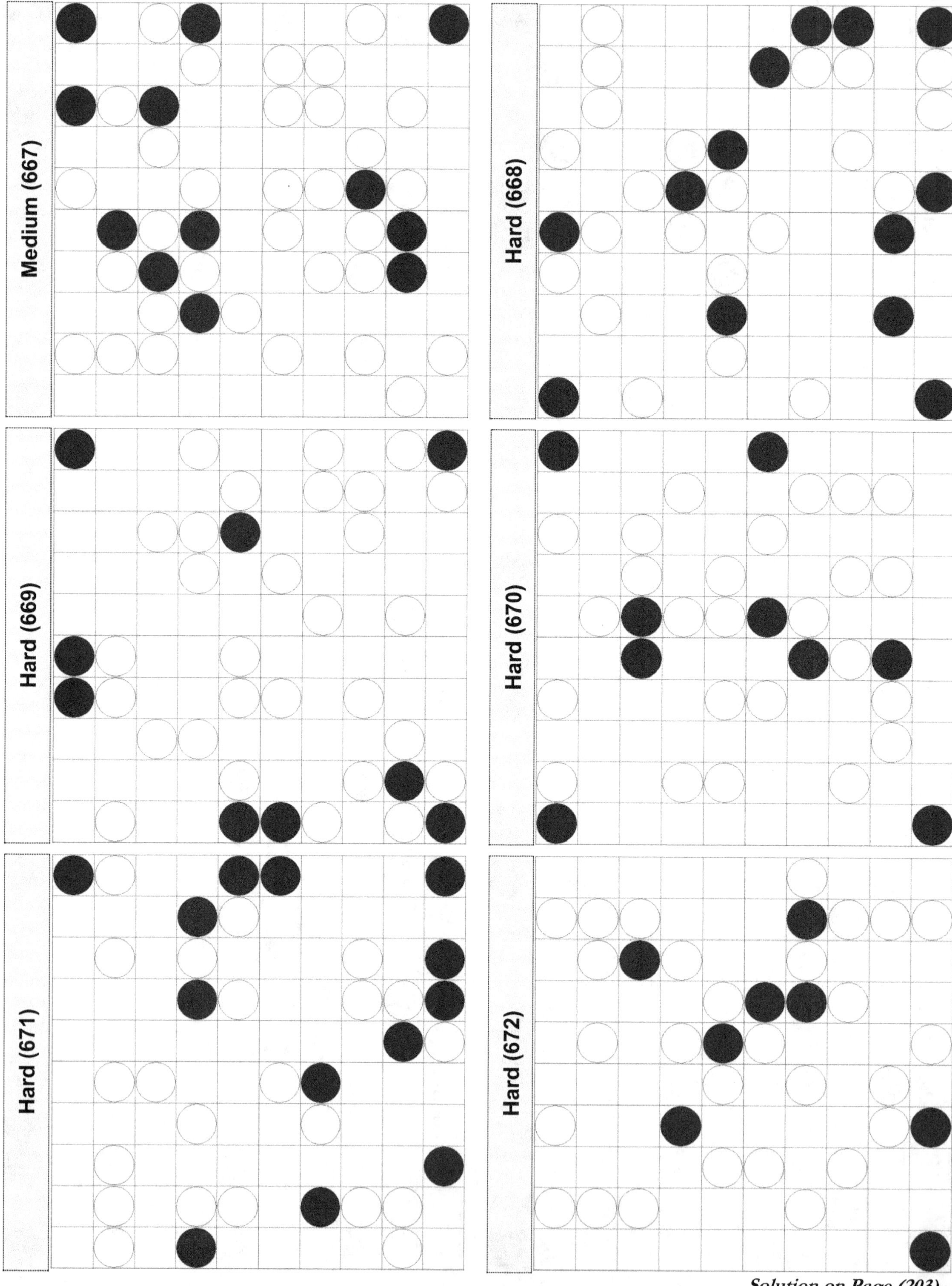

Medium (667)

Hard (668)

Hard (669)

Hard (670)

Hard (671)

Hard (672)

Solution on Page (203)

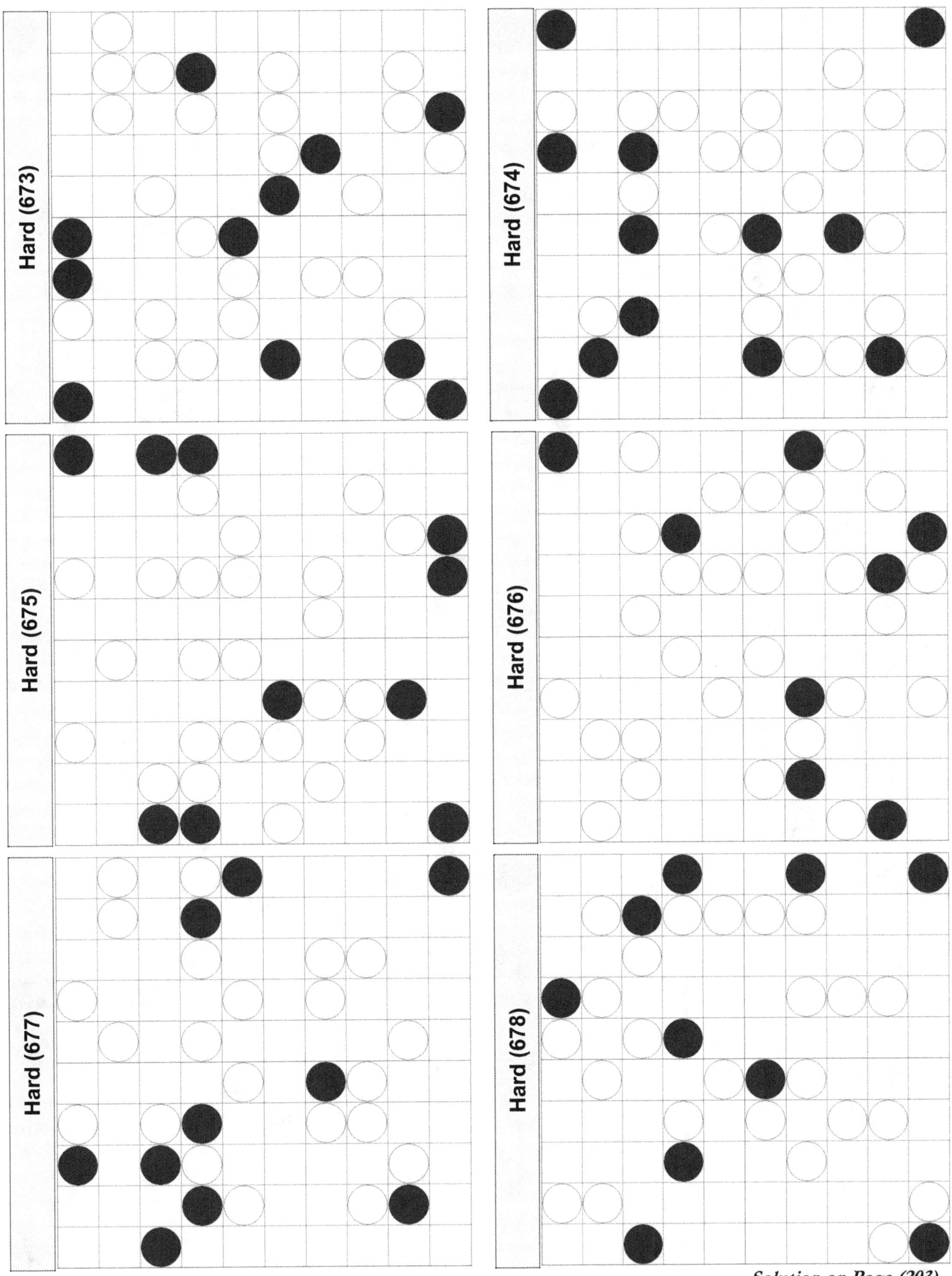

Solution on Page (203)

(115)

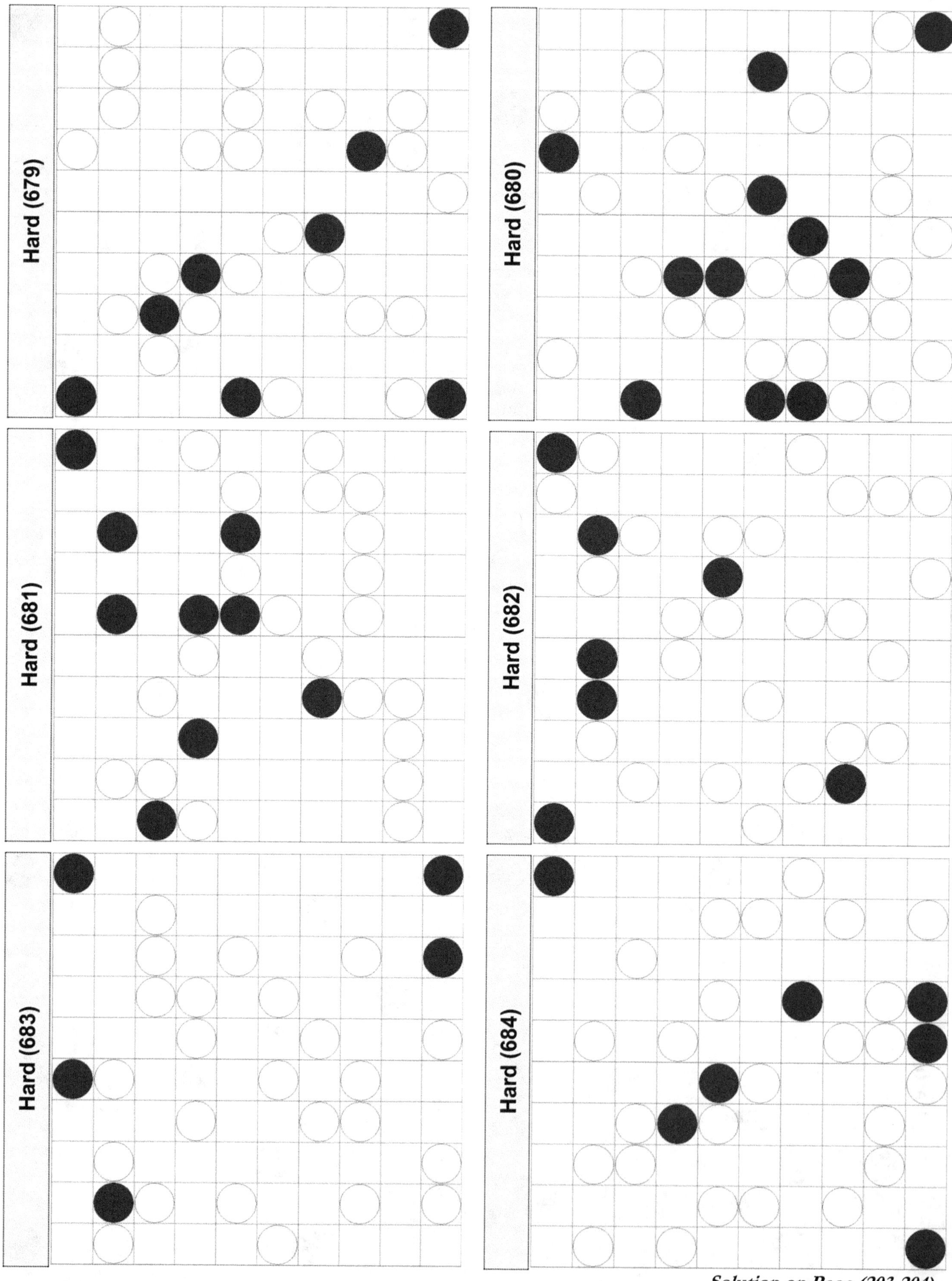

Solution on Page (203-204)

(116)

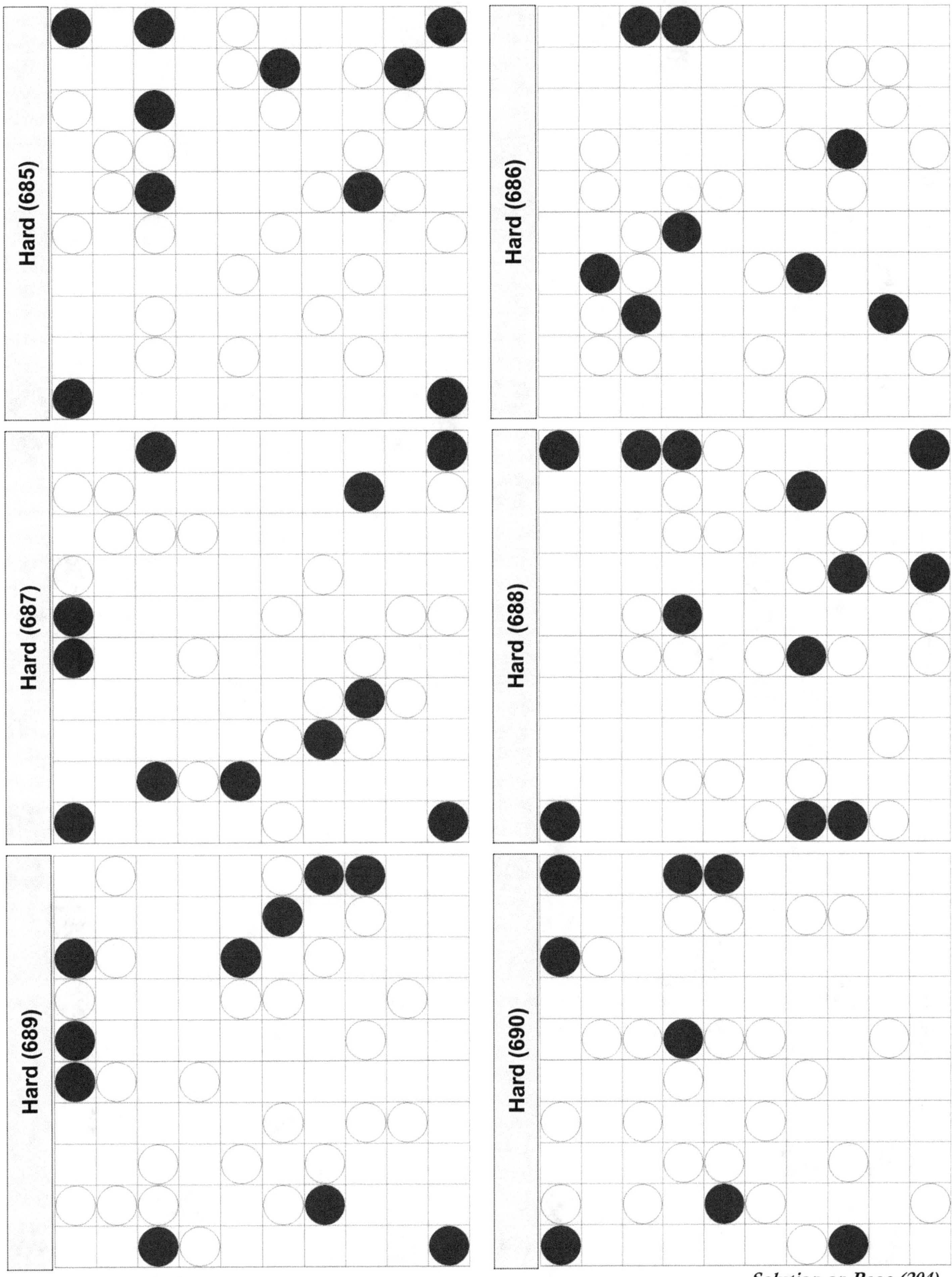

Solution on Page (204)

Hard (691)
Hard (692)
Hard (693)
Hard (694)
Hard (695)
Hard (696)
Solution on Page (204)

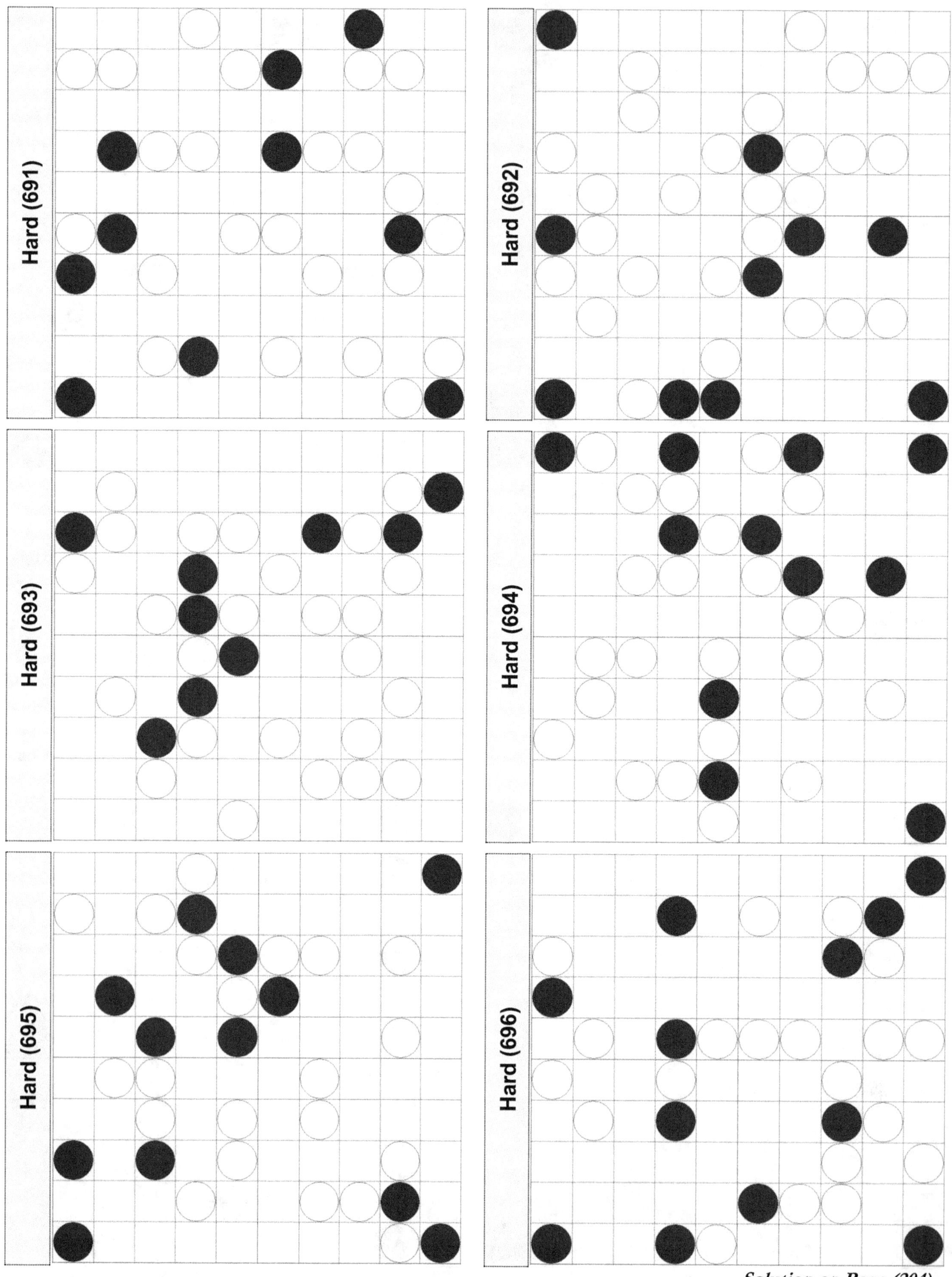

(118)

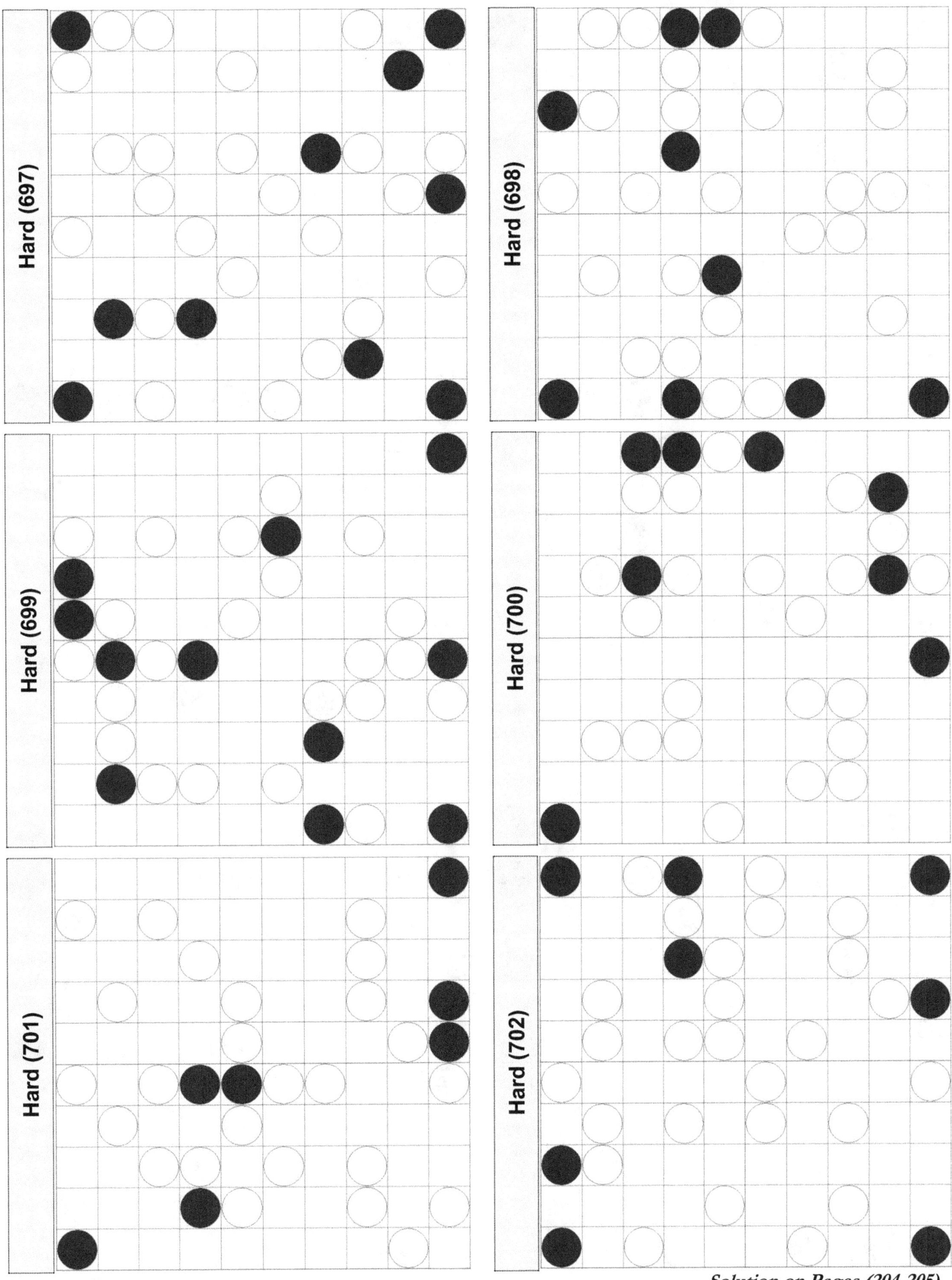

Solution on Pages (204-205)

(119)

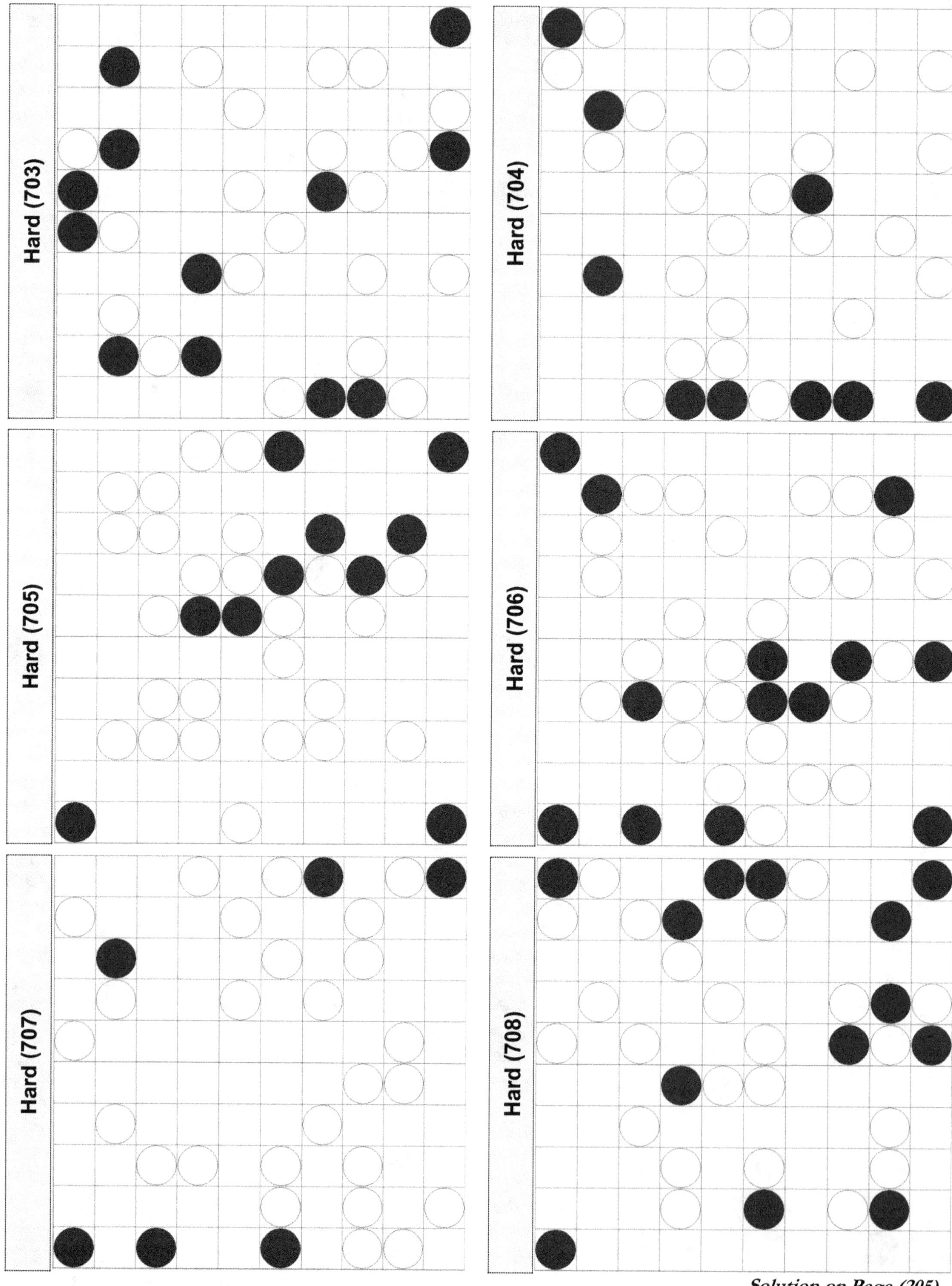

Hard (703)

Hard (704)

Hard (705)

Hard (706)

Hard (707)

Hard (708)

Solution on Page (205)

(120)

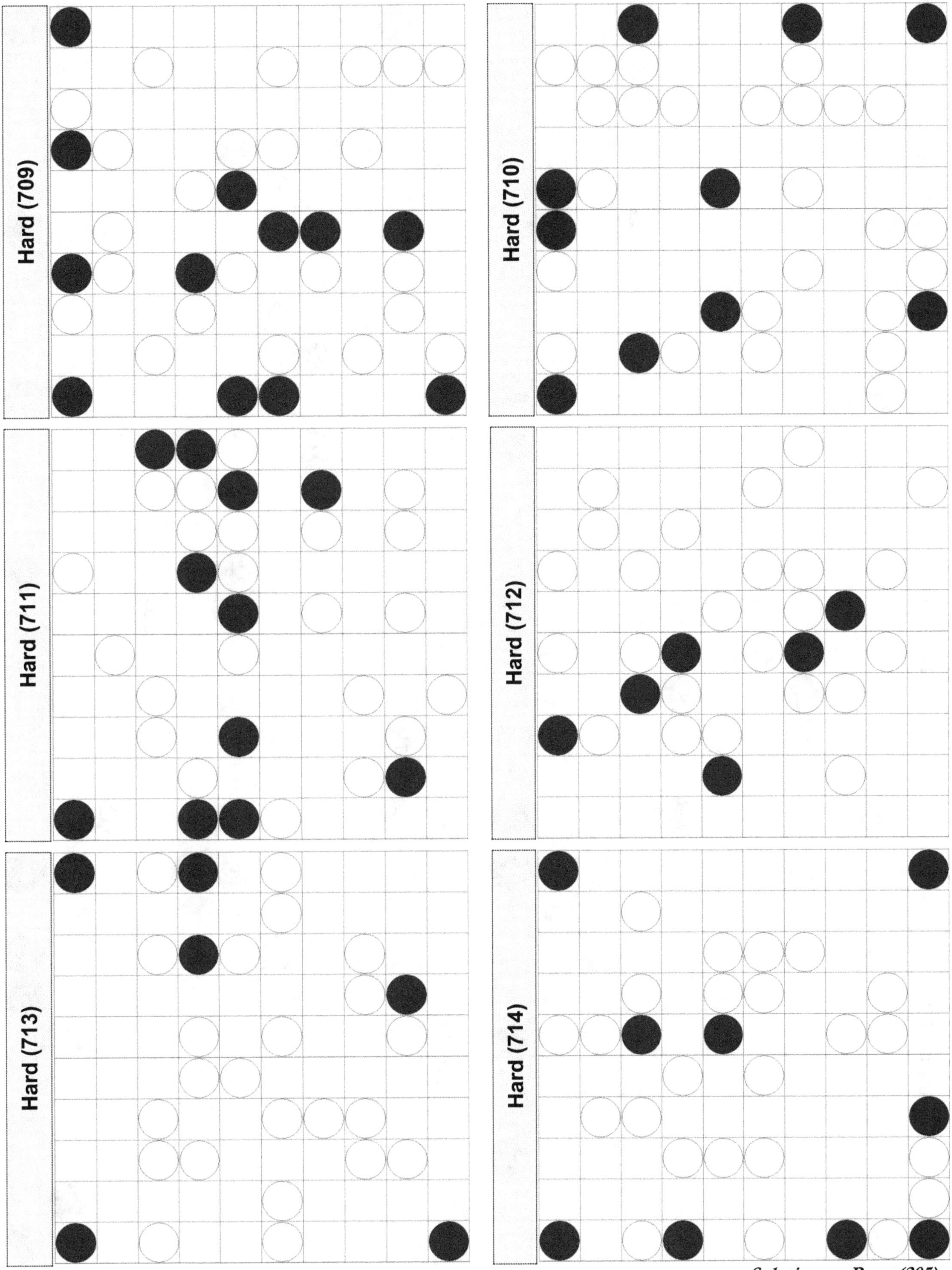

Solution on Page (205)

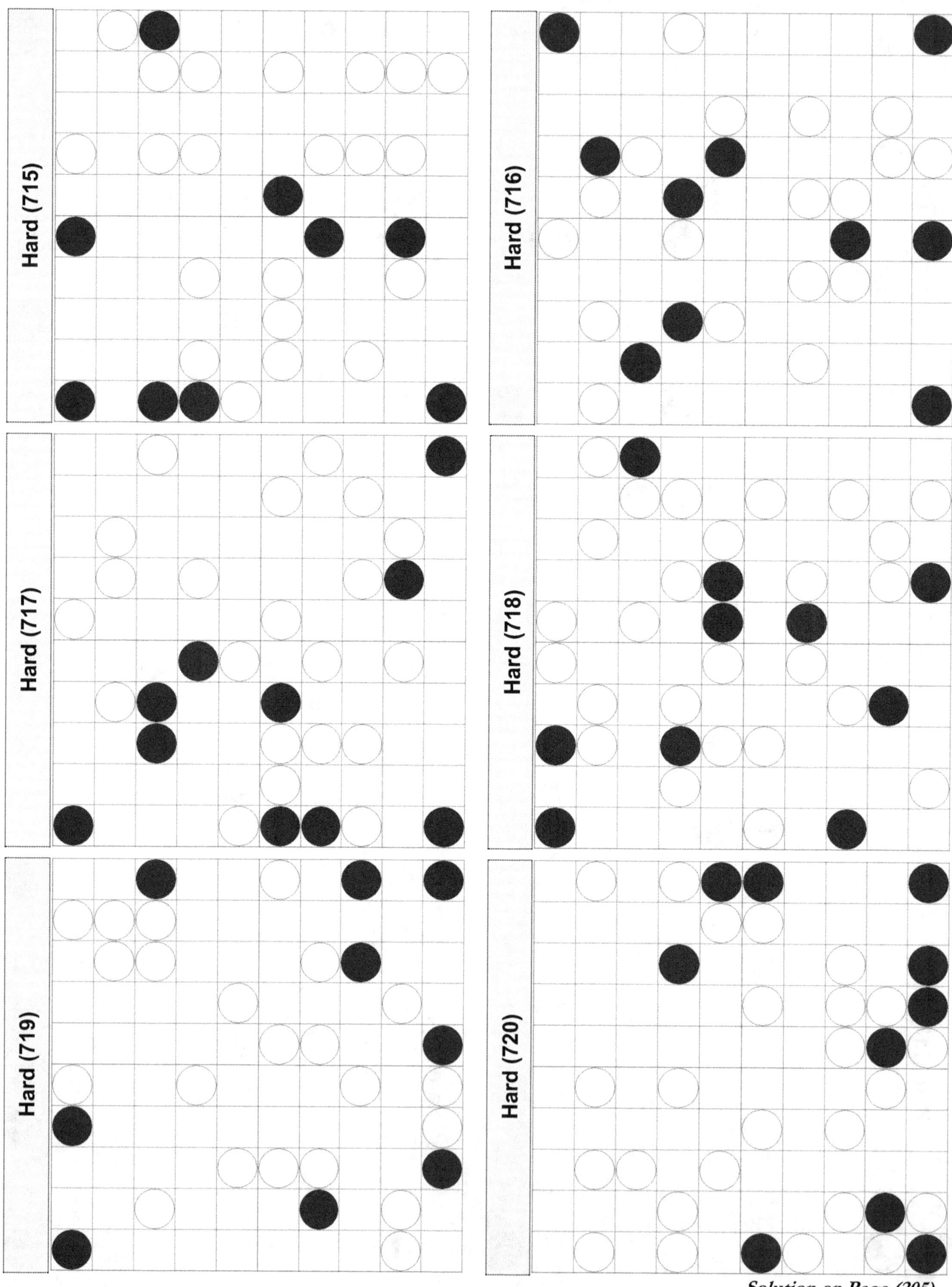

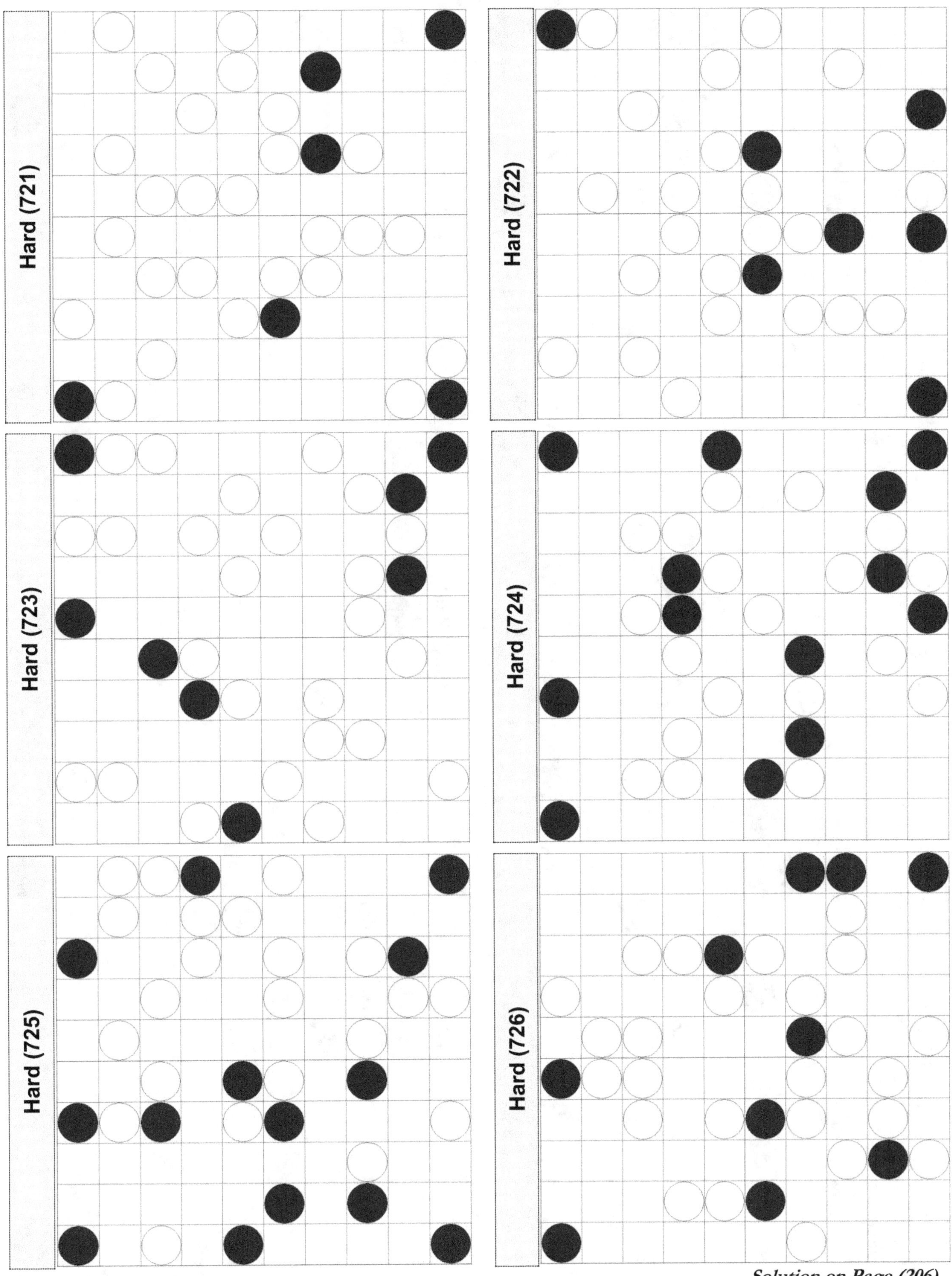

Solution on Page (206)

(123)

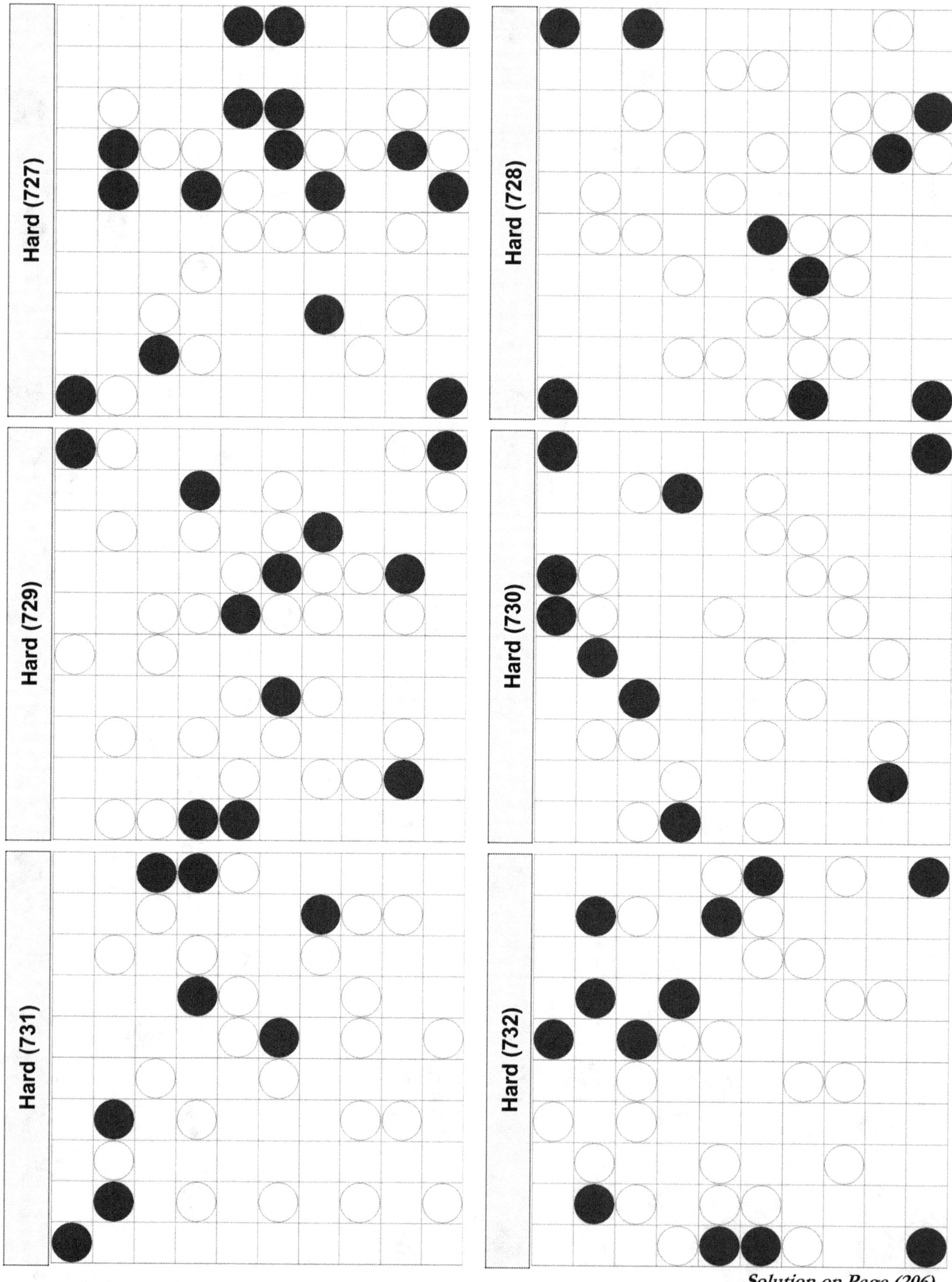

Solution on Page (206)

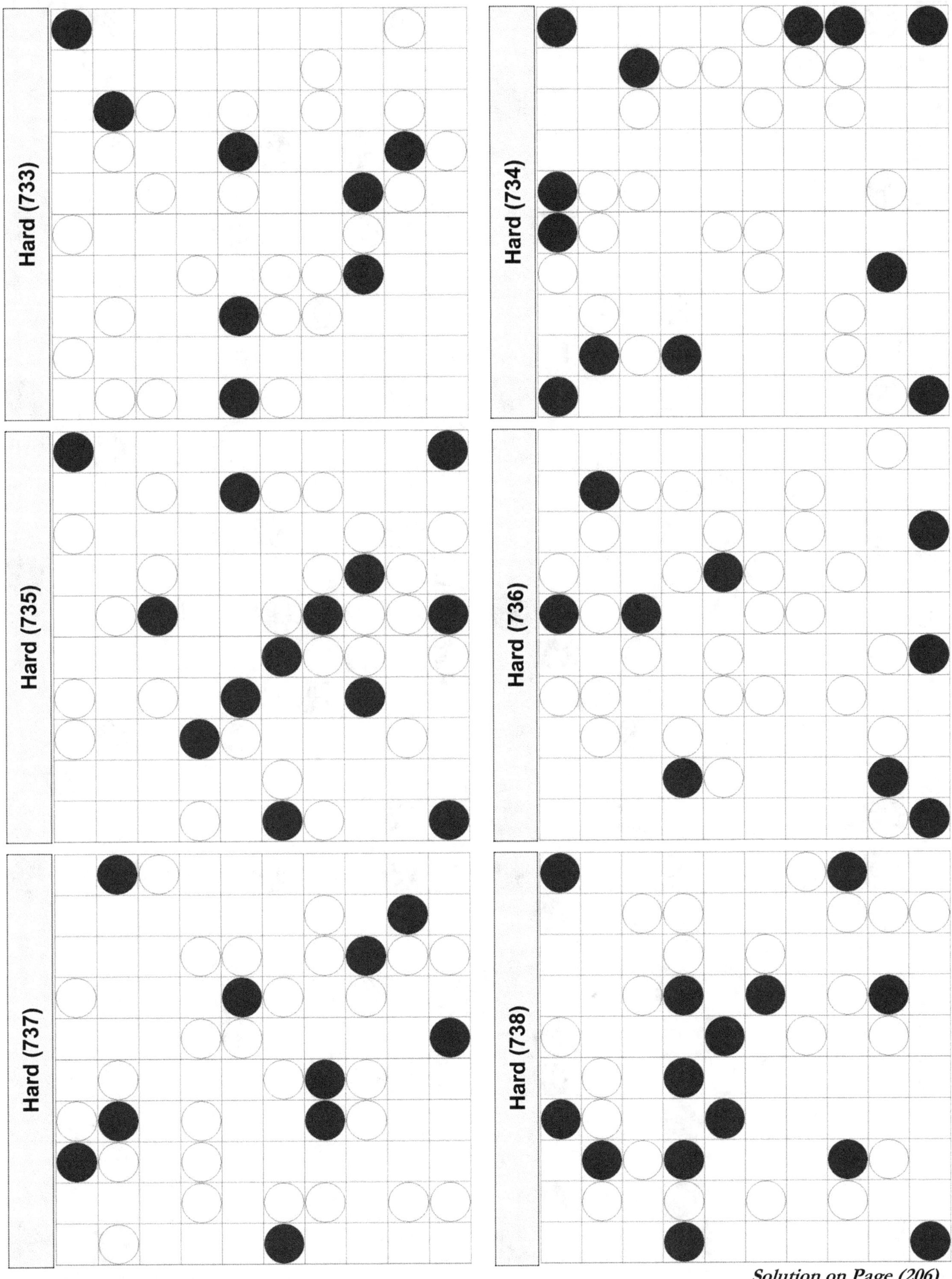

Hard (733)

Hard (734)

Hard (735)

Hard (736)

Hard (737)

Hard (738)

Solution on Page (206)

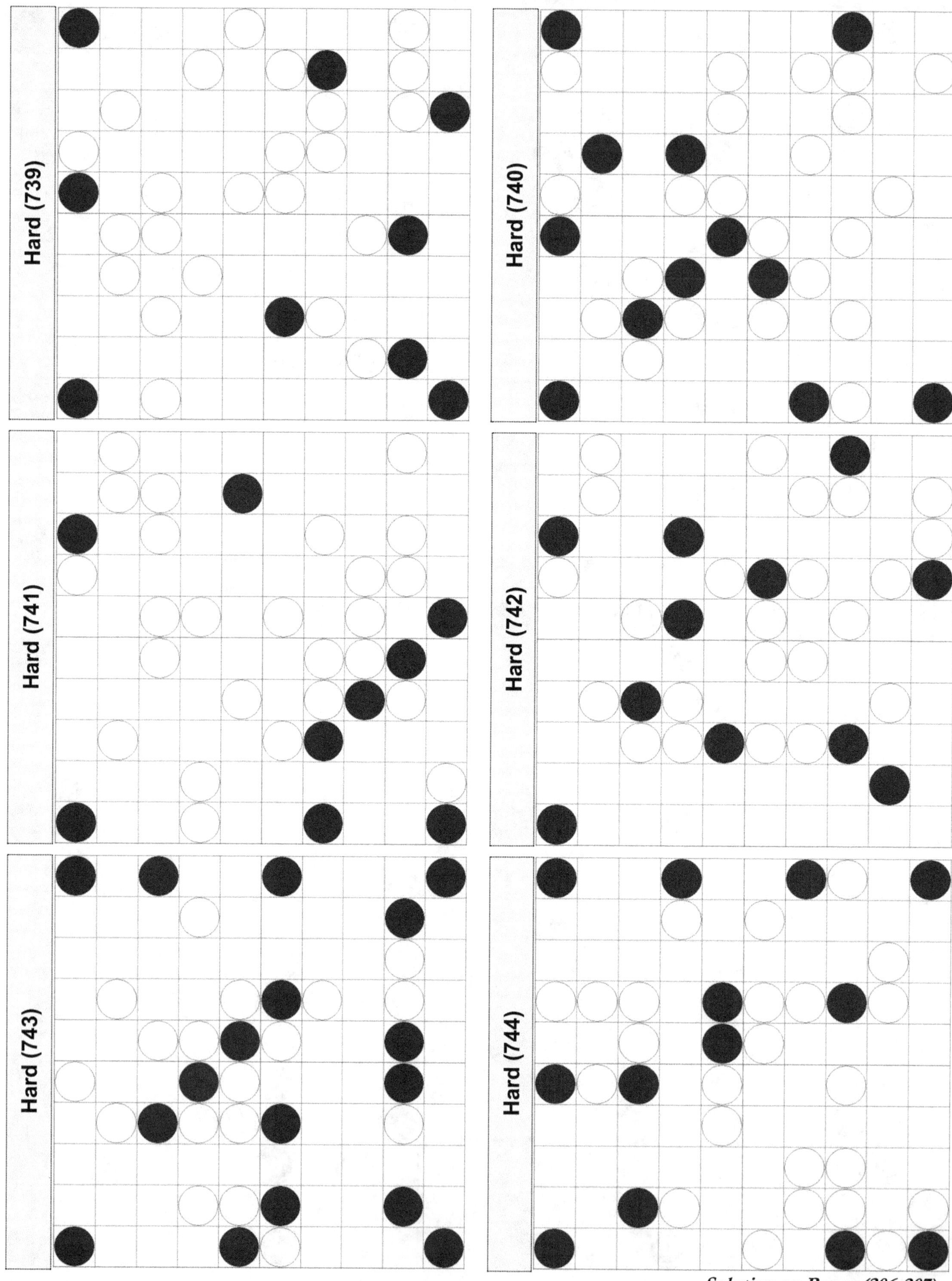

Solution on Pages (206-207)

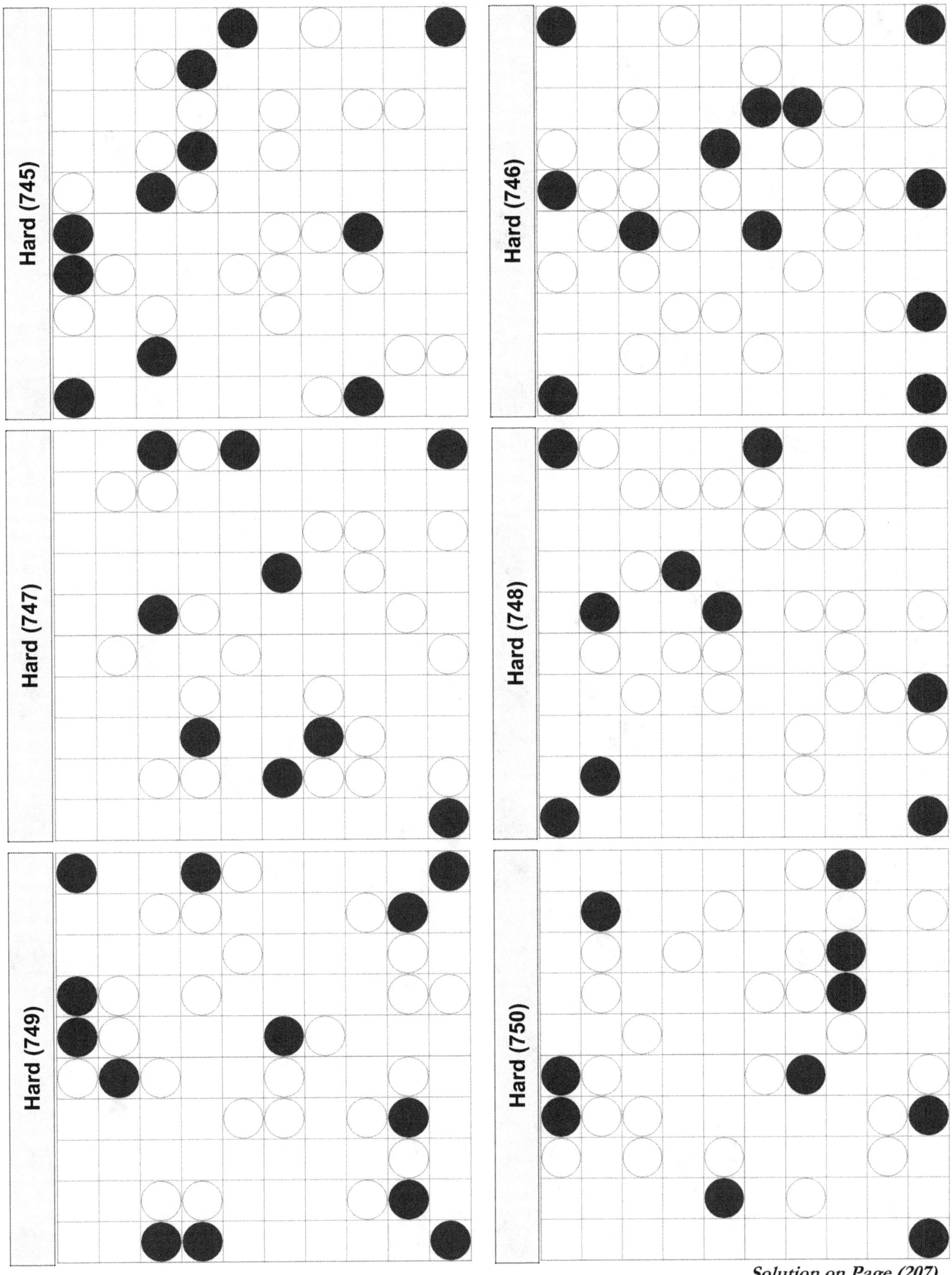

Solution on Page (207)

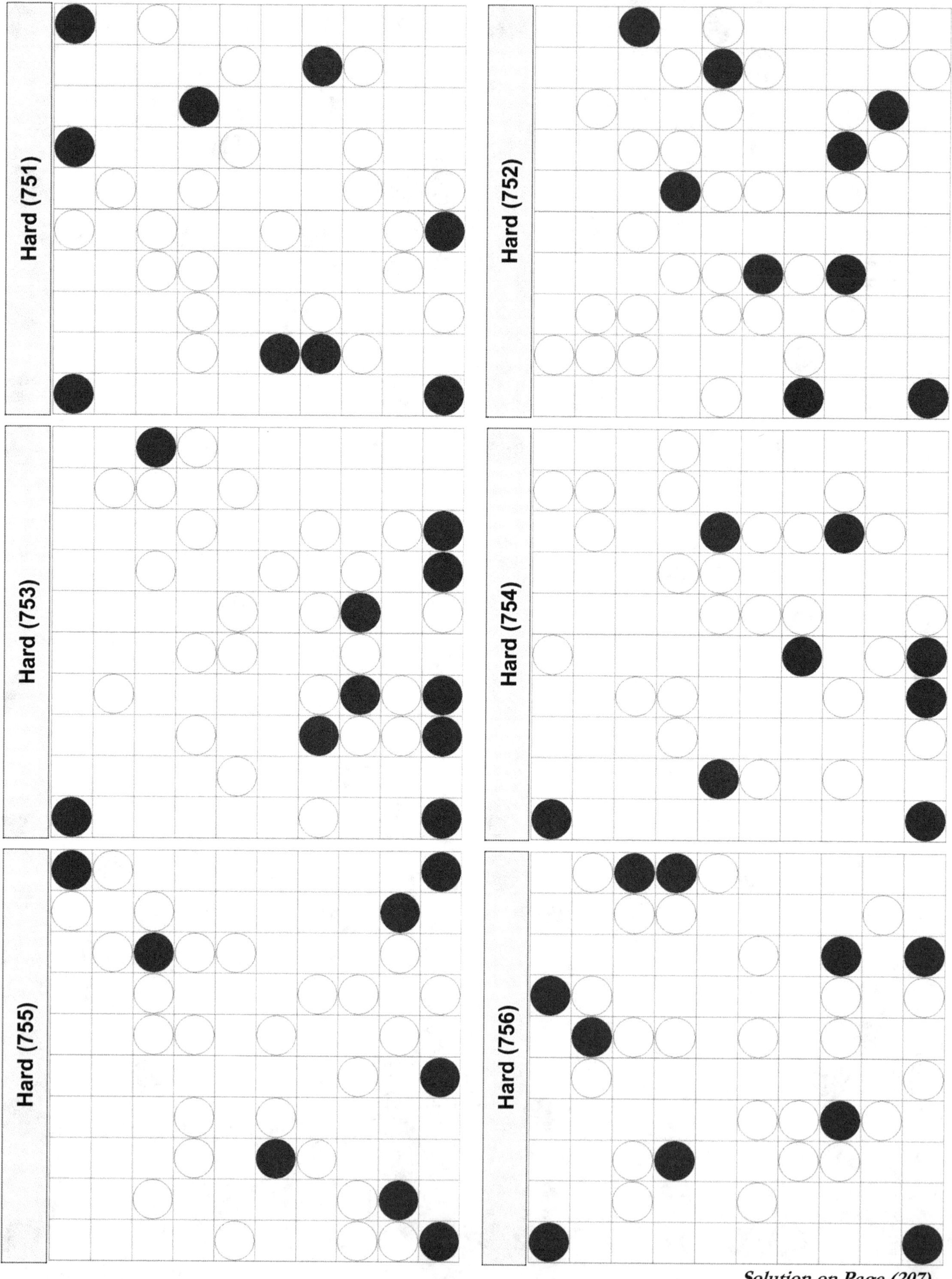

Hard (751)

Hard (752)

Hard (753)

Hard (754)

Hard (755)

Hard (756)

Solution on Page (207)

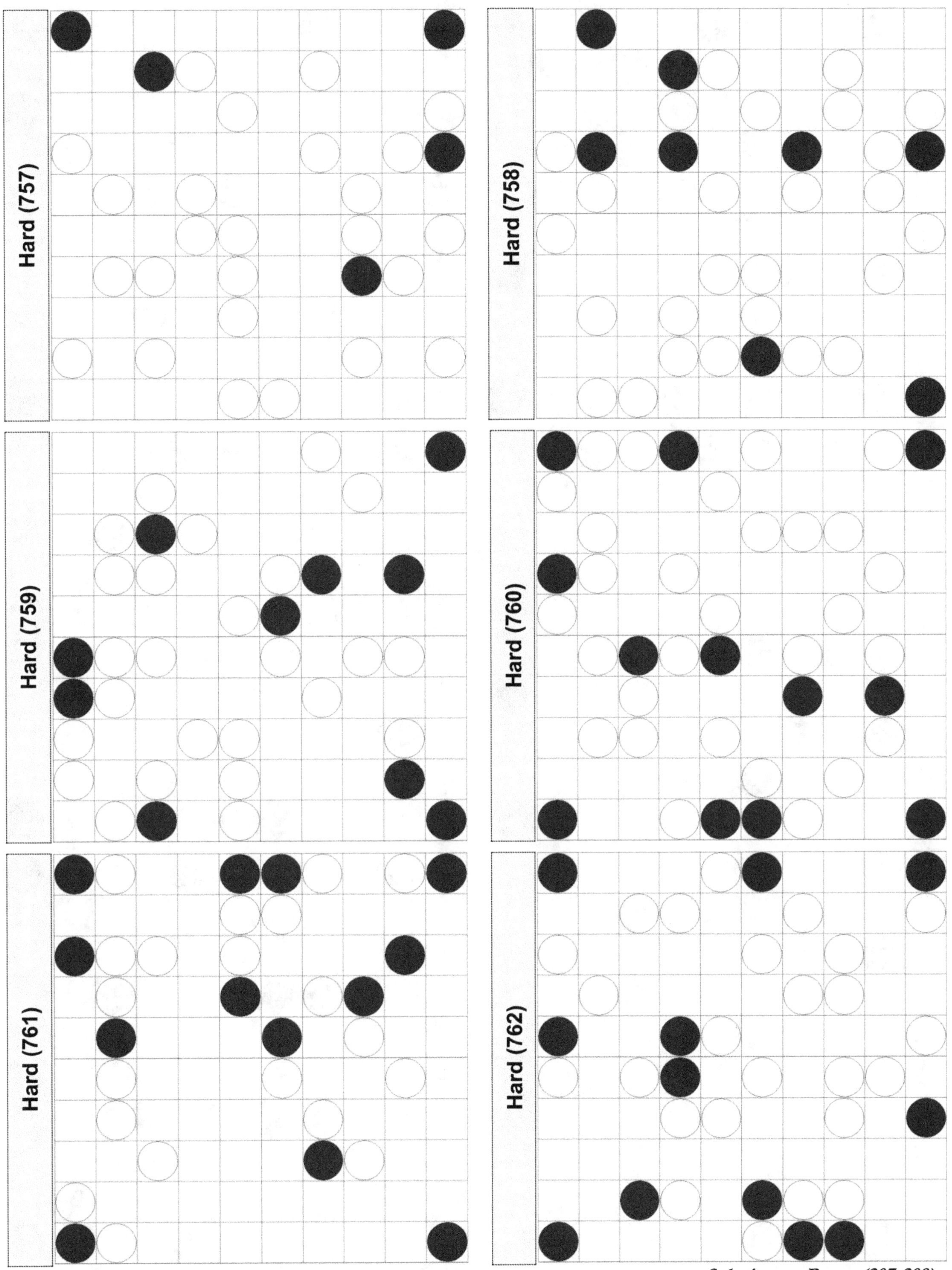

Solution on Pages (207-208)

(129)

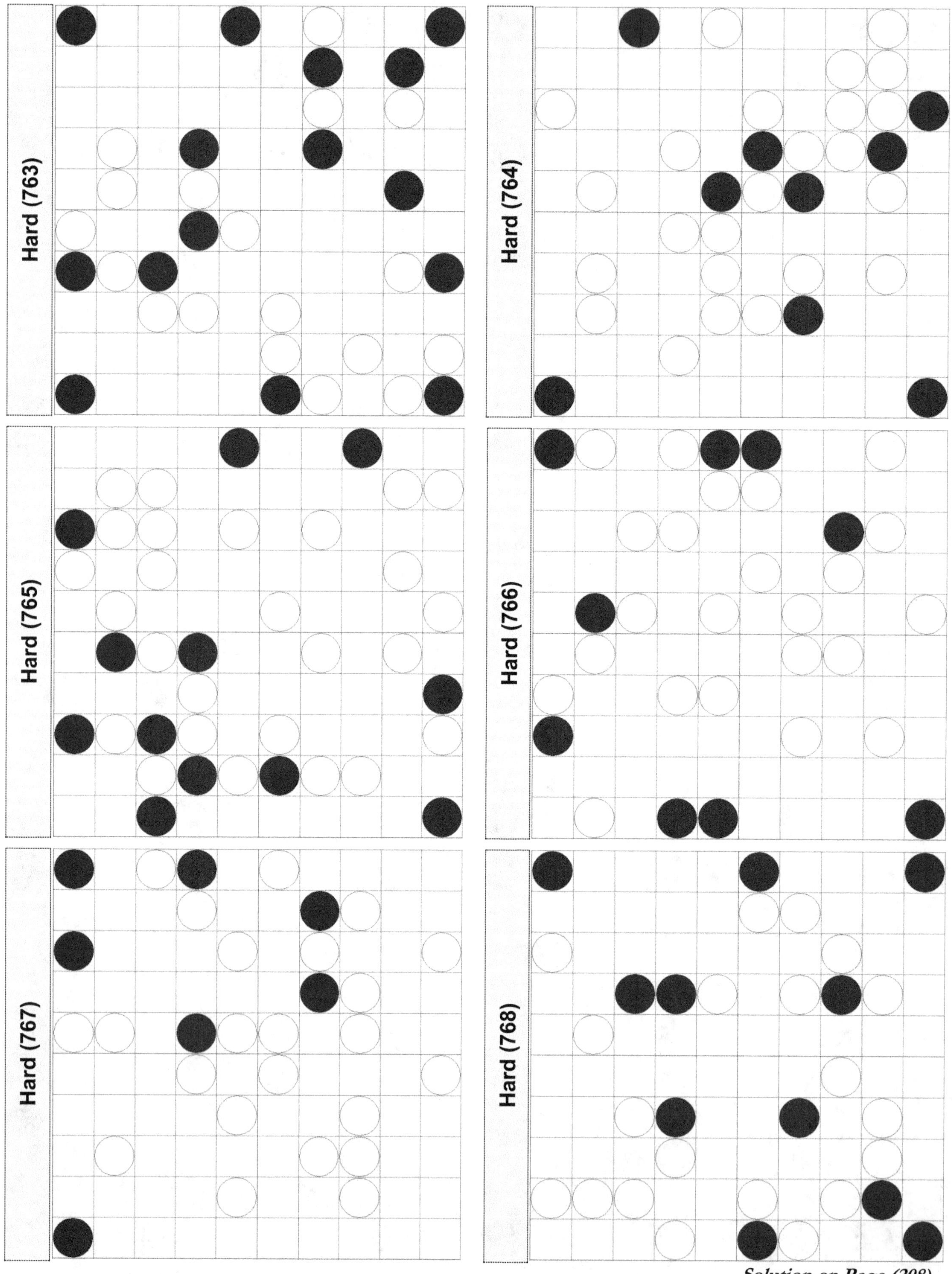

Solution on Page (208)

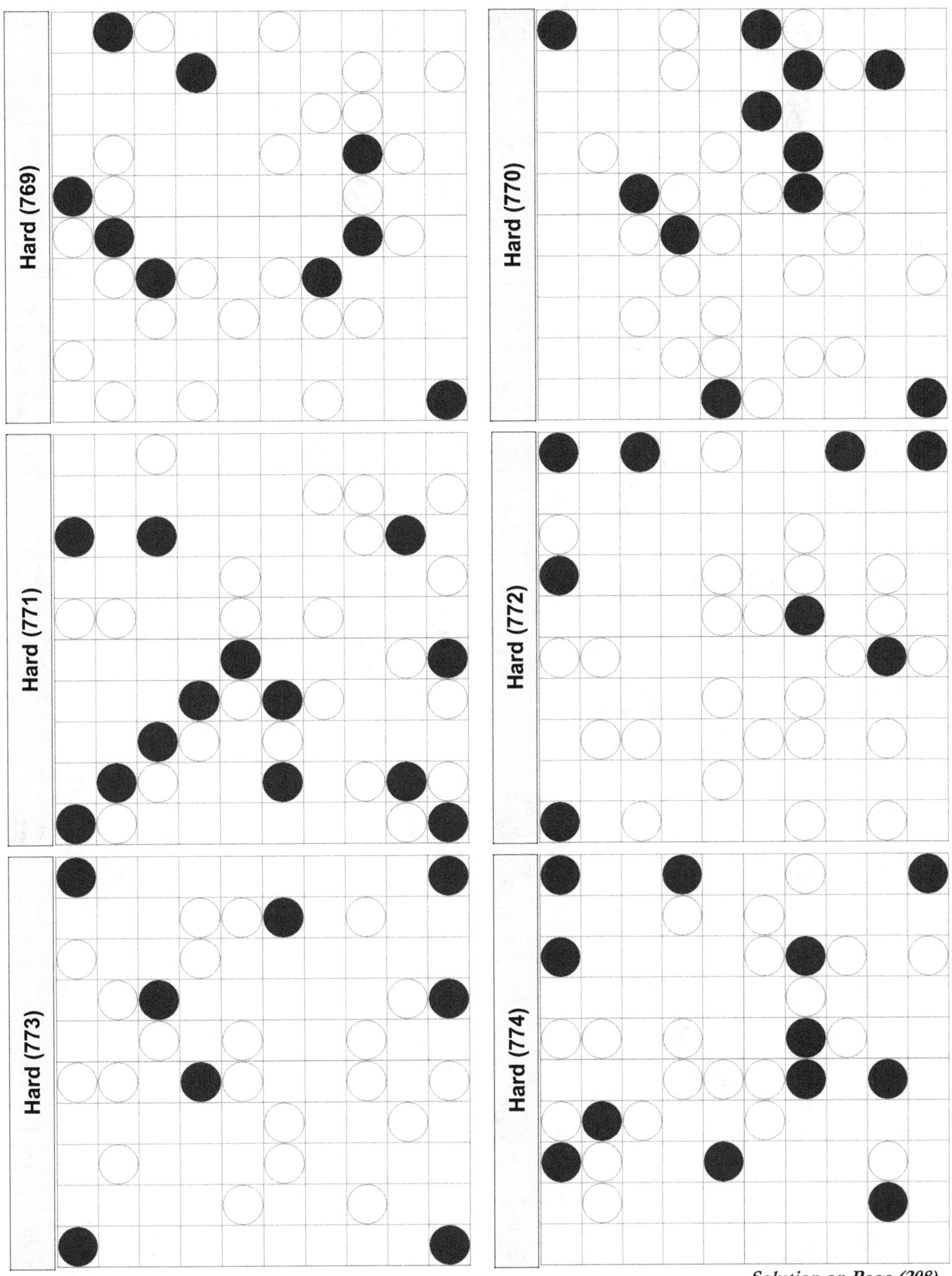

Hard (769)

Hard (770)

Hard (771)

Hard (772)

Hard (773)

Hard (774)

Solution on Page (208)

(131)

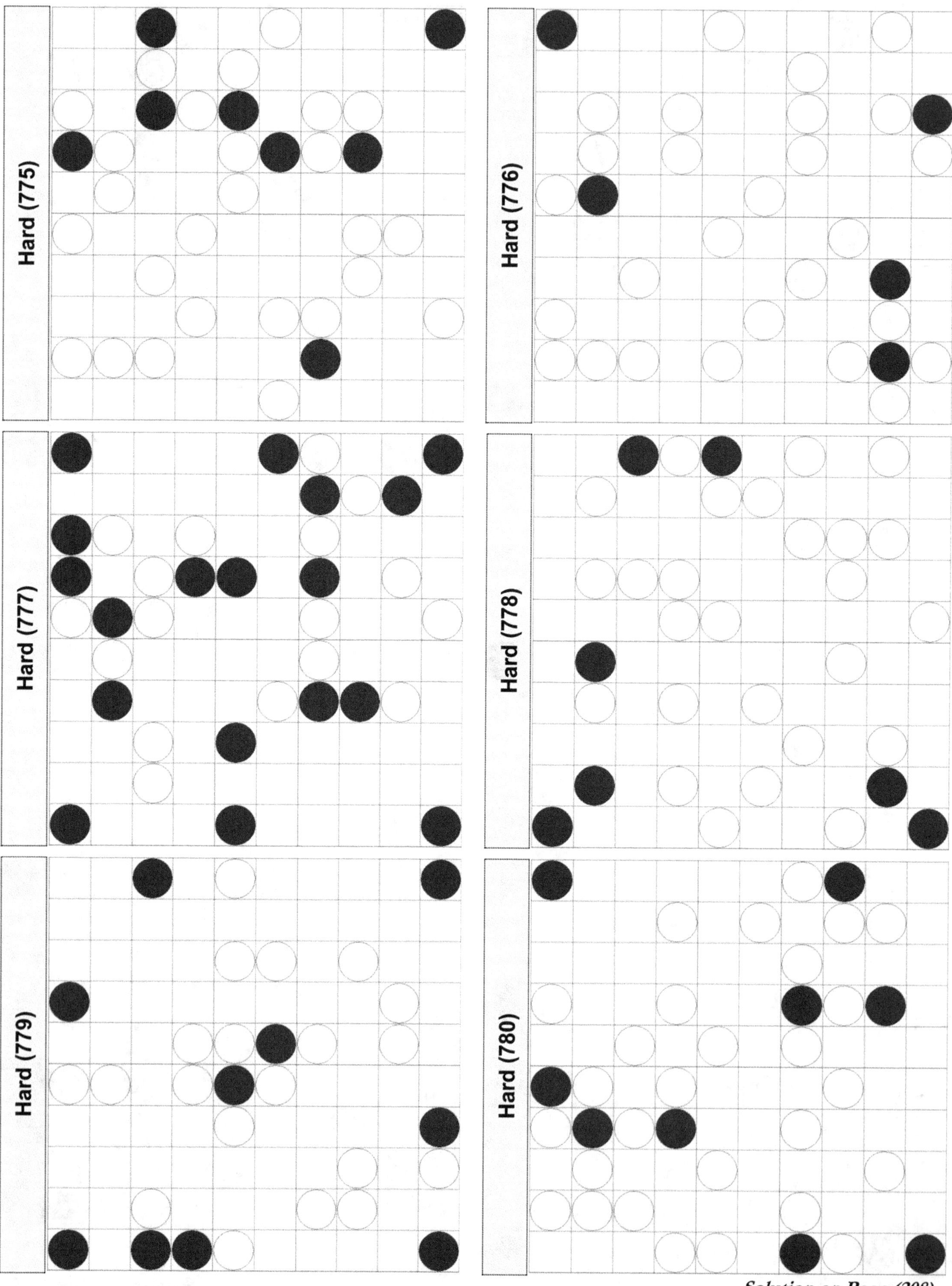

Hard (775)

Hard (776)

Hard (777)

Hard (778)

Hard (779)

Hard (780)

Solution on Page (208)

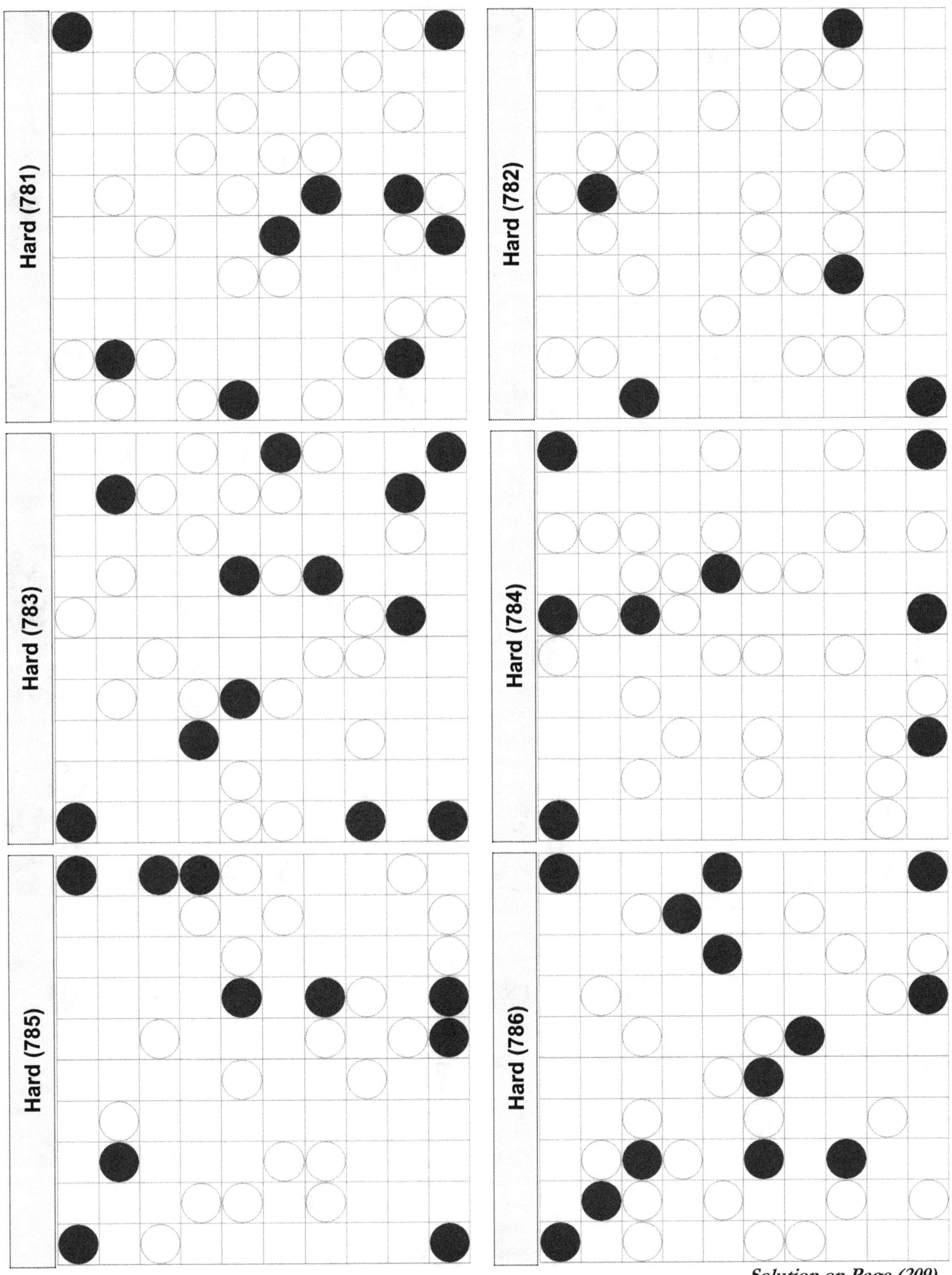

Solution on Page (209)

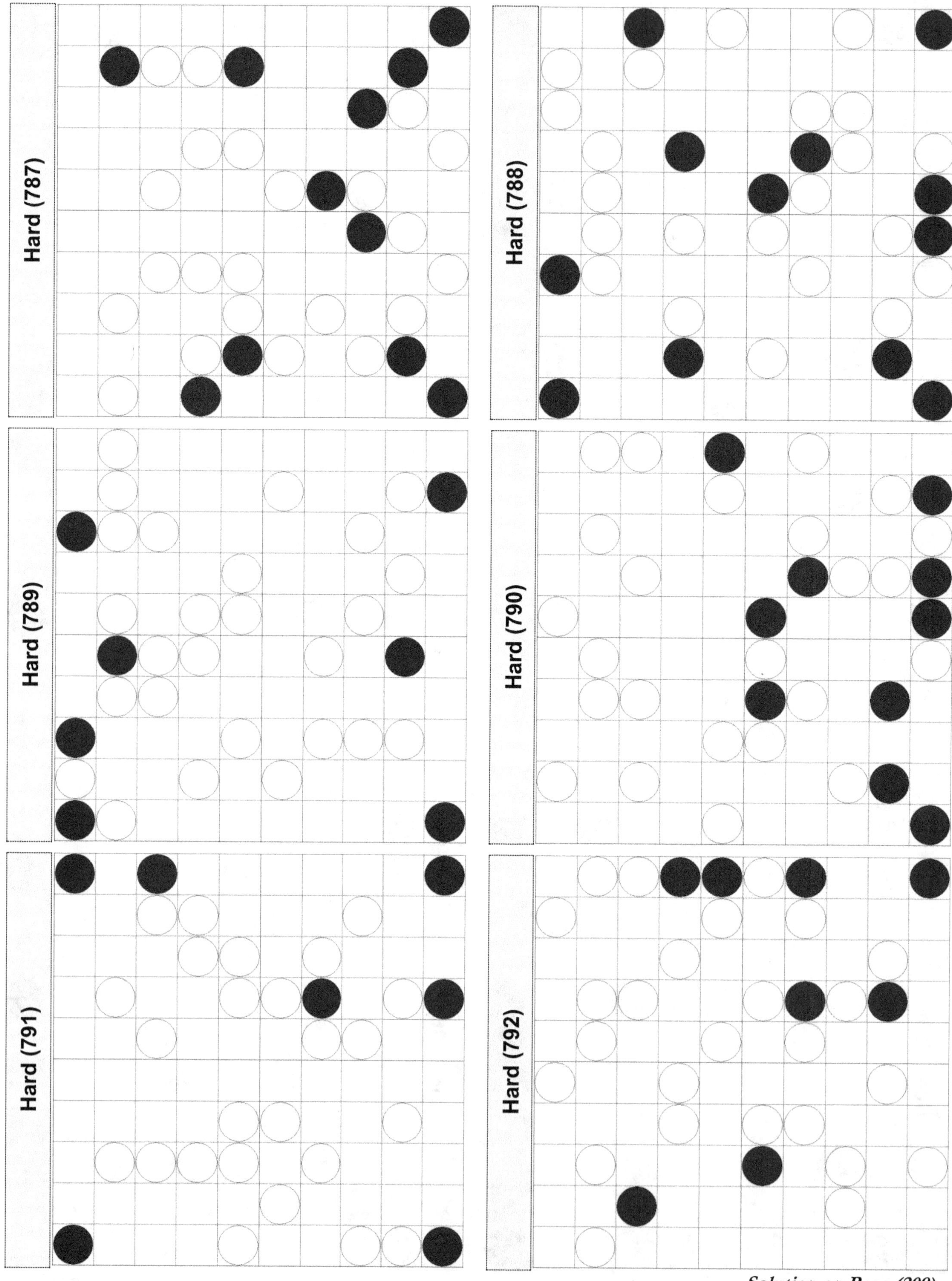

Solution on Page (209)

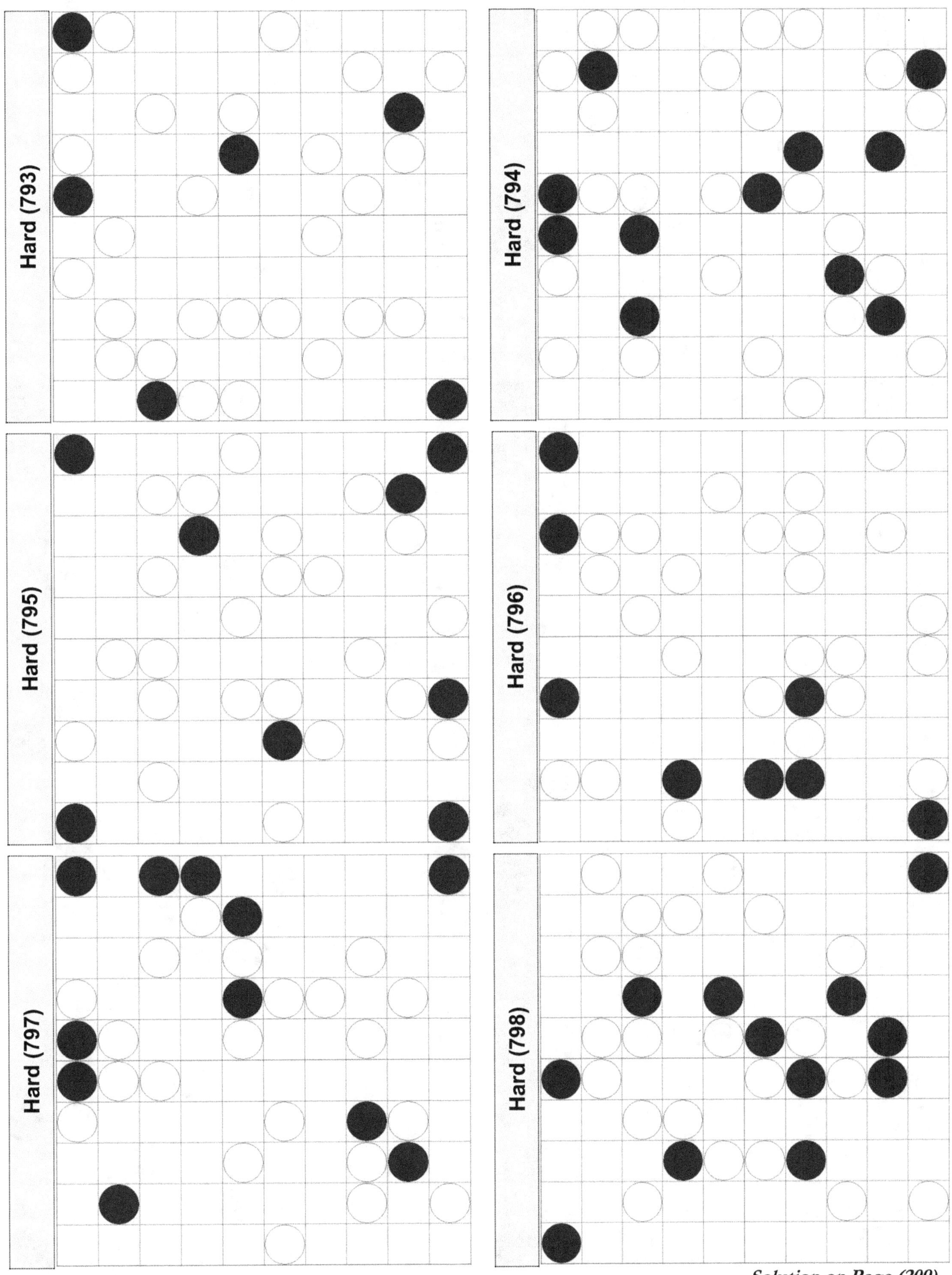

Solution on Page (209)

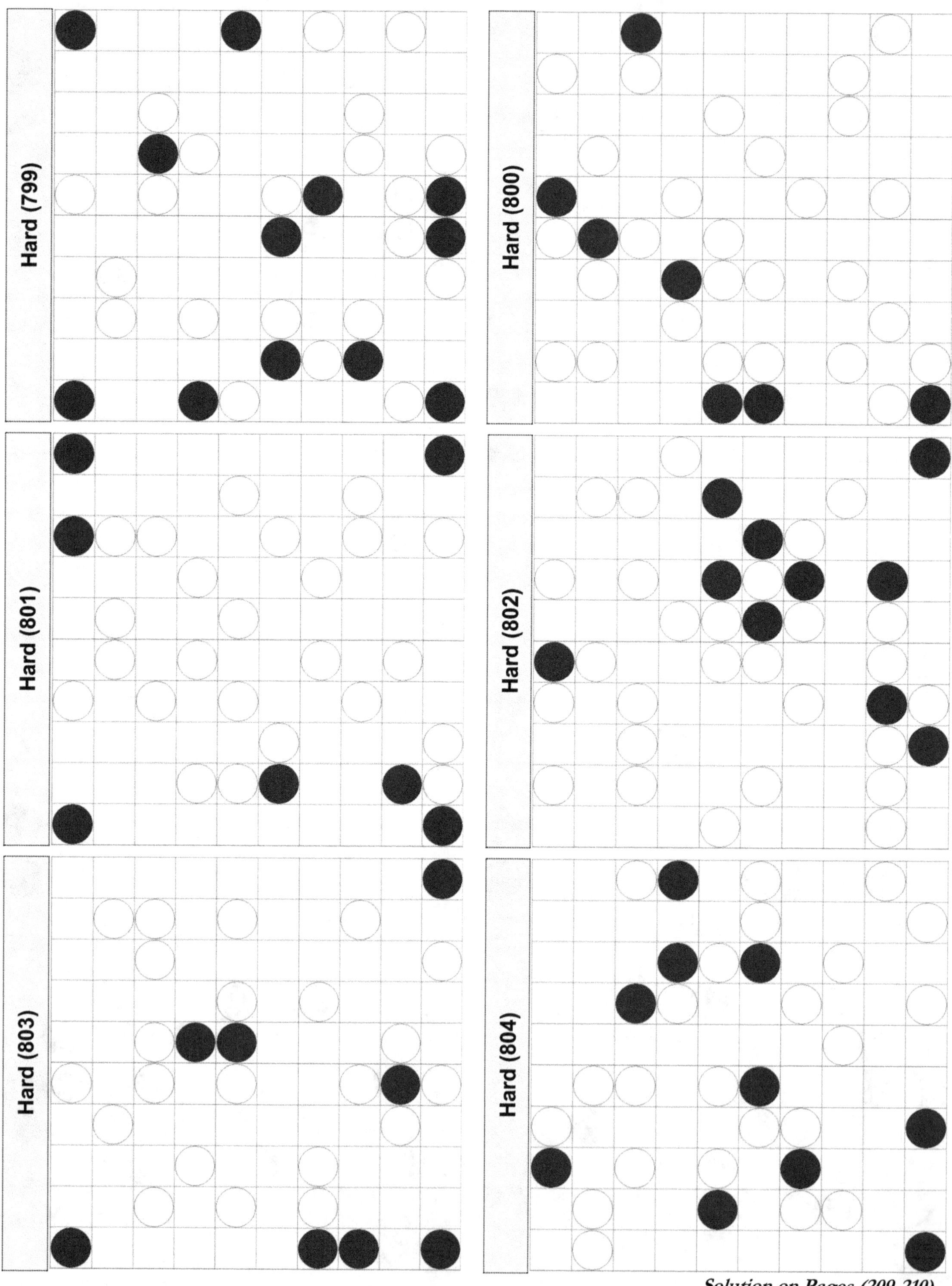

Solution on Pages (209-210)

(136)

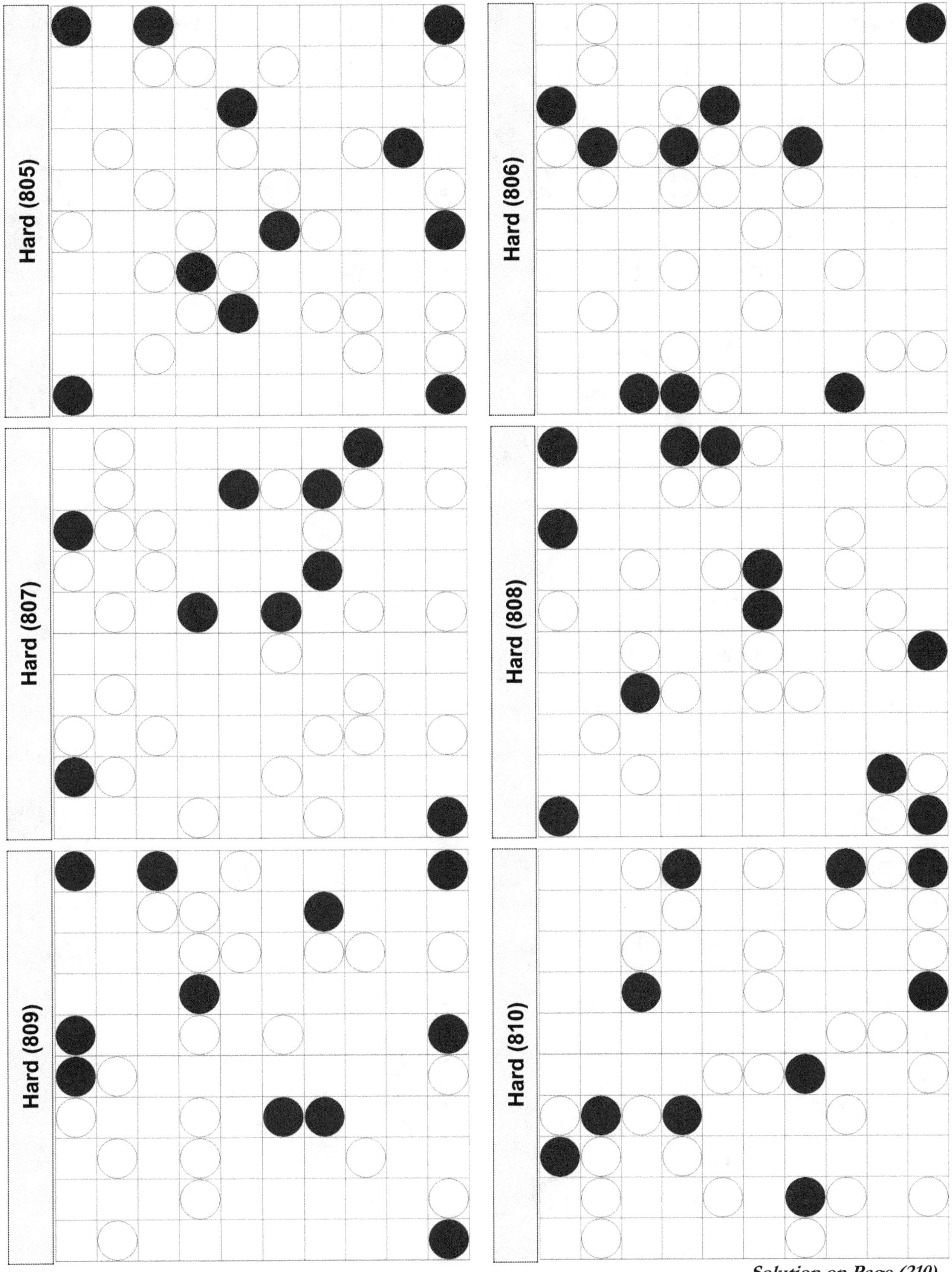

Solution on Page (210)

(137)

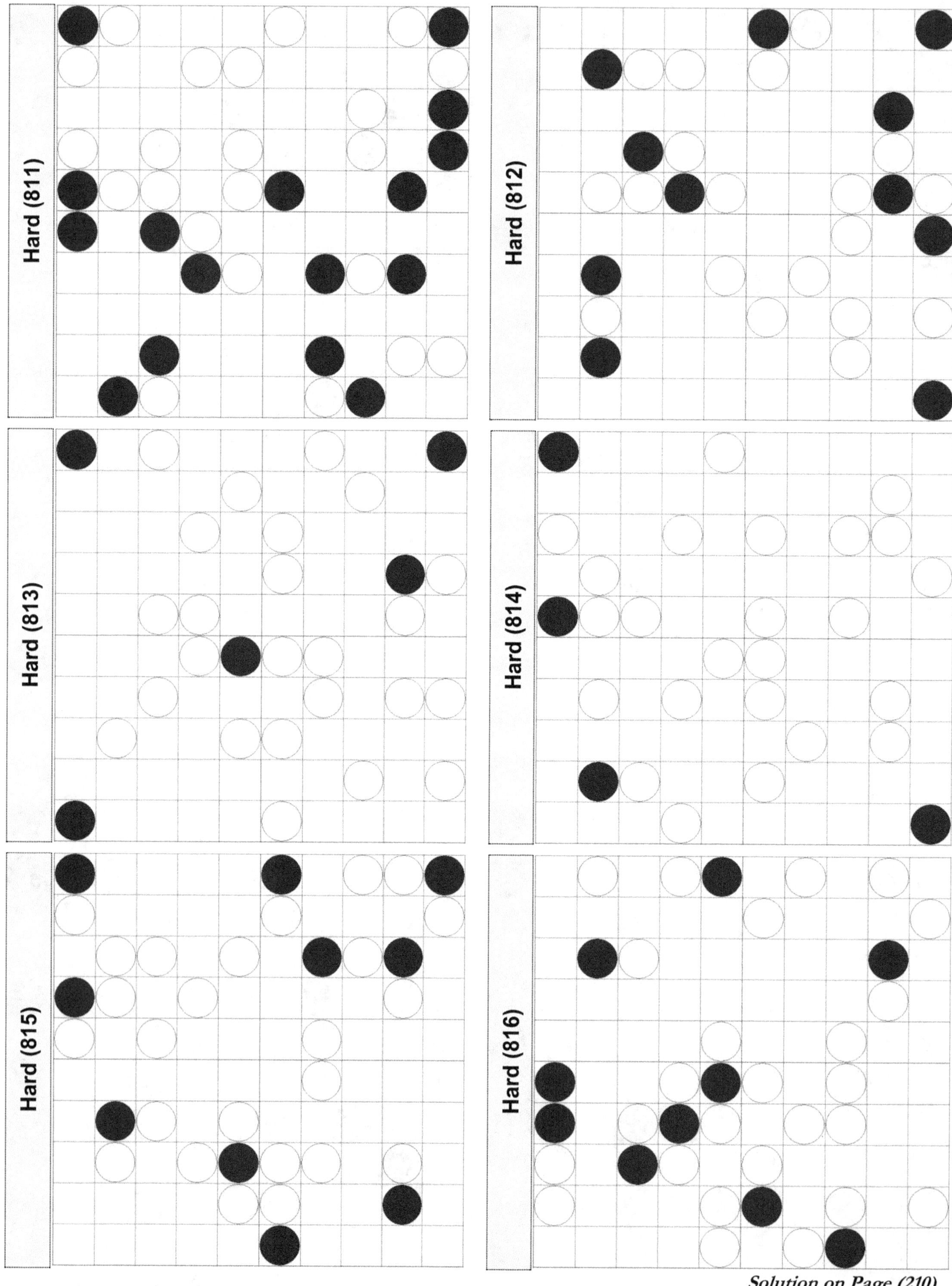

Solution on Page (210)

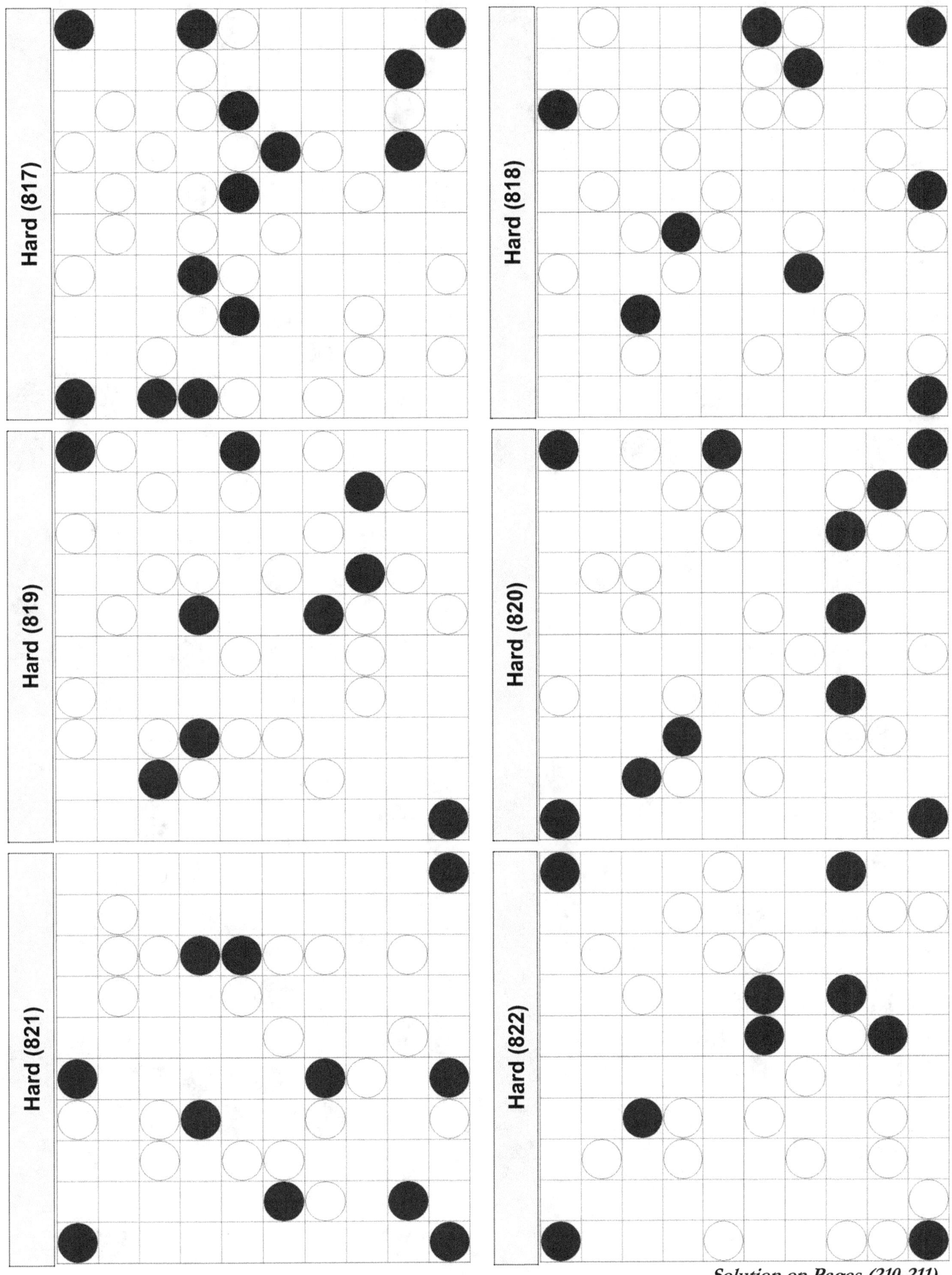

Solution on Pages (210-211)

(139)

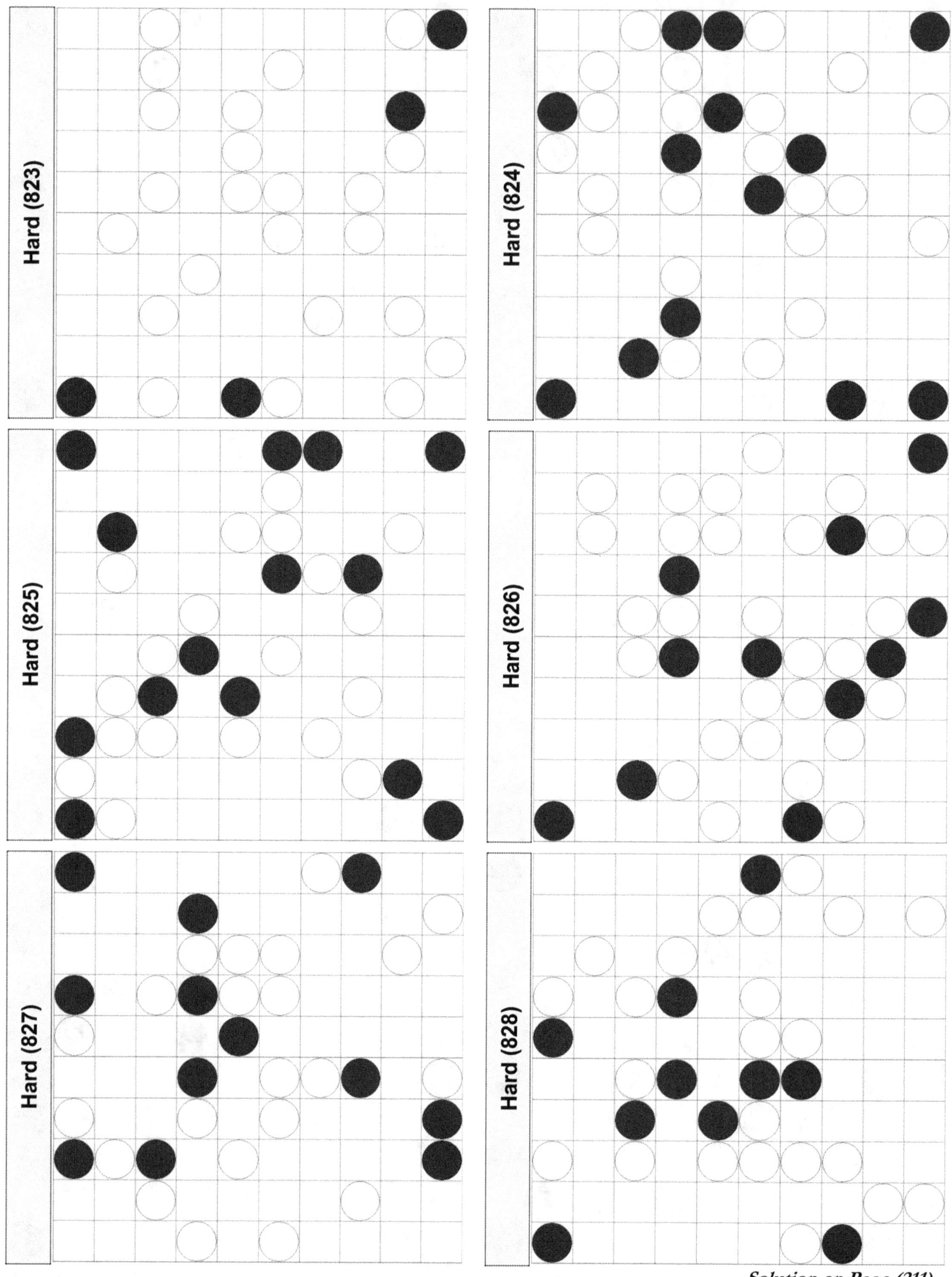

Hard (823)

Hard (824)

Hard (825)

Hard (826)

Hard (827)

Hard (828)

Solution on Page (211)

(140)

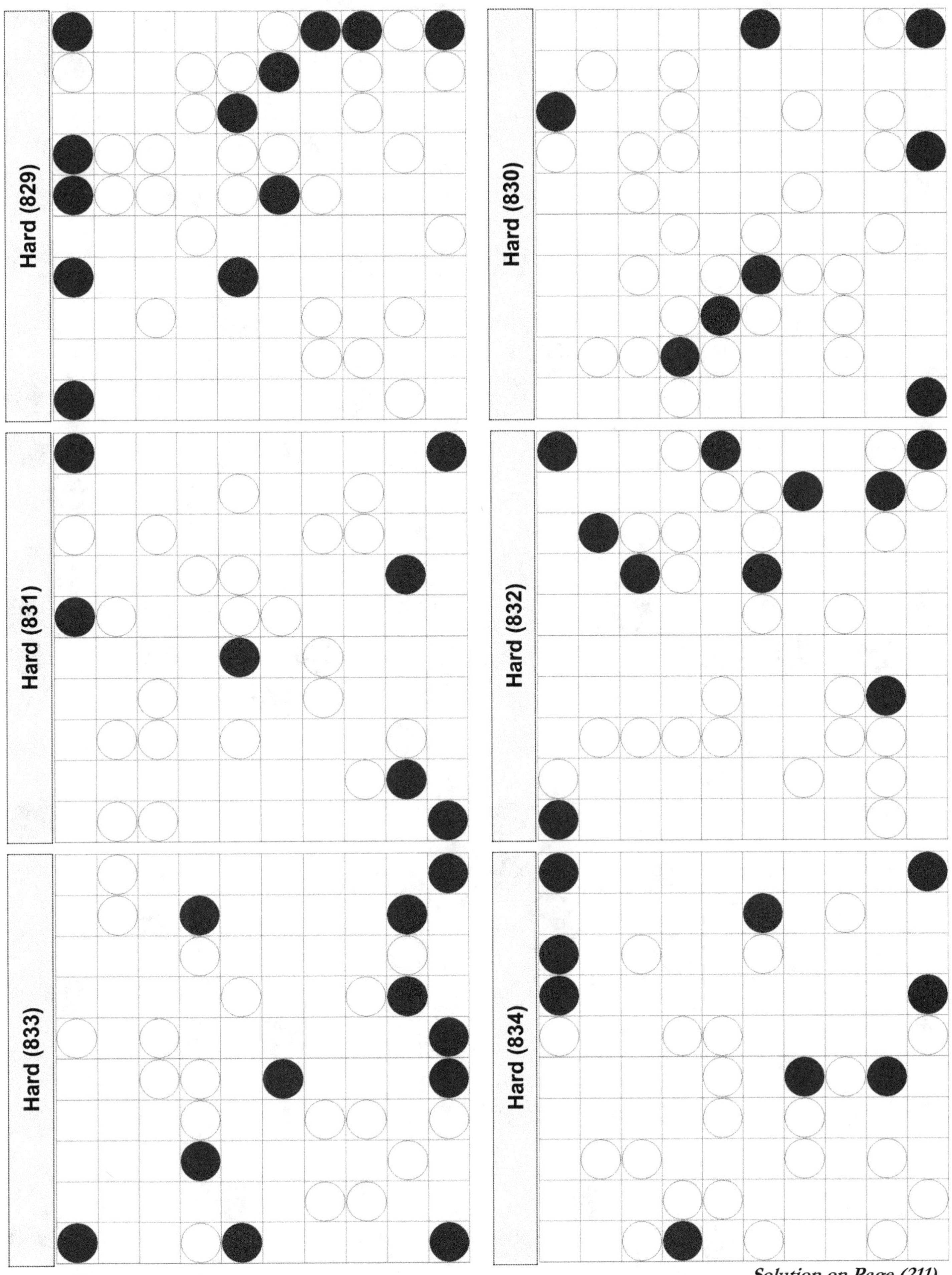

Hard (829)
Hard (830)
Hard (831)
Hard (832)
Hard (833)
Hard (834)
Solution on Page (211)

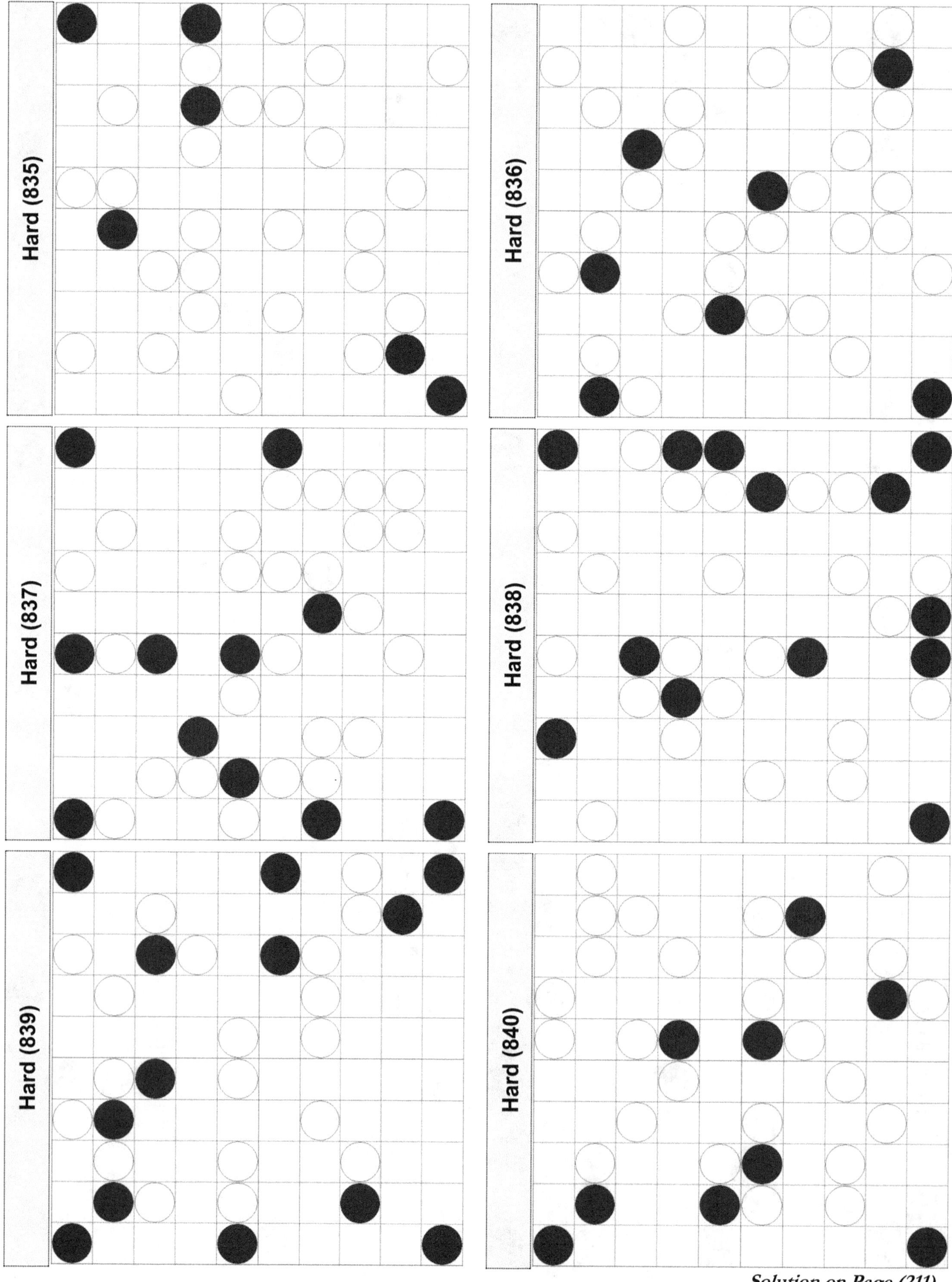

Hard (835)
Hard (836)
Hard (837)
Hard (838)
Hard (839)
Hard (840)
Solution on Page (211)

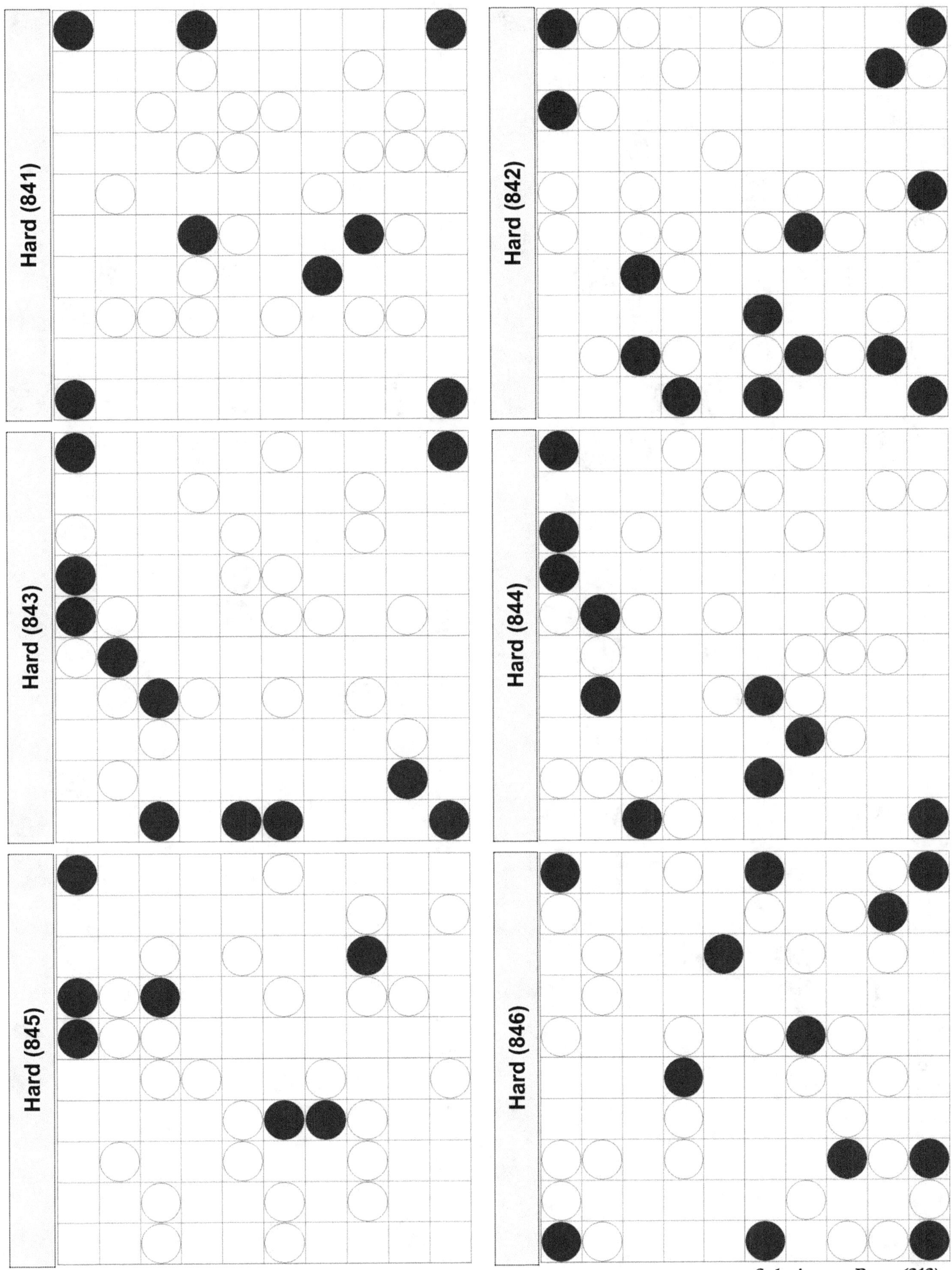

Solution on Page (212)

(143)

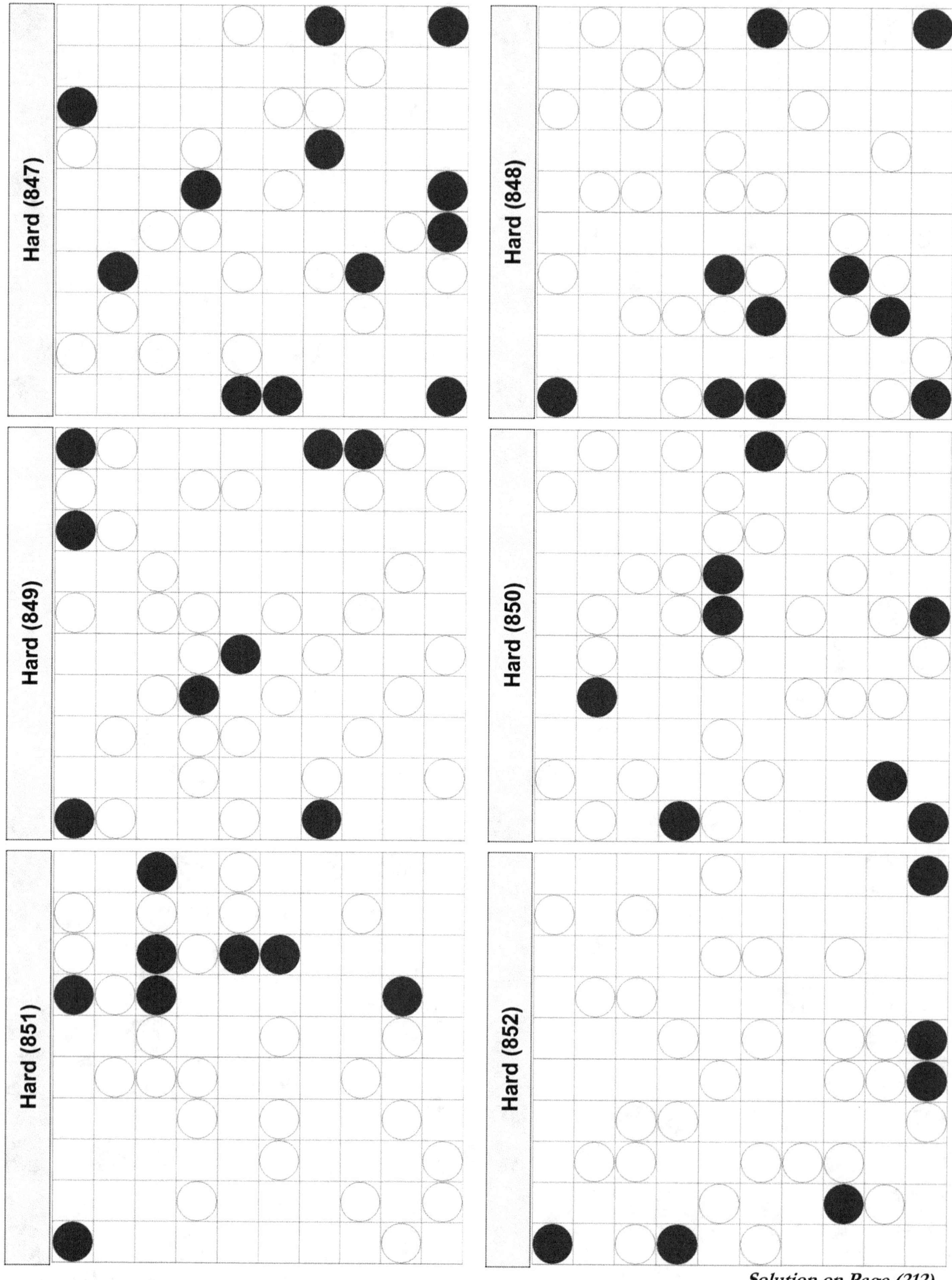

Solution on Page (212)

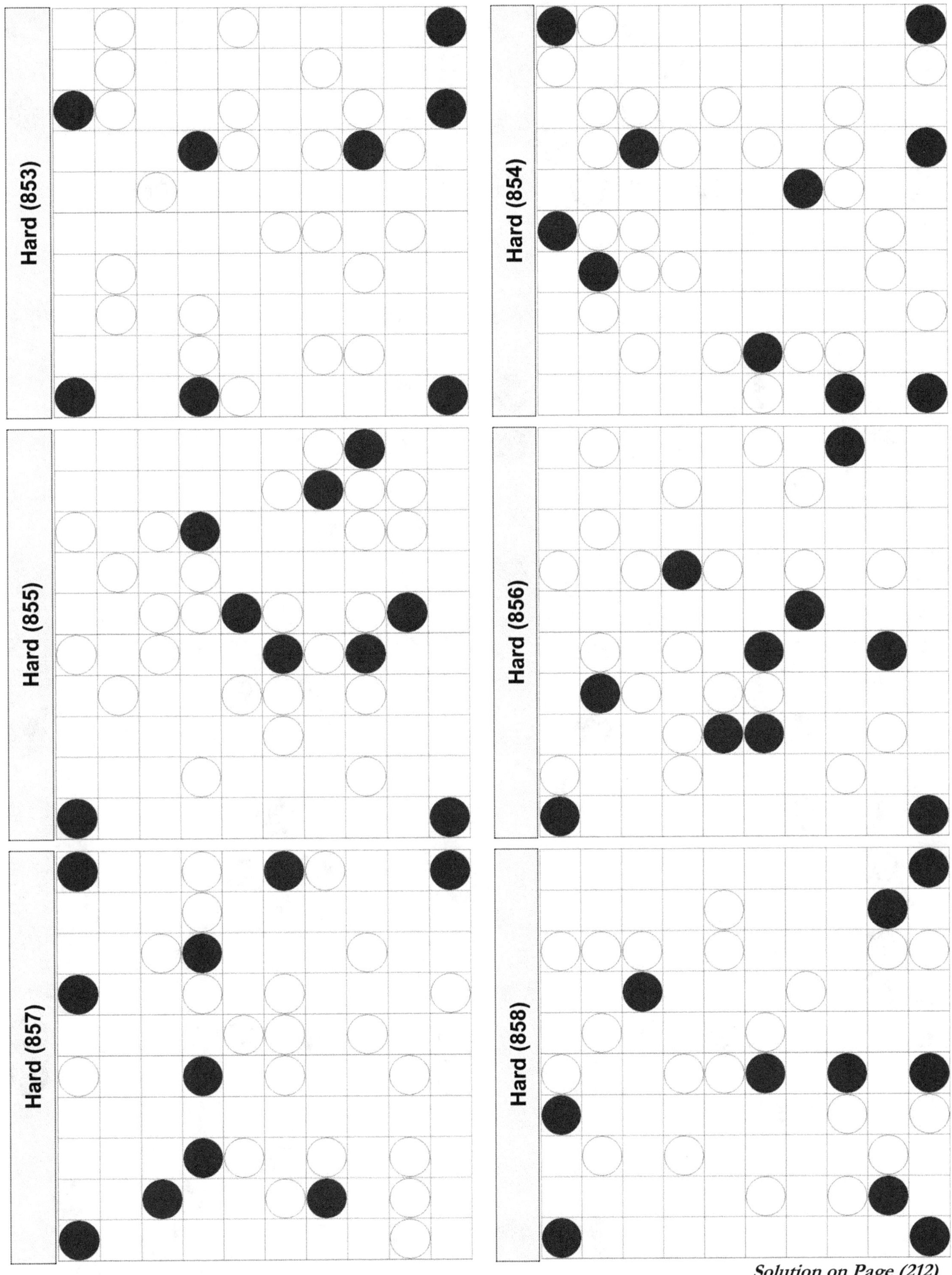

Hard (853)
Hard (854)
Hard (855)
Hard (856)
Hard (857)
Hard (858)
Solution on Page (212)

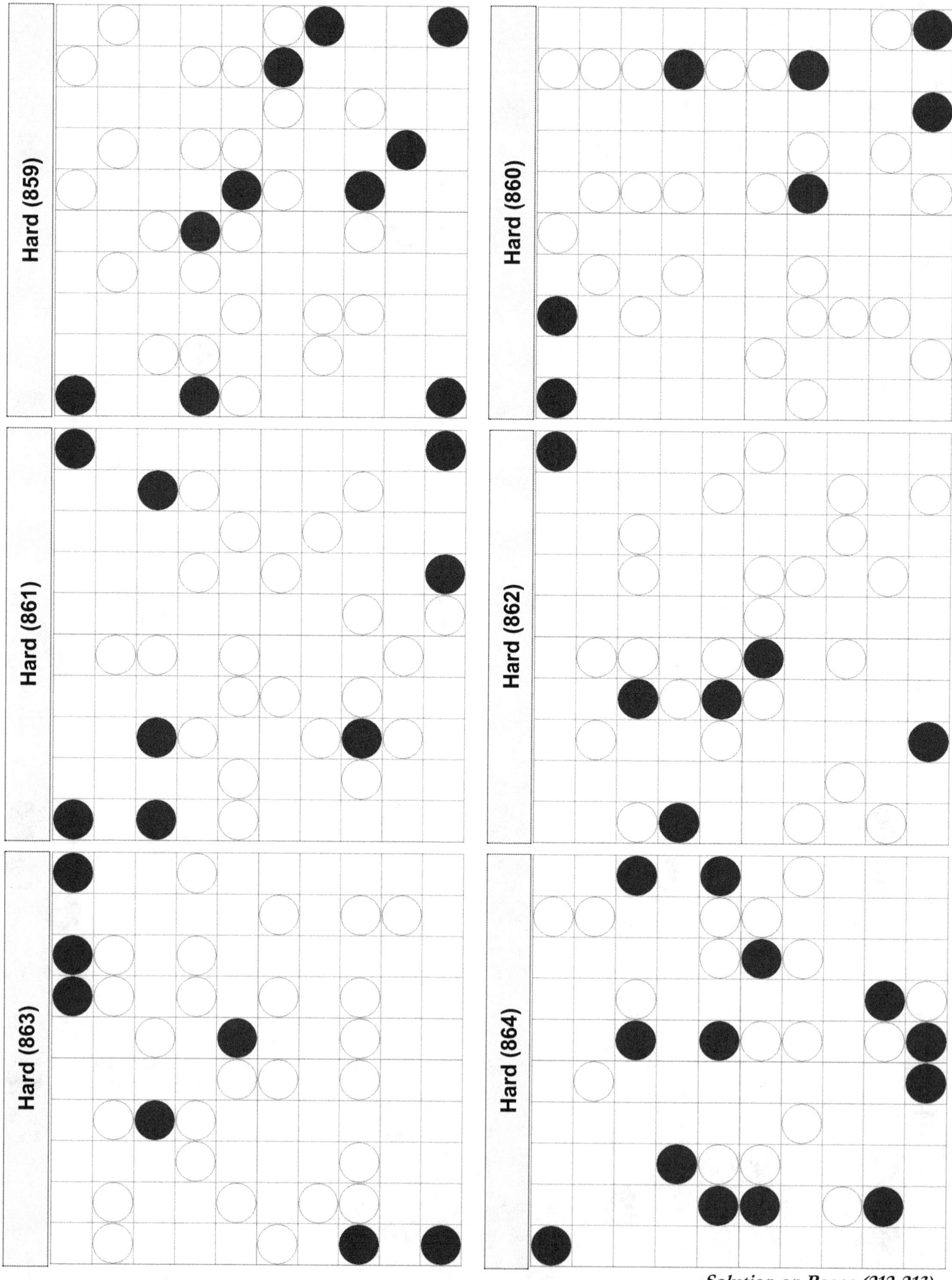

Solution on Pages (212-213)

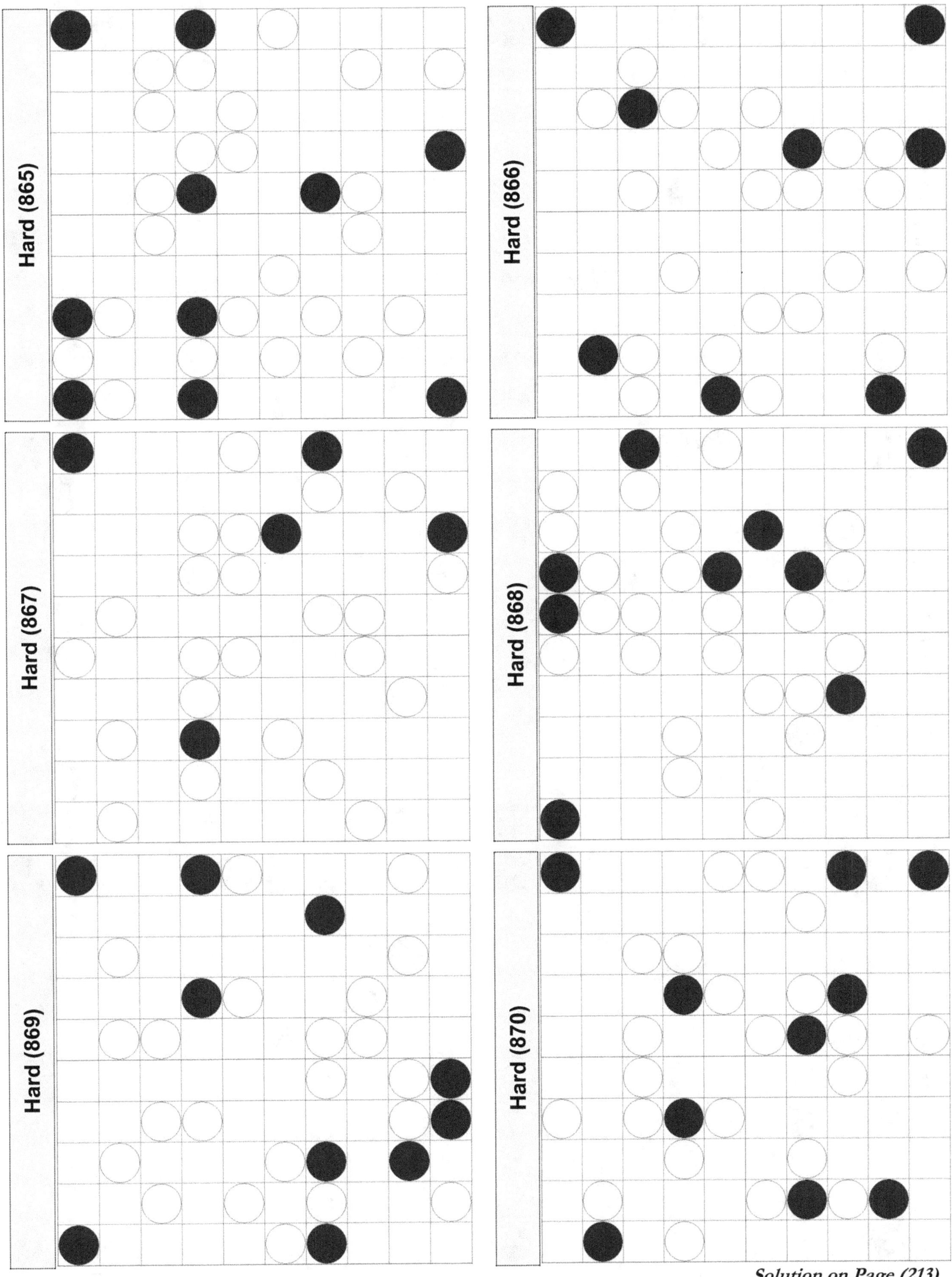

Solution on Page (213)

(147)

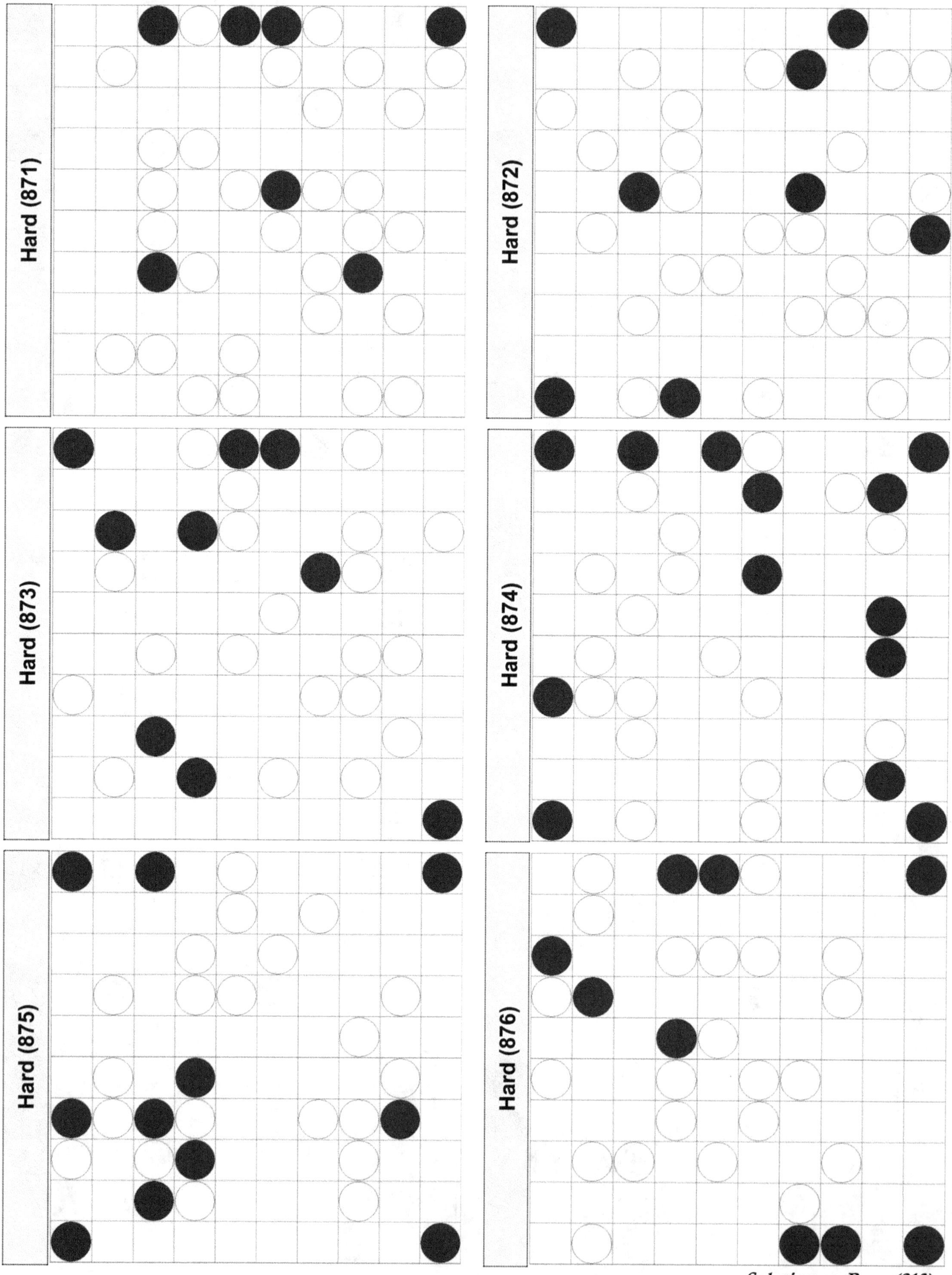

Hard (871)

Hard (872)

Hard (873)

Hard (874)

Hard (875)

Hard (876)

Solution on Page (213)

(148)

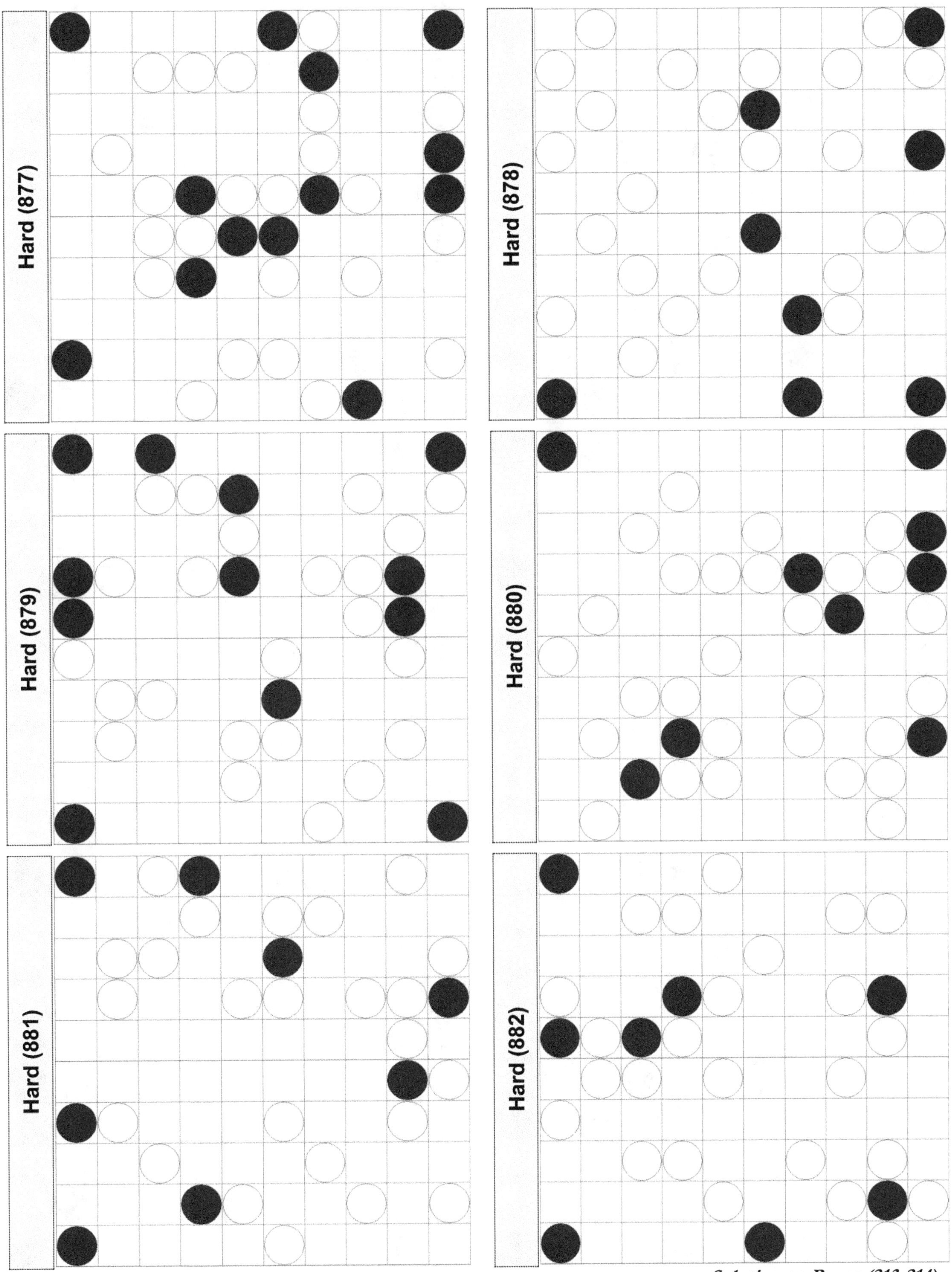

Solution on Pages (213-214)

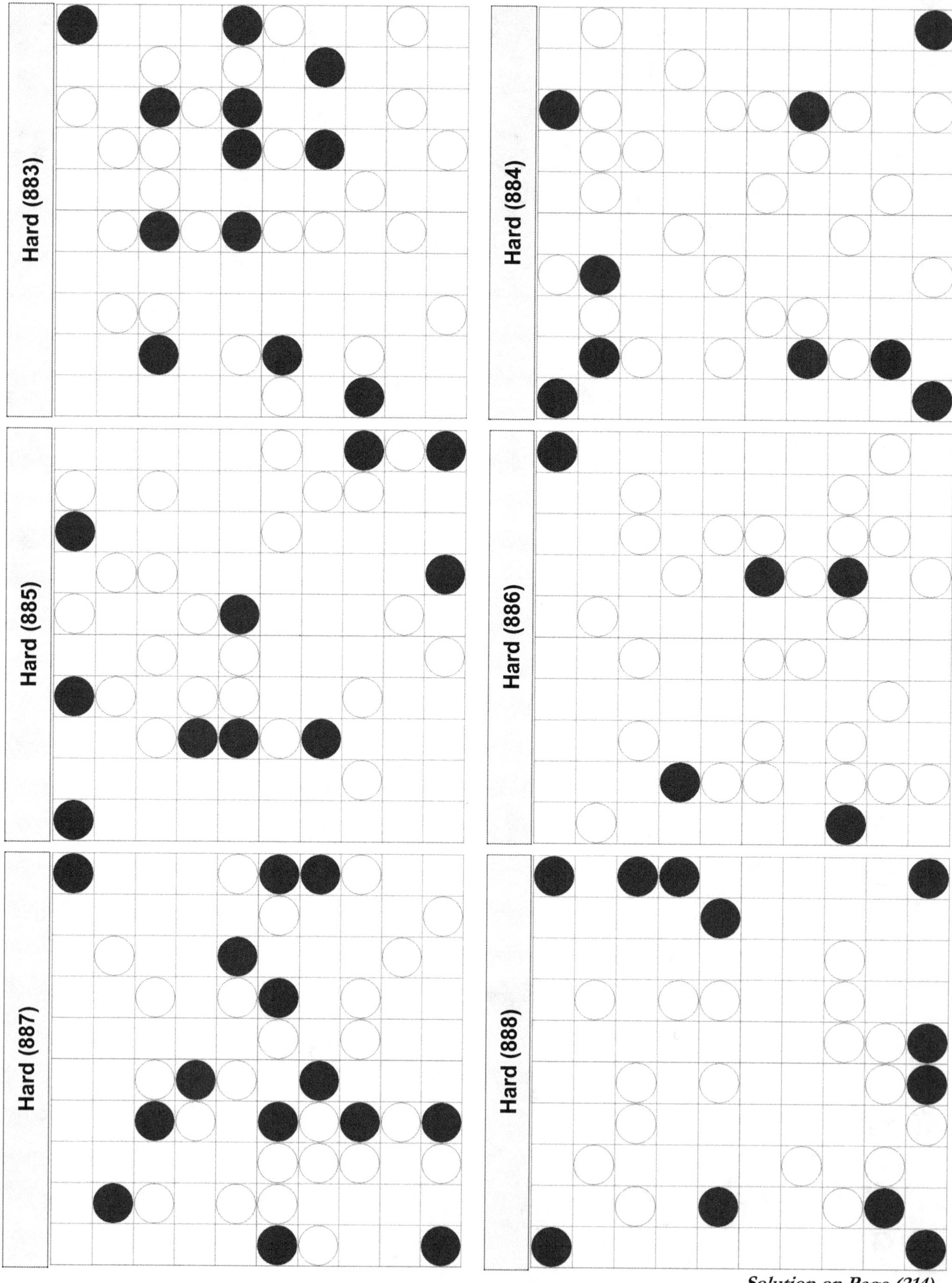

Hard (883)

Hard (884)

Hard (885)

Hard (886)

Hard (887)

Hard (888)

Solution on Page (214)

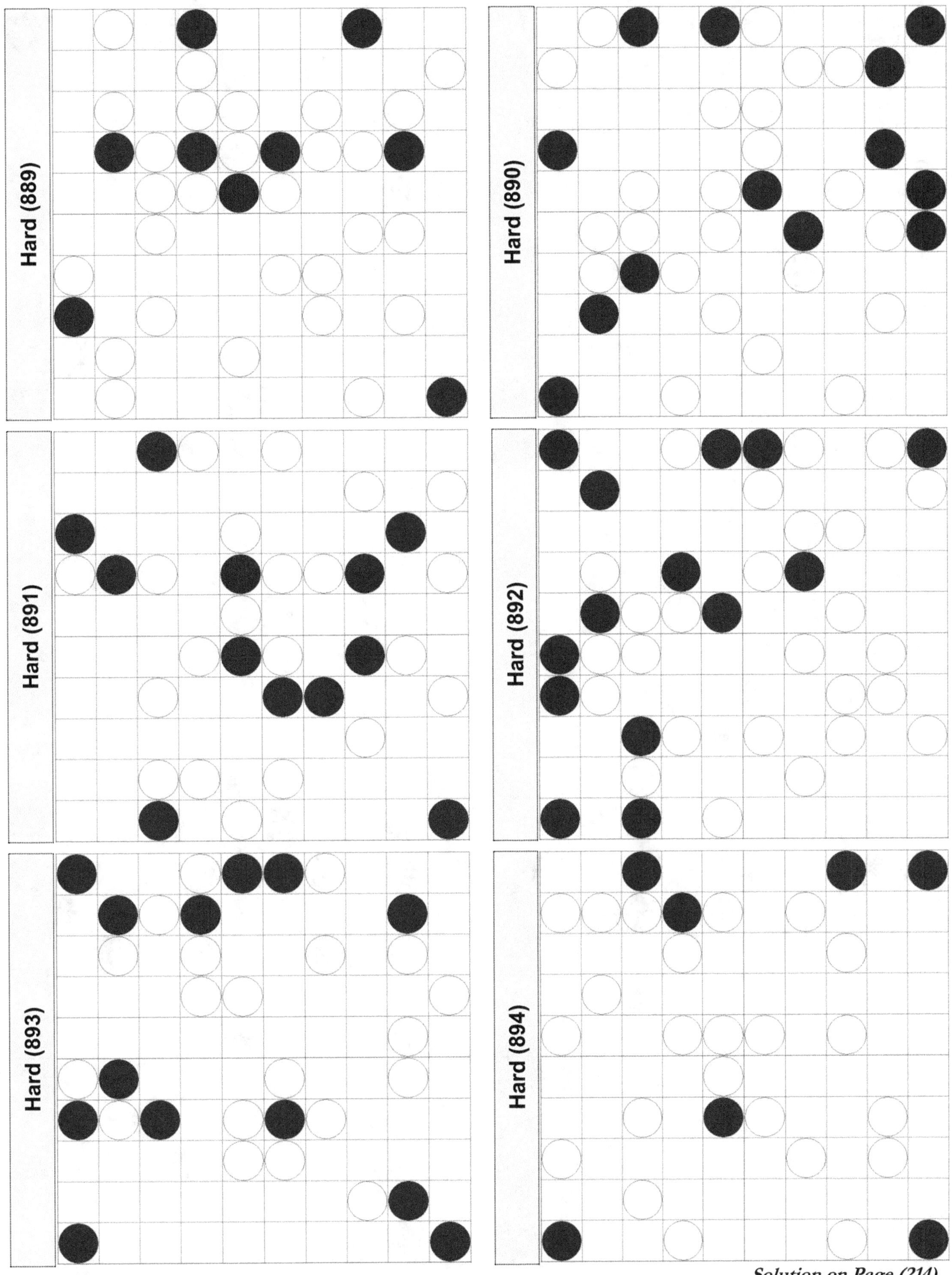

Solution on Page (214)

(151)

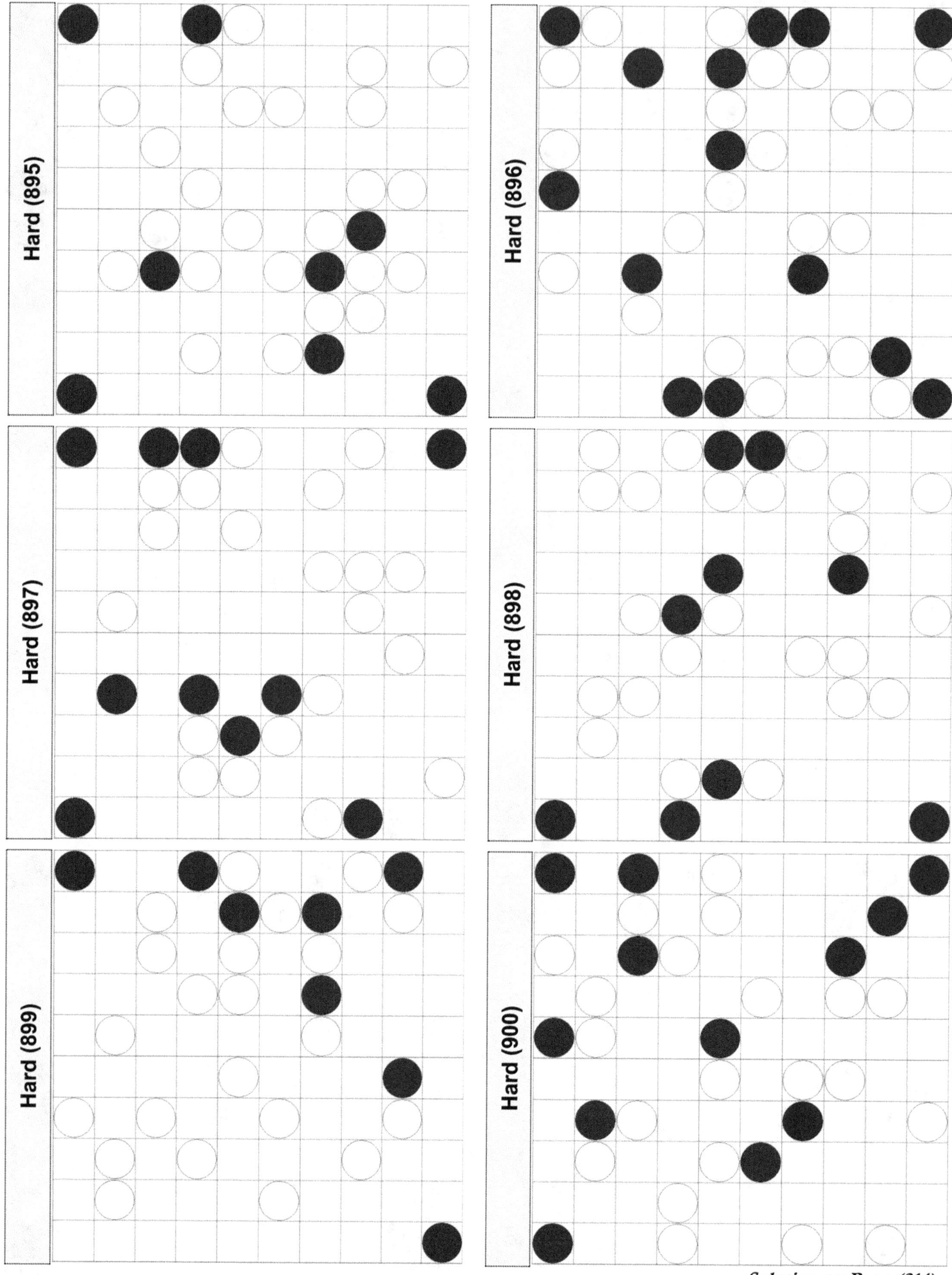

Solution on Page (214)

(152)

Hard (901)
Hard (902)
Hard (903)
Hard (904)
Hard (905)
Hard (906)
Solution on Page (215)

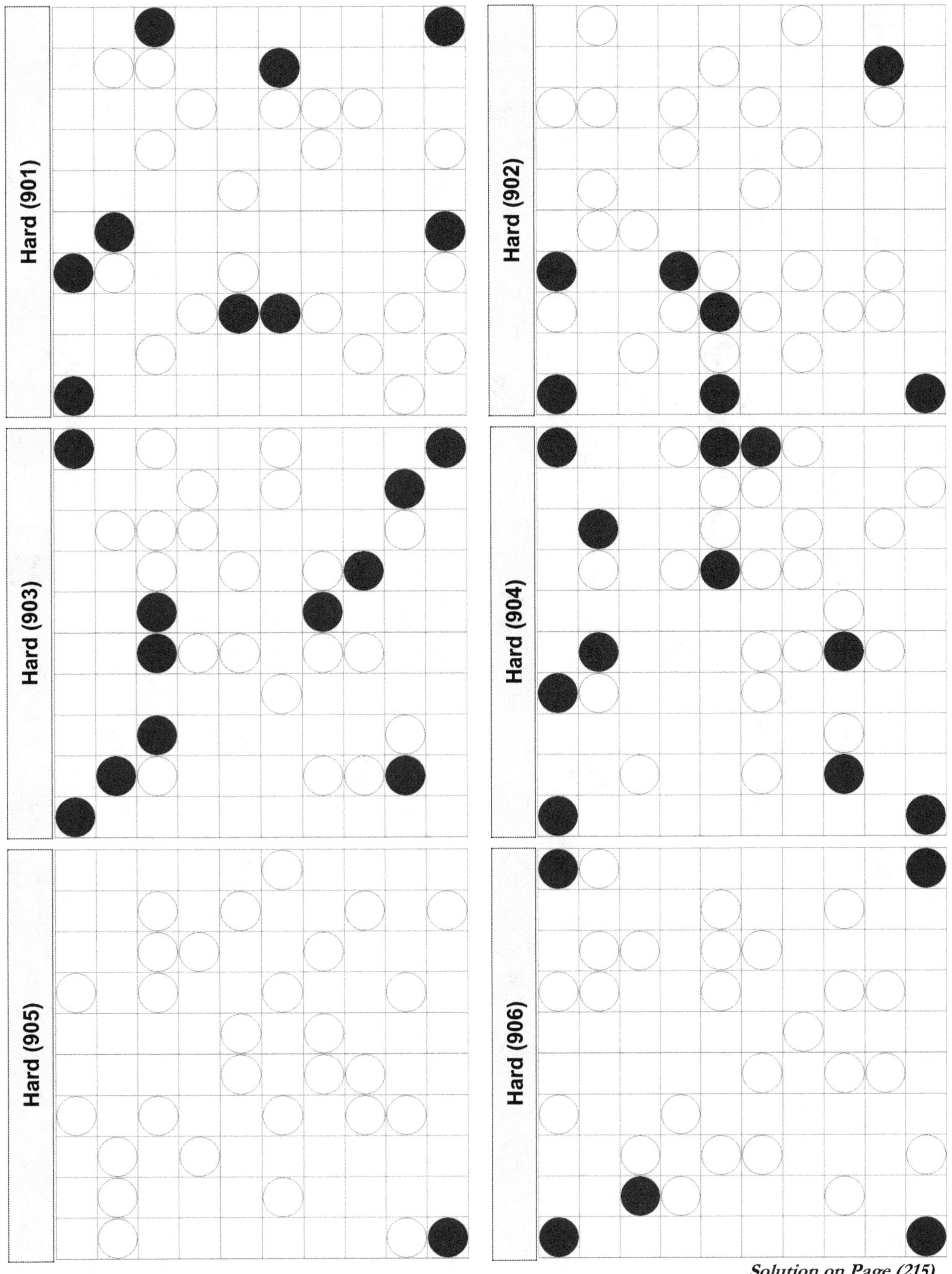

(153)

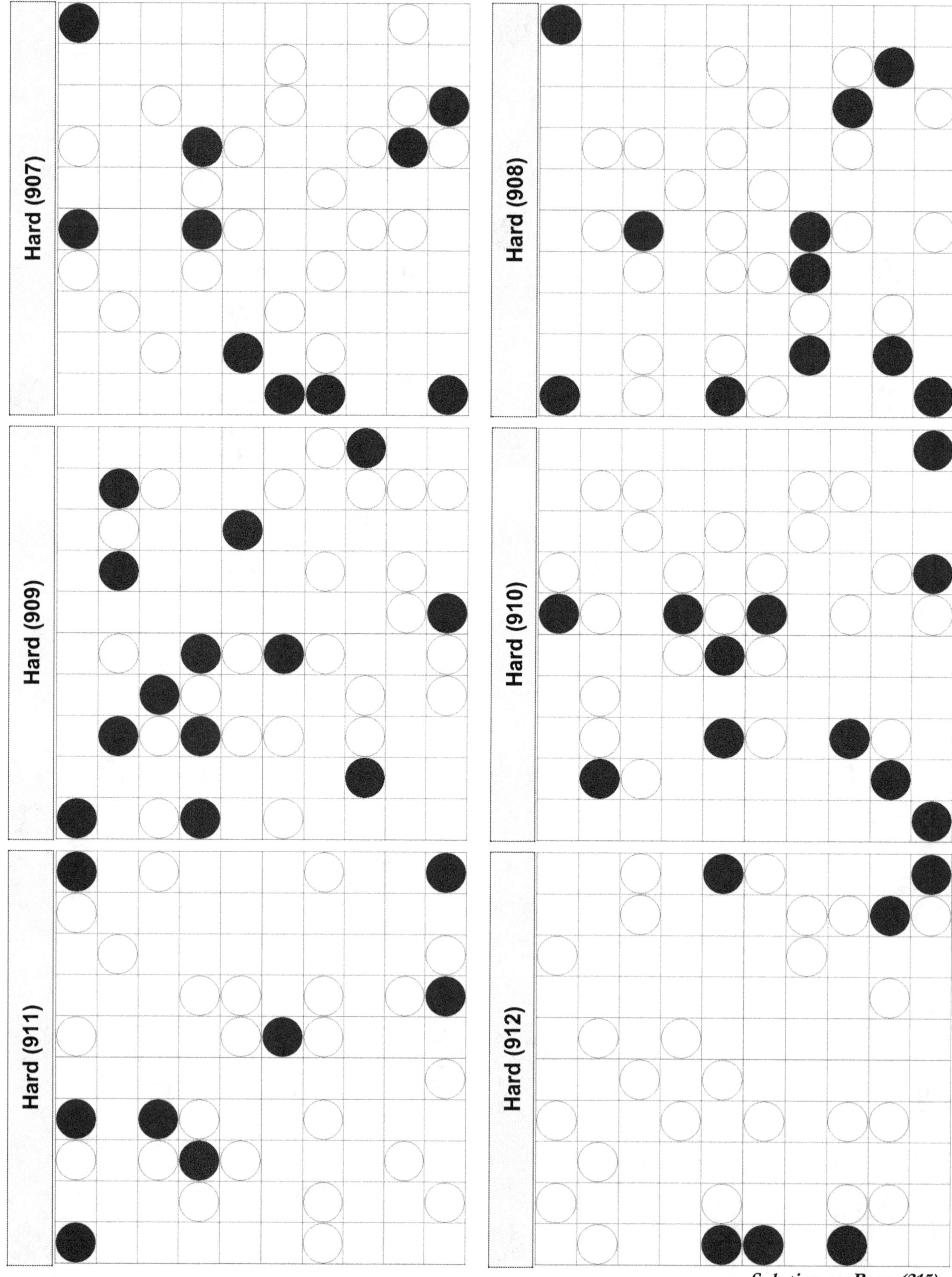

Solution on Page (215)

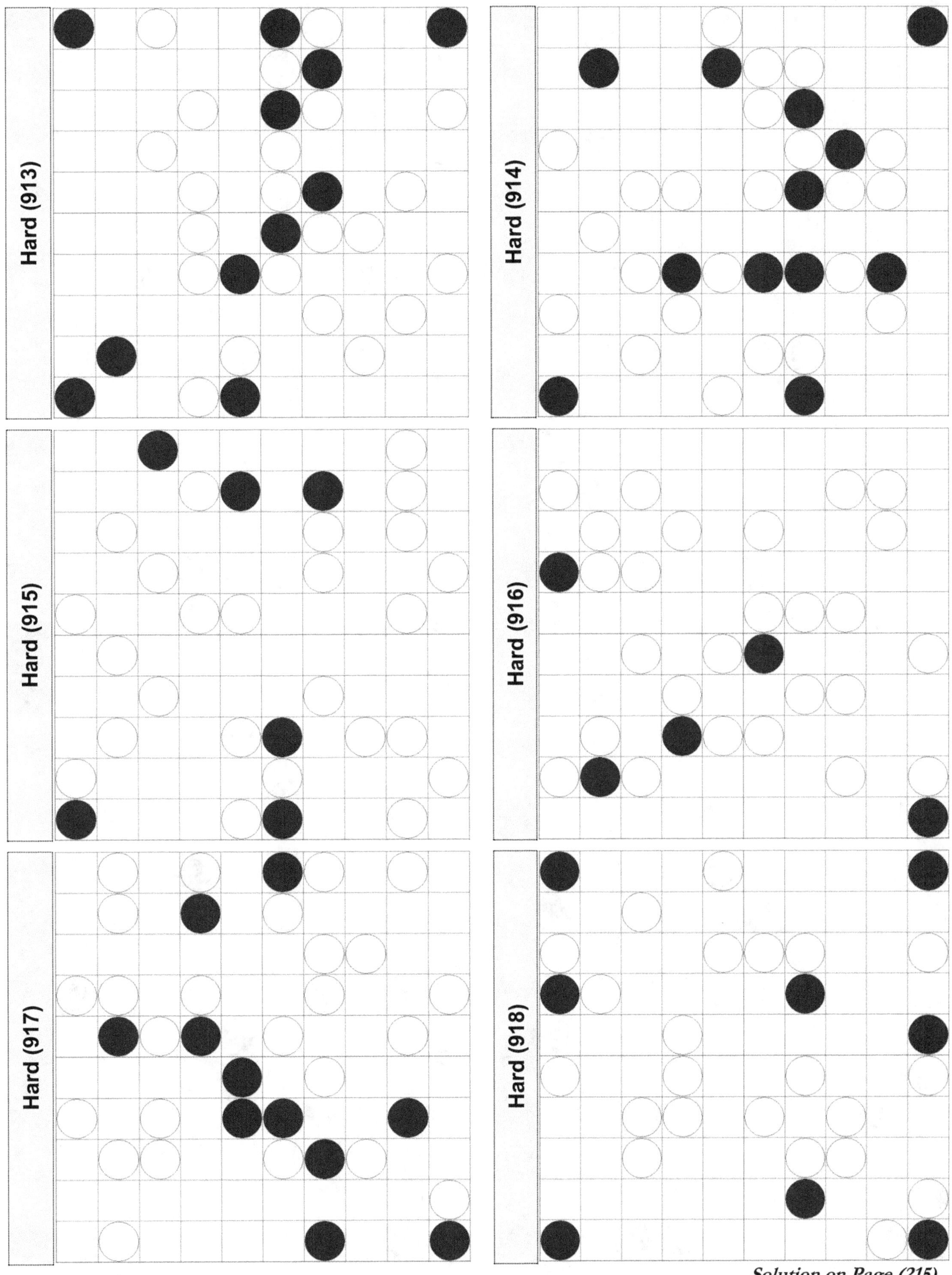

Hard (913)

Hard (914)

Hard (915)

Hard (916)

Hard (917)

Hard (918)

Solution on Page (215)

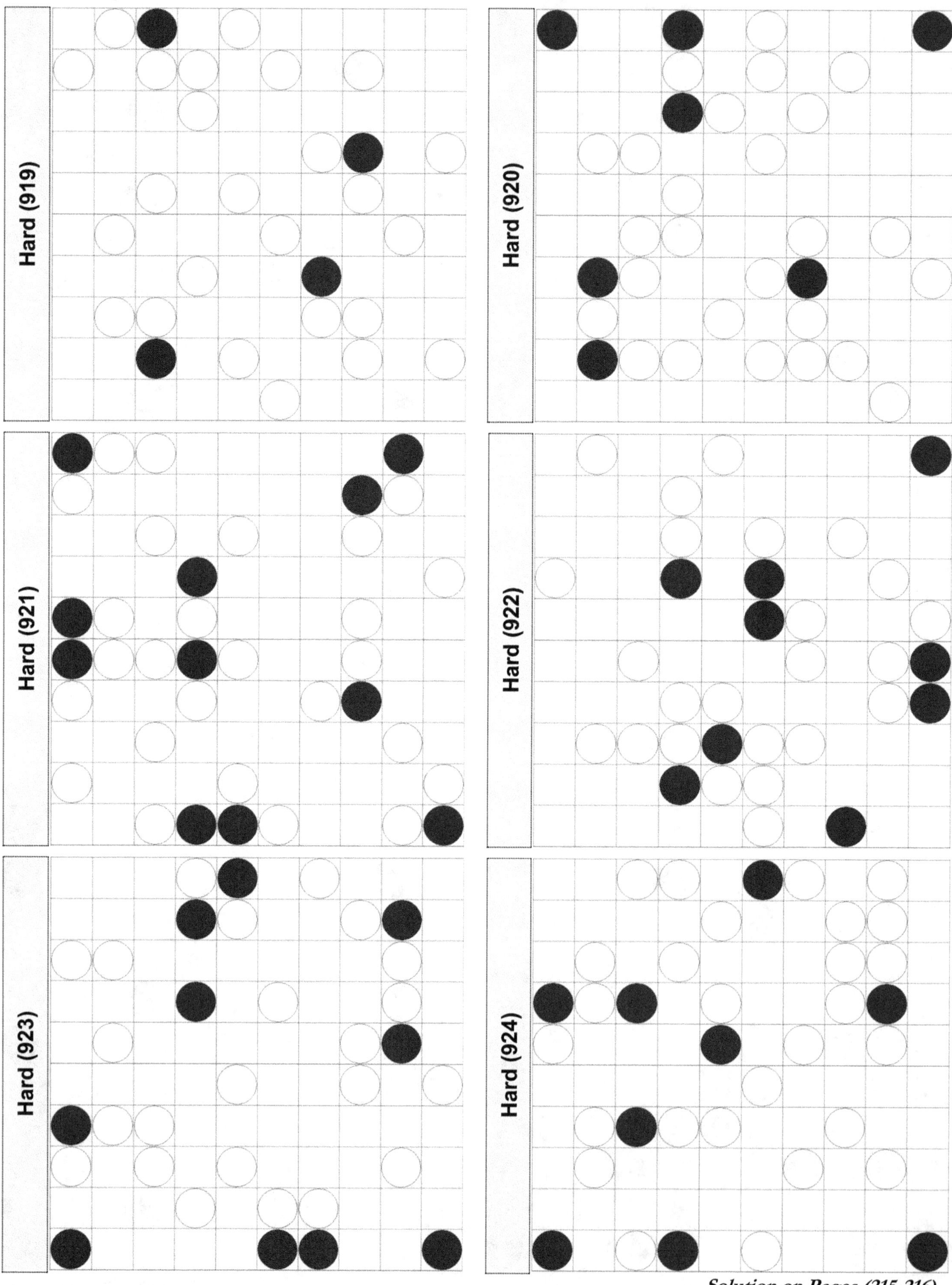

Solution on Pages (215-216)

(156)

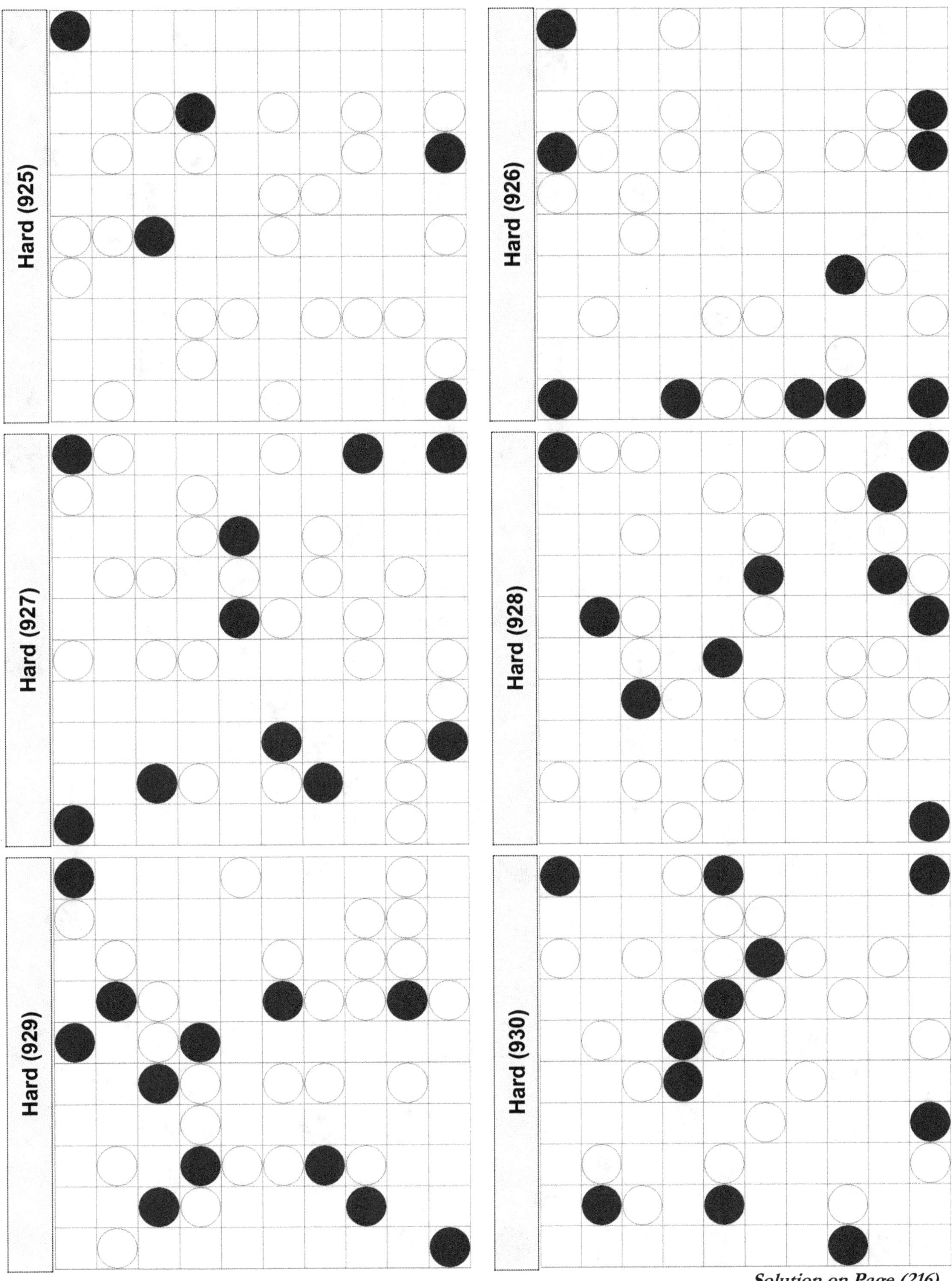

Hard (925)
Hard (926)
Hard (927)
Hard (928)
Hard (929)
Hard (930)
Solution on Page (216)

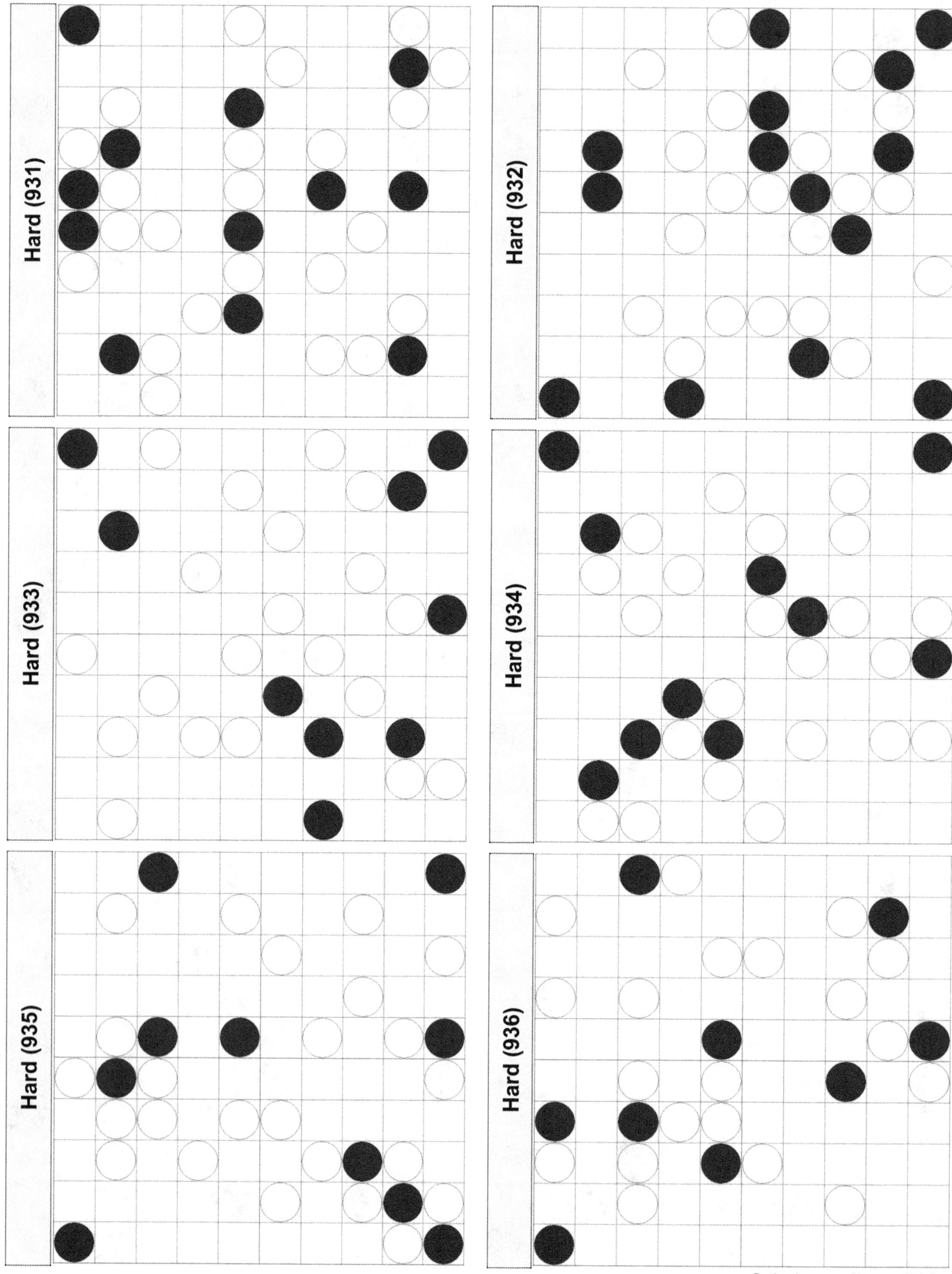

Solution on Page (216)

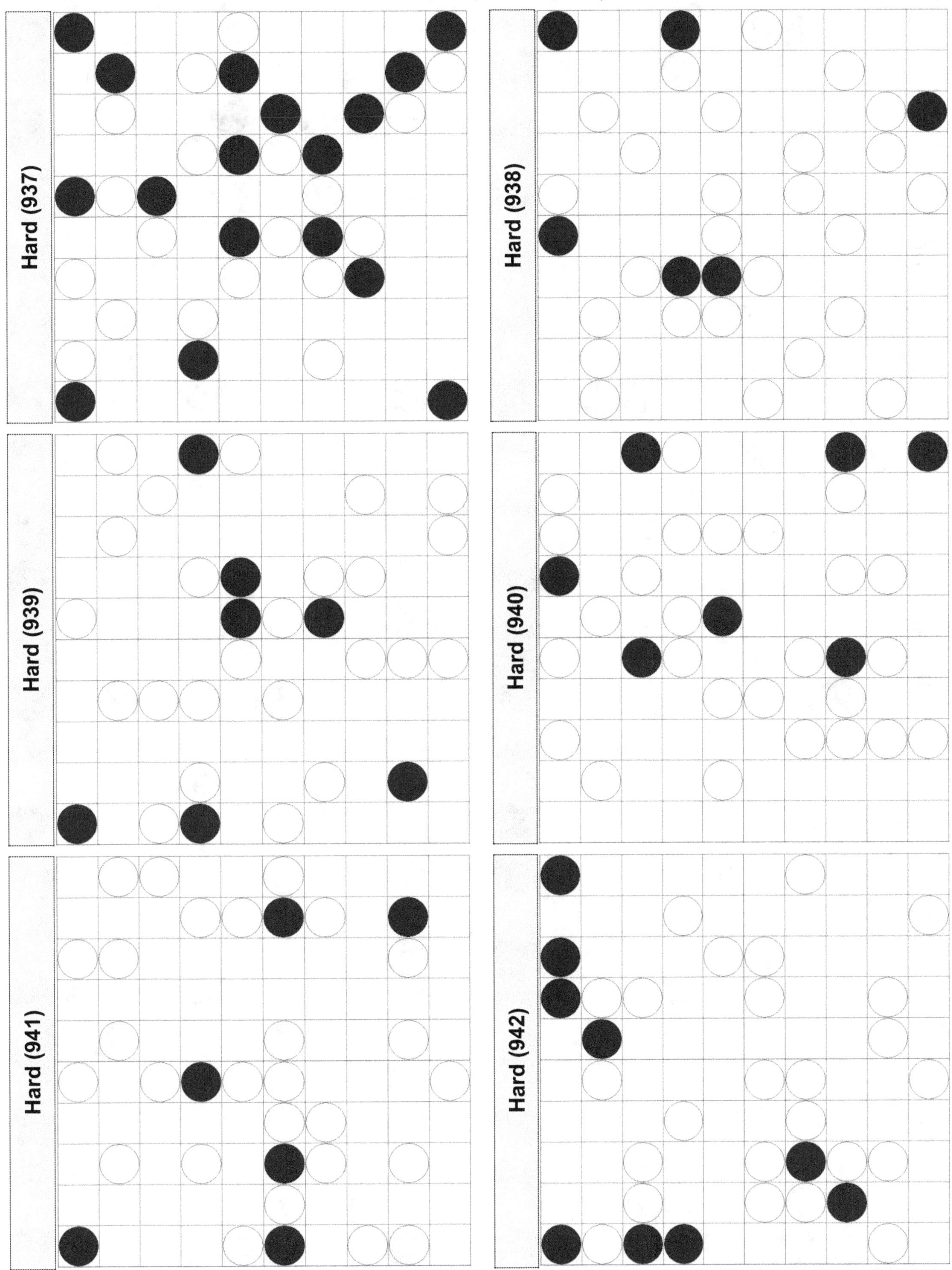

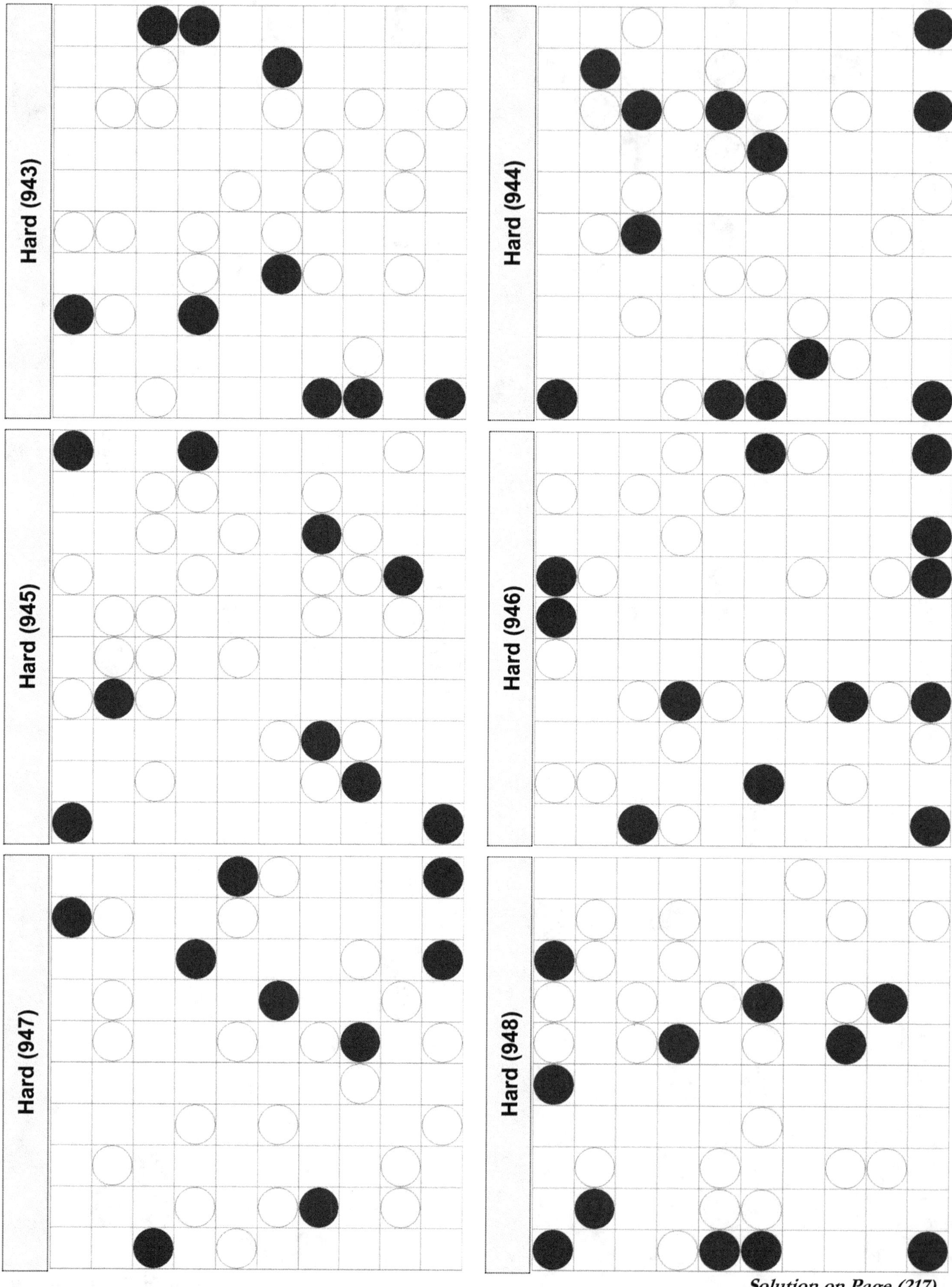

Solution on Page (217)

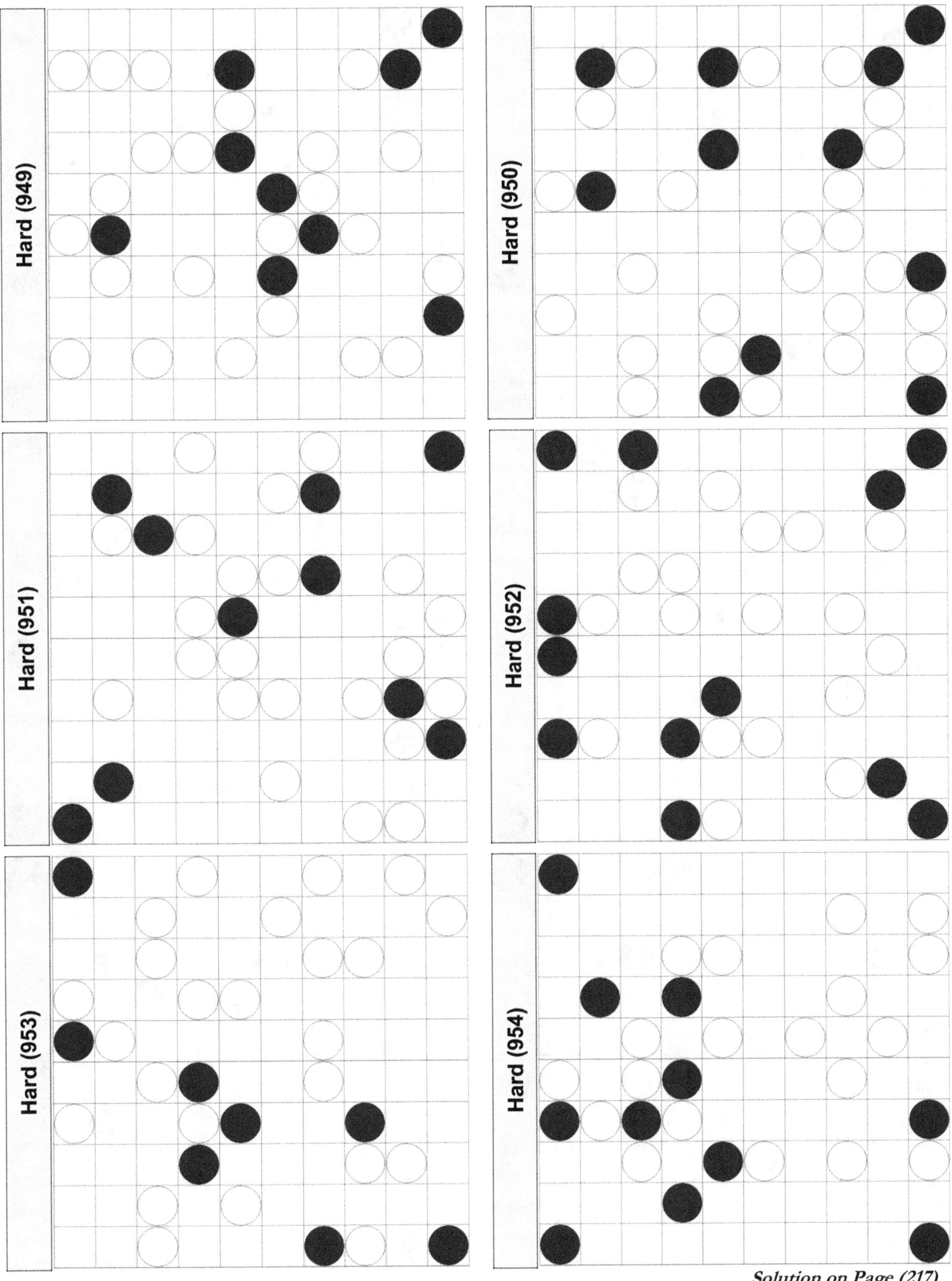

Hard (949)

Hard (950)

Hard (951)

Hard (952)

Hard (953)

Hard (954)

Solution on Page (217)

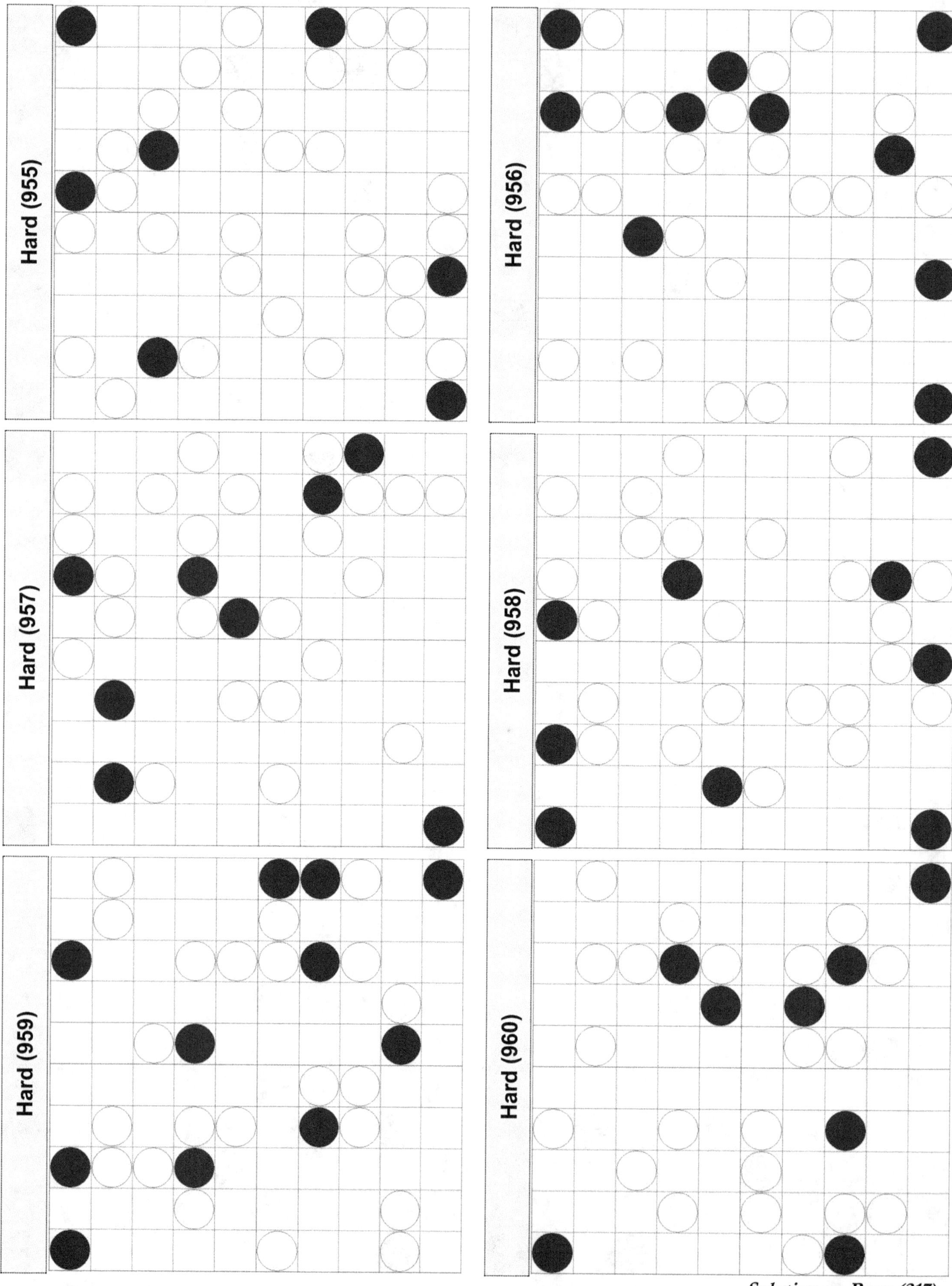

Solution on Page (217)

(162)

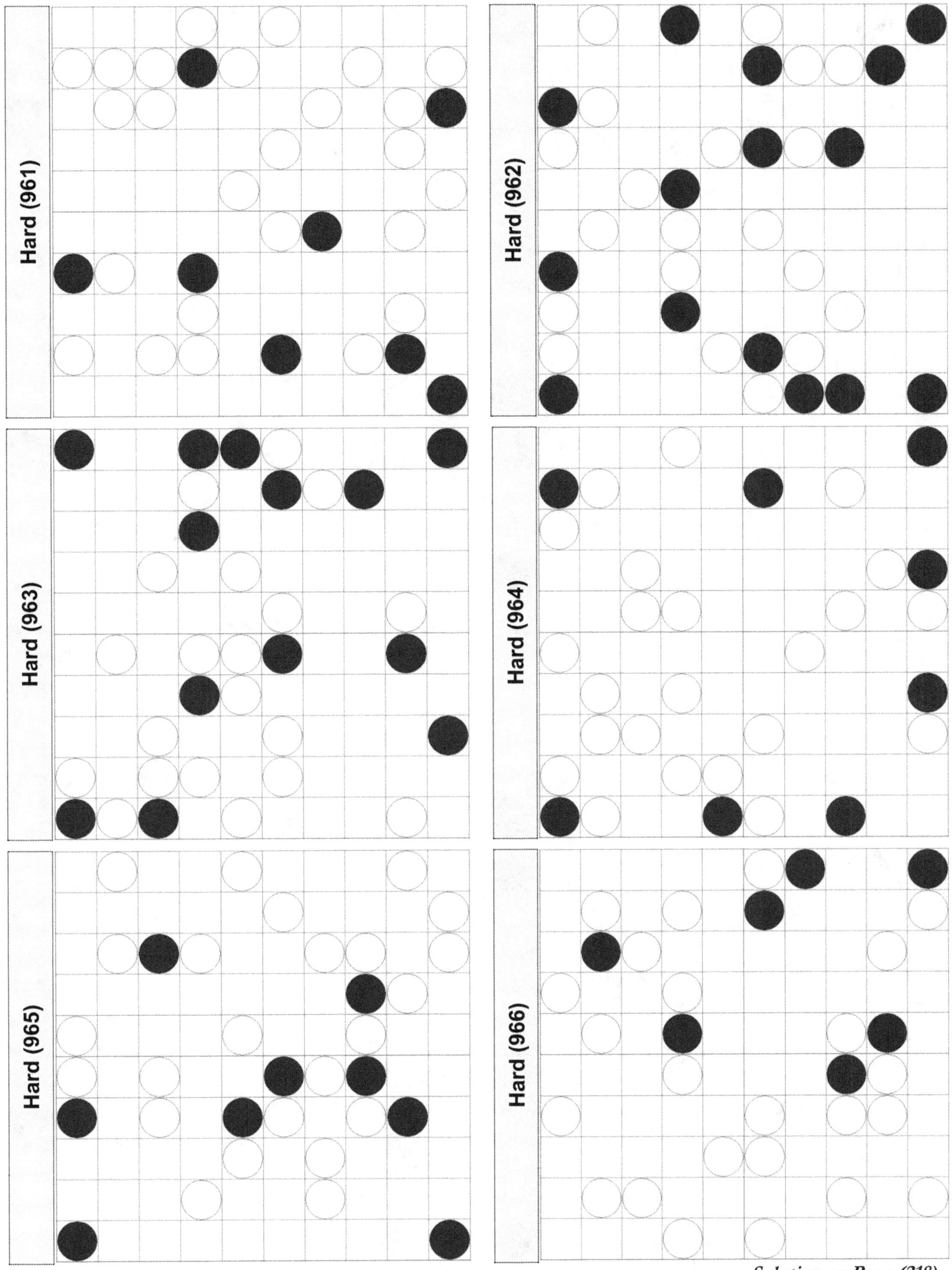

Solution on Page (218)

(163)

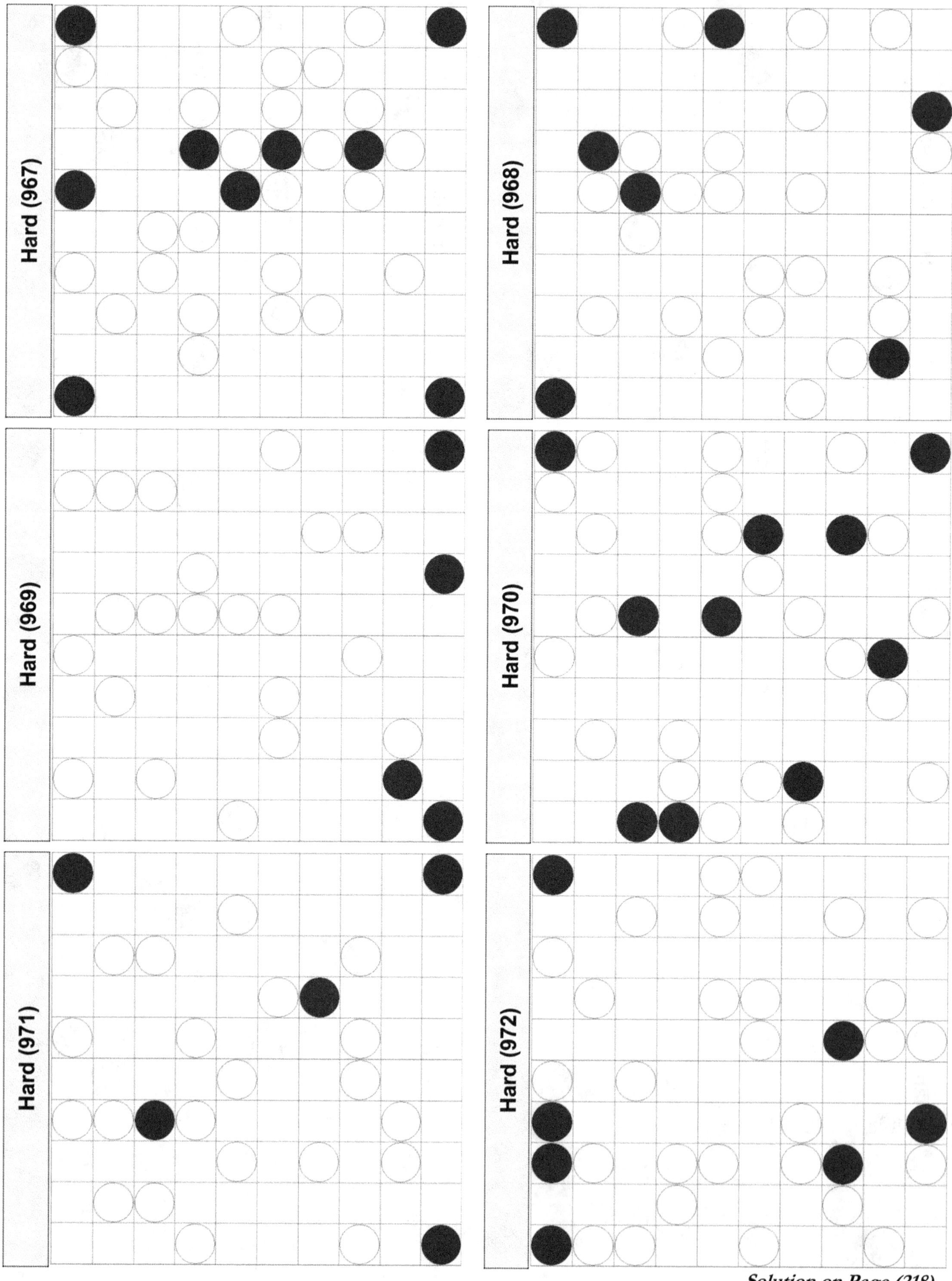

Solution on Page (218)

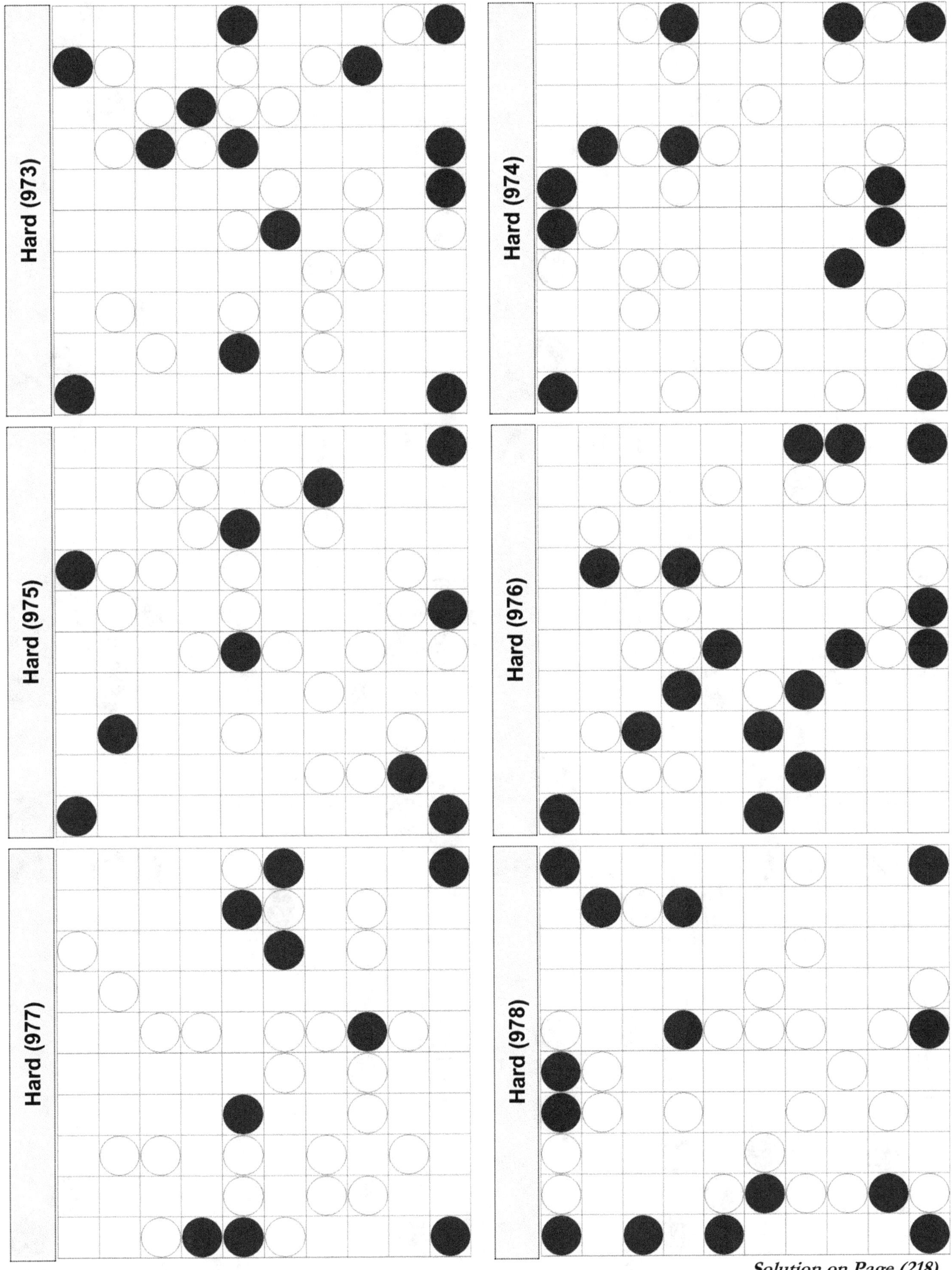

Solution on Page (218)

(165)

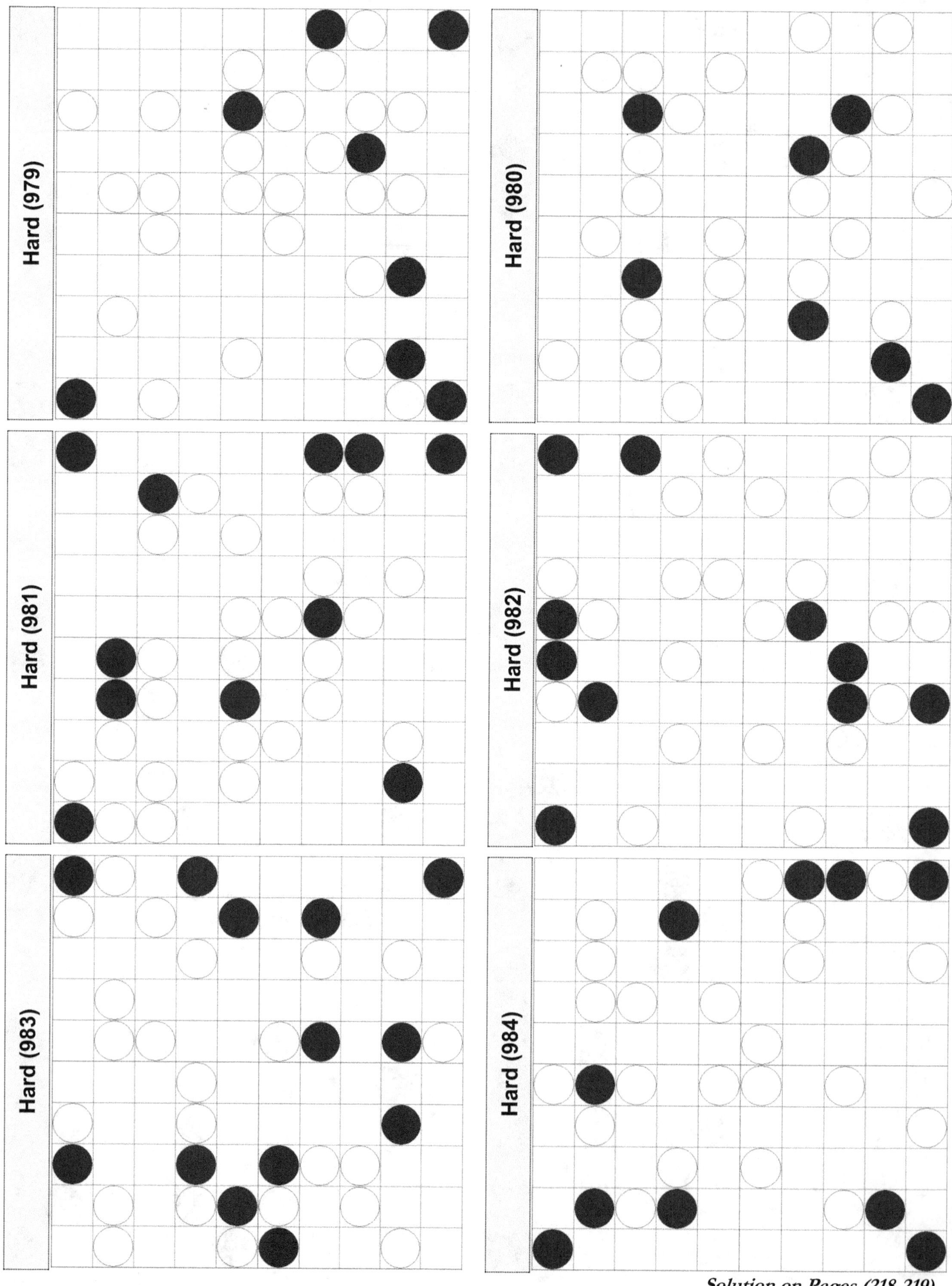

Solution on Pages (218-219)

(166)

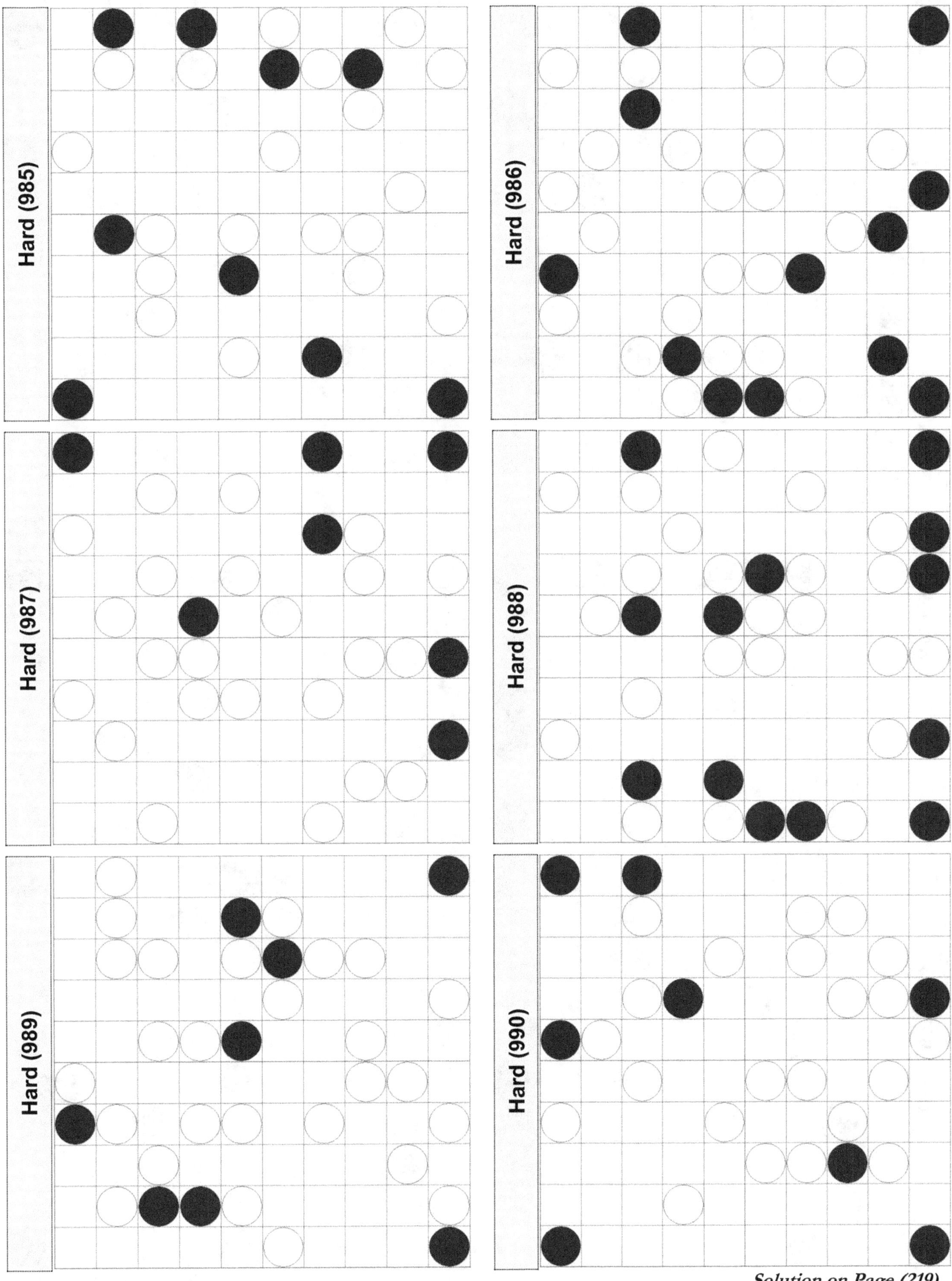

Solution on Page (219)

(167)

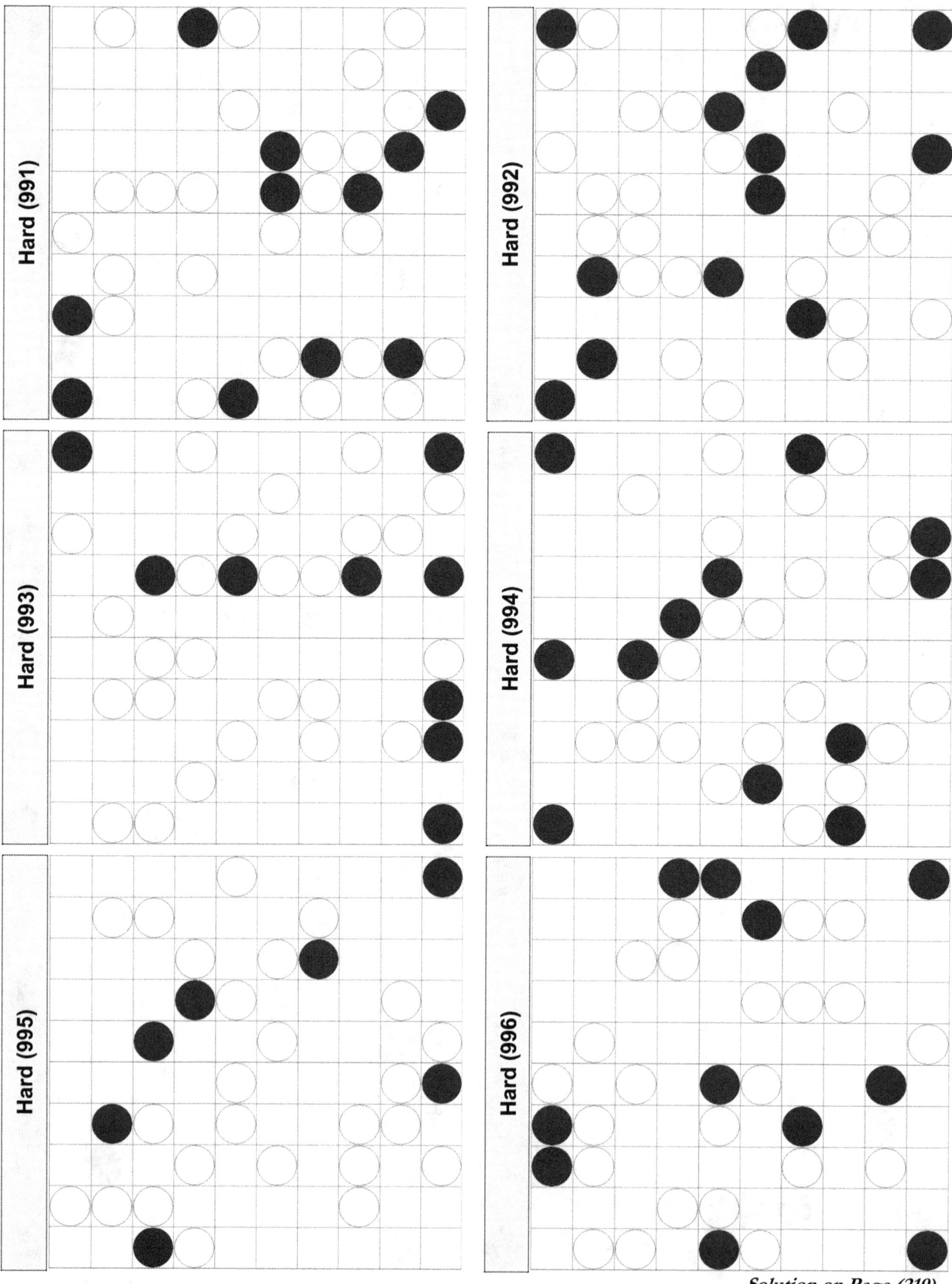

Hard (991)
Hard (992)
Hard (993)
Hard (994)
Hard (995)
Hard (996)
Solution on Page (219)

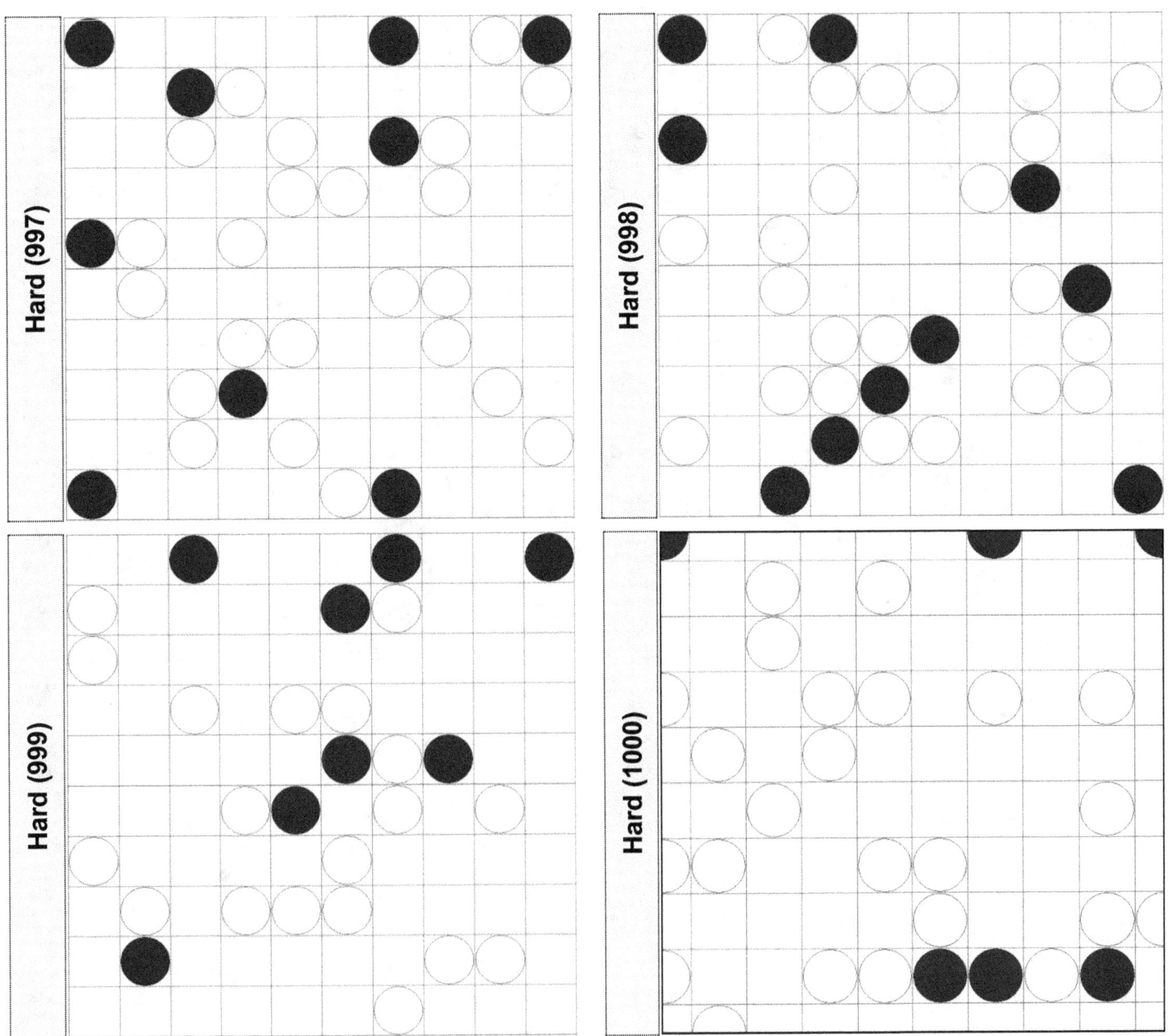

Solution on Page (219)

::::: Puzzle (1) :::::
::::: Puzzle (2) :::::
::::: Puzzle (3) :::::
::::: Puzzle (4) :::::
::::: Puzzle (5) :::::
::::: Puzzle (6) :::::
::::: Puzzle (7) :::::
::::: Puzzle (8) :::::
::::: Puzzle (9) :::::
::::: Puzzle (10) :::::
::::: Puzzle (11) :::::
::::: Puzzle (12) :::::
::::: Puzzle (13) :::::
::::: Puzzle (14) :::::
::::: Puzzle (15) :::::
::::: Puzzle (16) :::::
::::: Puzzle (17) :::::
::::: Puzzle (18) :::::
::::: Puzzle (19) :::::
::::: Puzzle (20) :::::

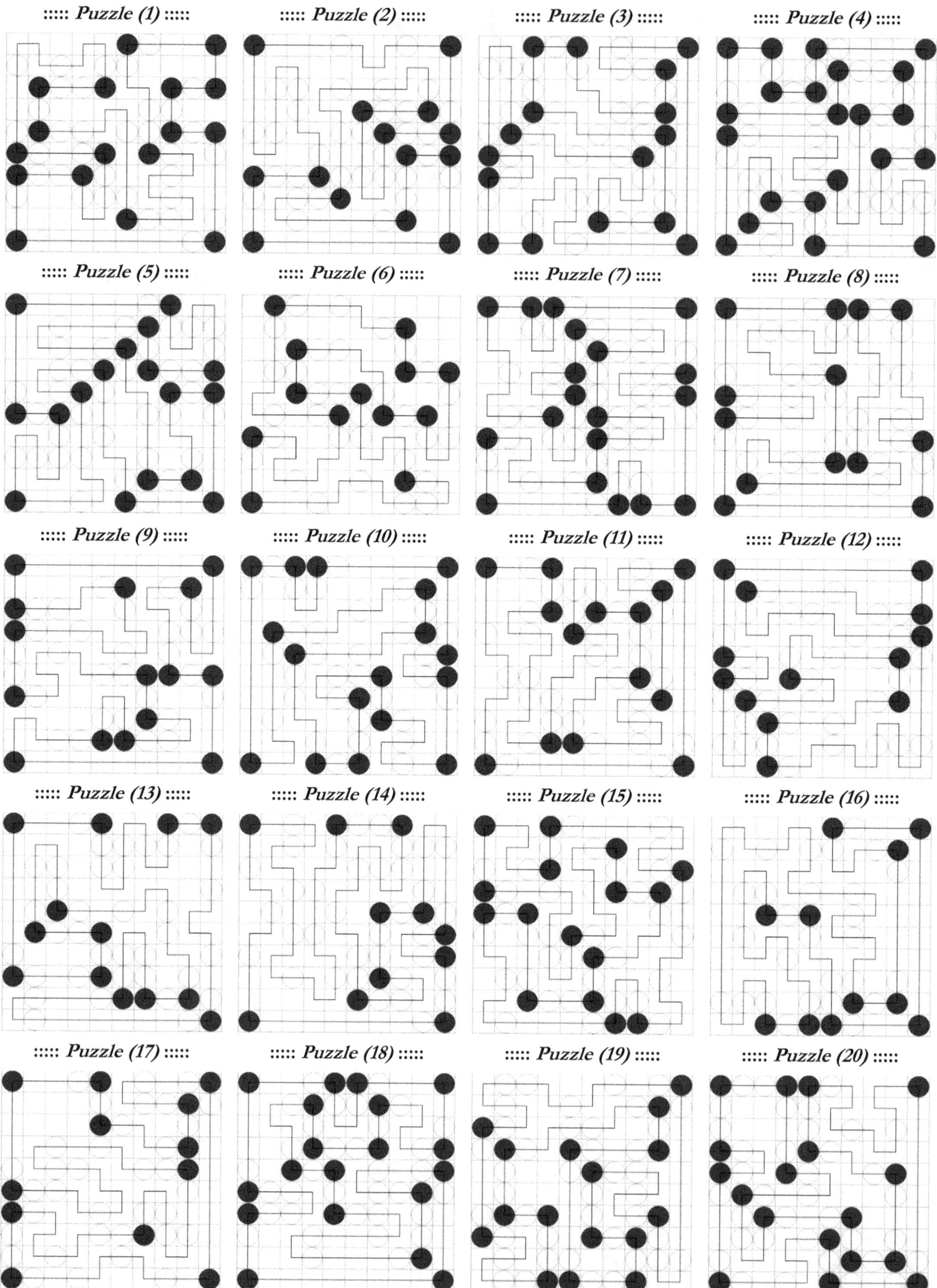

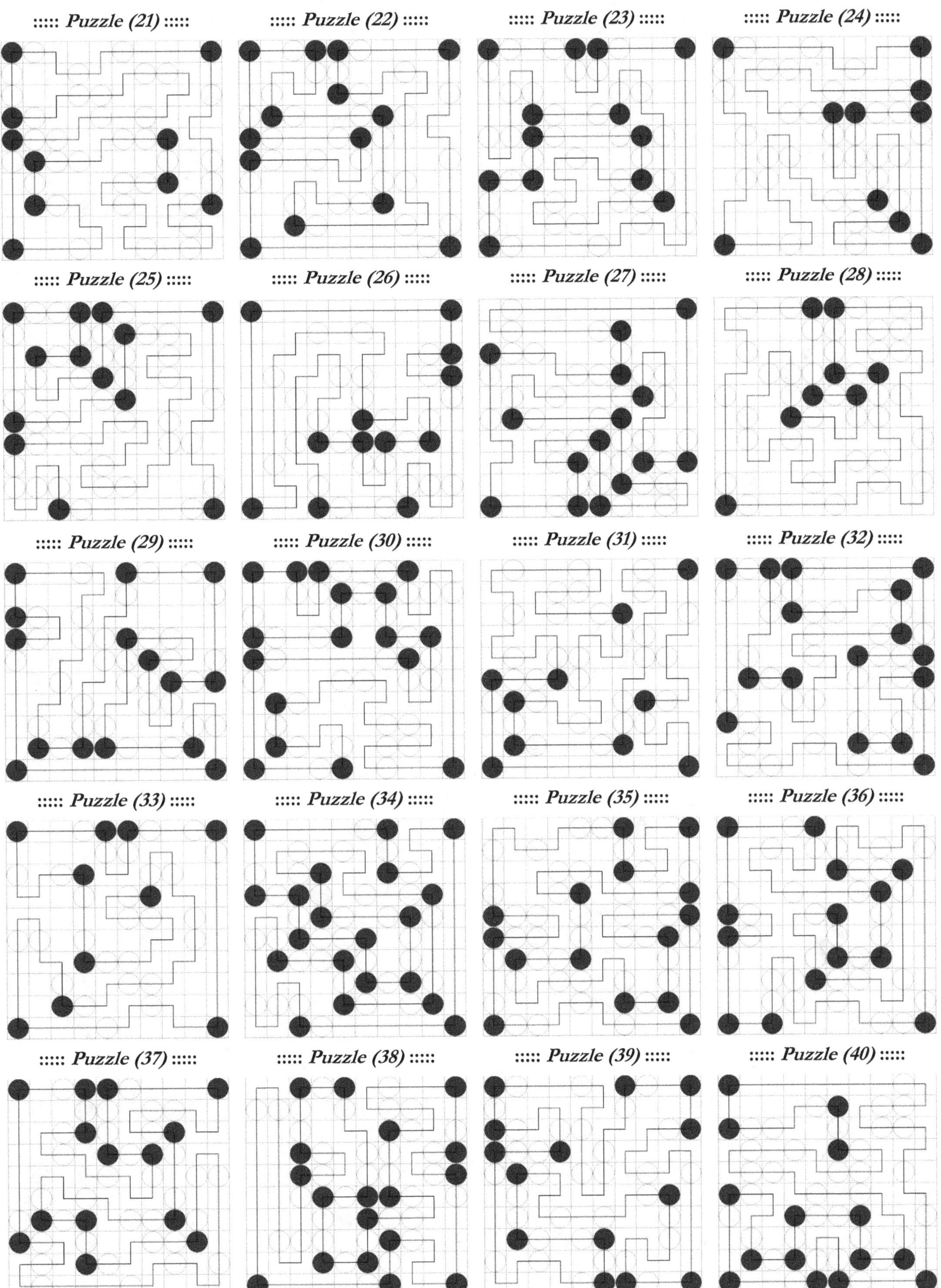

::::: Puzzle (21) :::::
::::: Puzzle (22) :::::
::::: Puzzle (23) :::::
::::: Puzzle (24) :::::
::::: Puzzle (25) :::::
::::: Puzzle (26) :::::
::::: Puzzle (27) :::::
::::: Puzzle (28) :::::
::::: Puzzle (29) :::::
::::: Puzzle (30) :::::
::::: Puzzle (31) :::::
::::: Puzzle (32) :::::
::::: Puzzle (33) :::::
::::: Puzzle (34) :::::
::::: Puzzle (35) :::::
::::: Puzzle (36) :::::
::::: Puzzle (37) :::::
::::: Puzzle (38) :::::
::::: Puzzle (39) :::::
::::: Puzzle (40) :::::

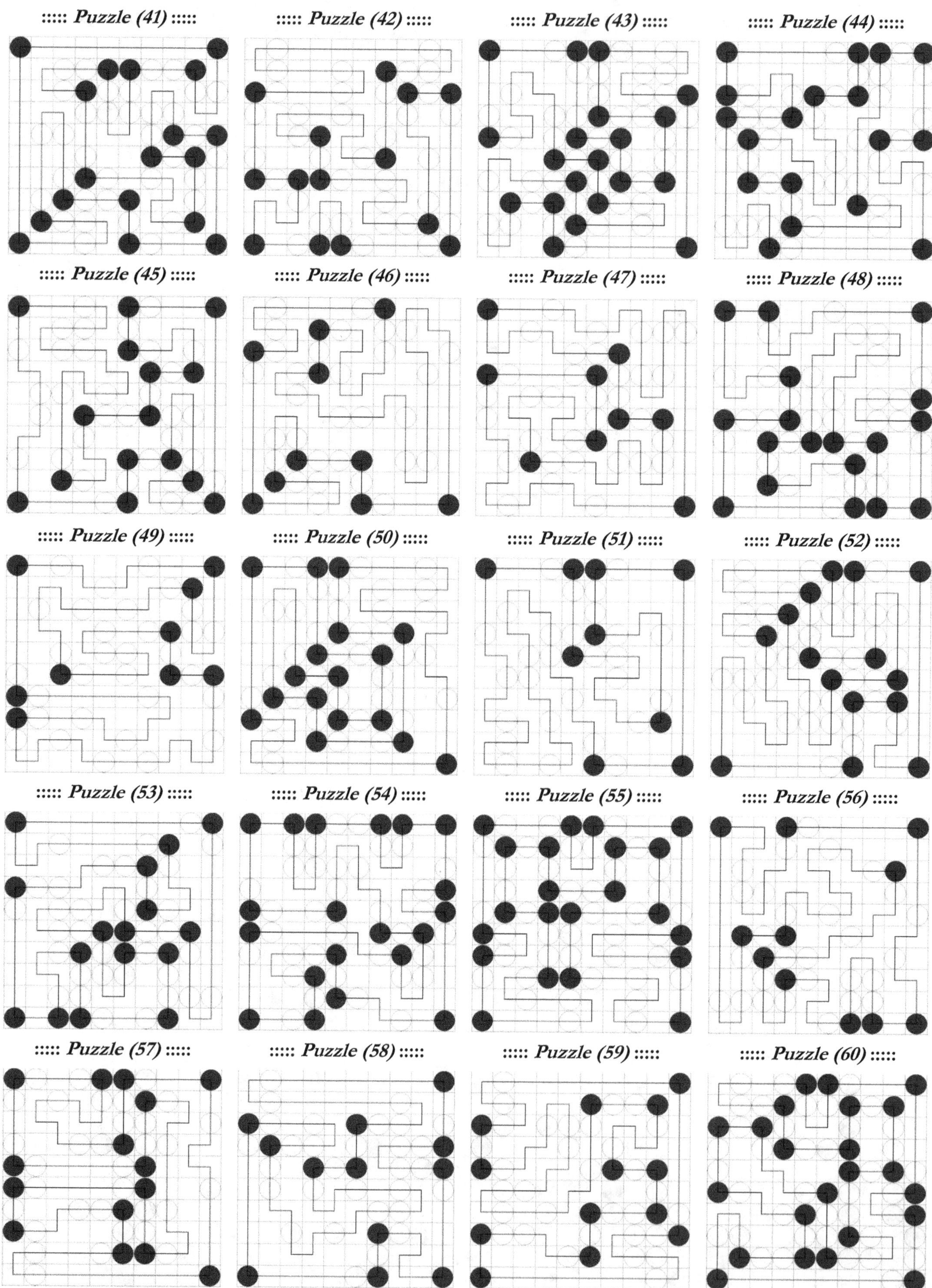

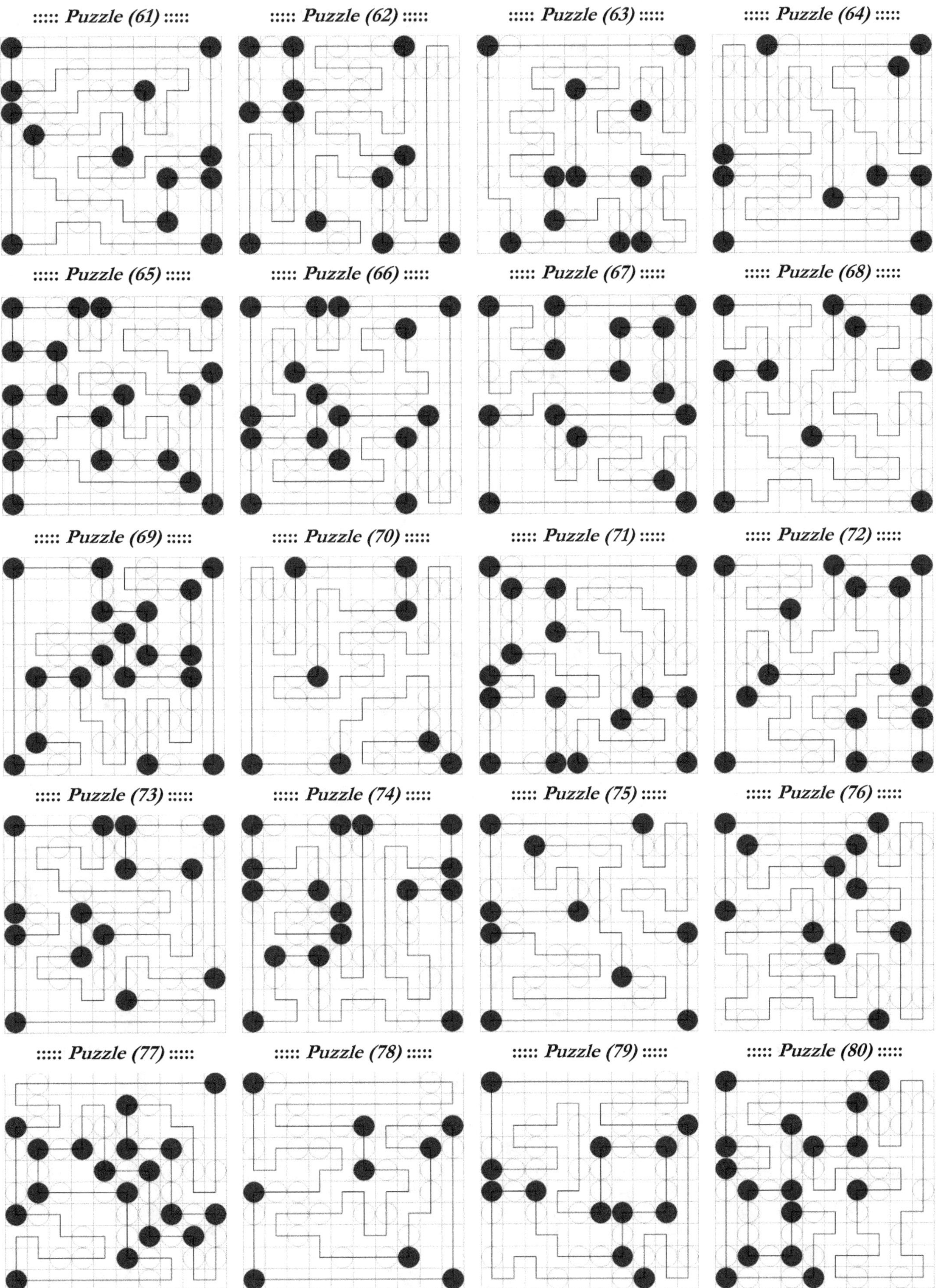

::::: *Puzzle (61)* ::::: ::::: *Puzzle (62)* ::::: ::::: *Puzzle (63)* ::::: ::::: *Puzzle (64)* :::::

::::: *Puzzle (65)* ::::: ::::: *Puzzle (66)* ::::: ::::: *Puzzle (67)* ::::: ::::: *Puzzle (68)* :::::

::::: *Puzzle (69)* ::::: ::::: *Puzzle (70)* ::::: ::::: *Puzzle (71)* ::::: ::::: *Puzzle (72)* :::::

::::: *Puzzle (73)* ::::: ::::: *Puzzle (74)* ::::: ::::: *Puzzle (75)* ::::: ::::: *Puzzle (76)* :::::

::::: *Puzzle (77)* ::::: ::::: *Puzzle (78)* ::::: ::::: *Puzzle (79)* ::::: ::::: *Puzzle (80)* :::::

(173)

::::: *Puzzle (81)* :::::	::::: *Puzzle (82)* :::::	::::: *Puzzle (83)* :::::	::::: *Puzzle (84)* :::::
::::: *Puzzle (85)* :::::	::::: *Puzzle (86)* :::::	::::: *Puzzle (87)* :::::	::::: *Puzzle (88)* :::::
::::: *Puzzle (89)* :::::	::::: *Puzzle (90)* :::::	::::: *Puzzle (91)* :::::	::::: *Puzzle (92)* :::::
::::: *Puzzle (93)* :::::	::::: *Puzzle (94)* :::::	::::: *Puzzle (95)* :::::	::::: *Puzzle (96)* :::::
::::: *Puzzle (97)* :::::	::::: *Puzzle (98)* :::::	::::: *Puzzle (99)* :::::	::::: *Puzzle (100)* :::::

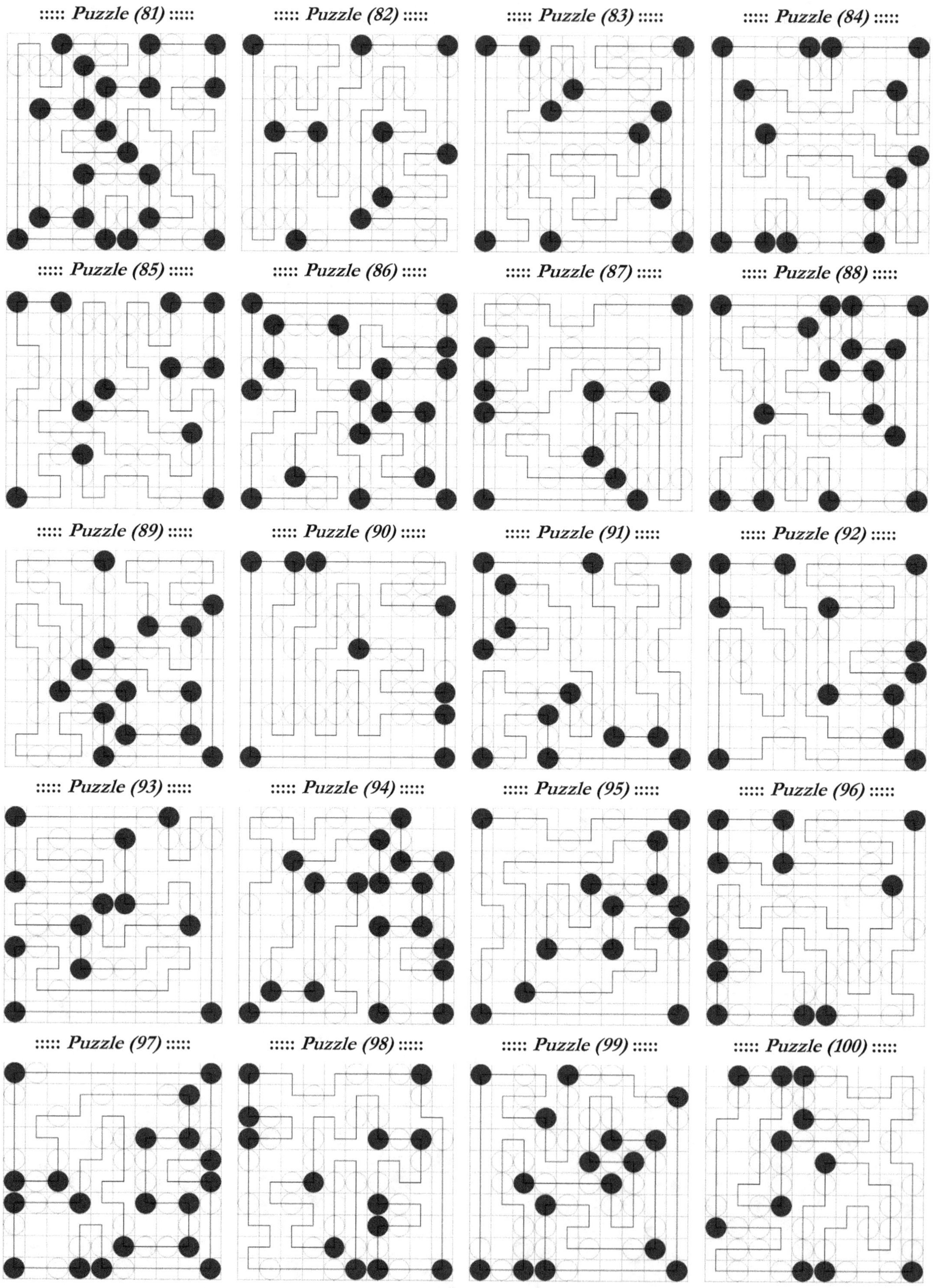

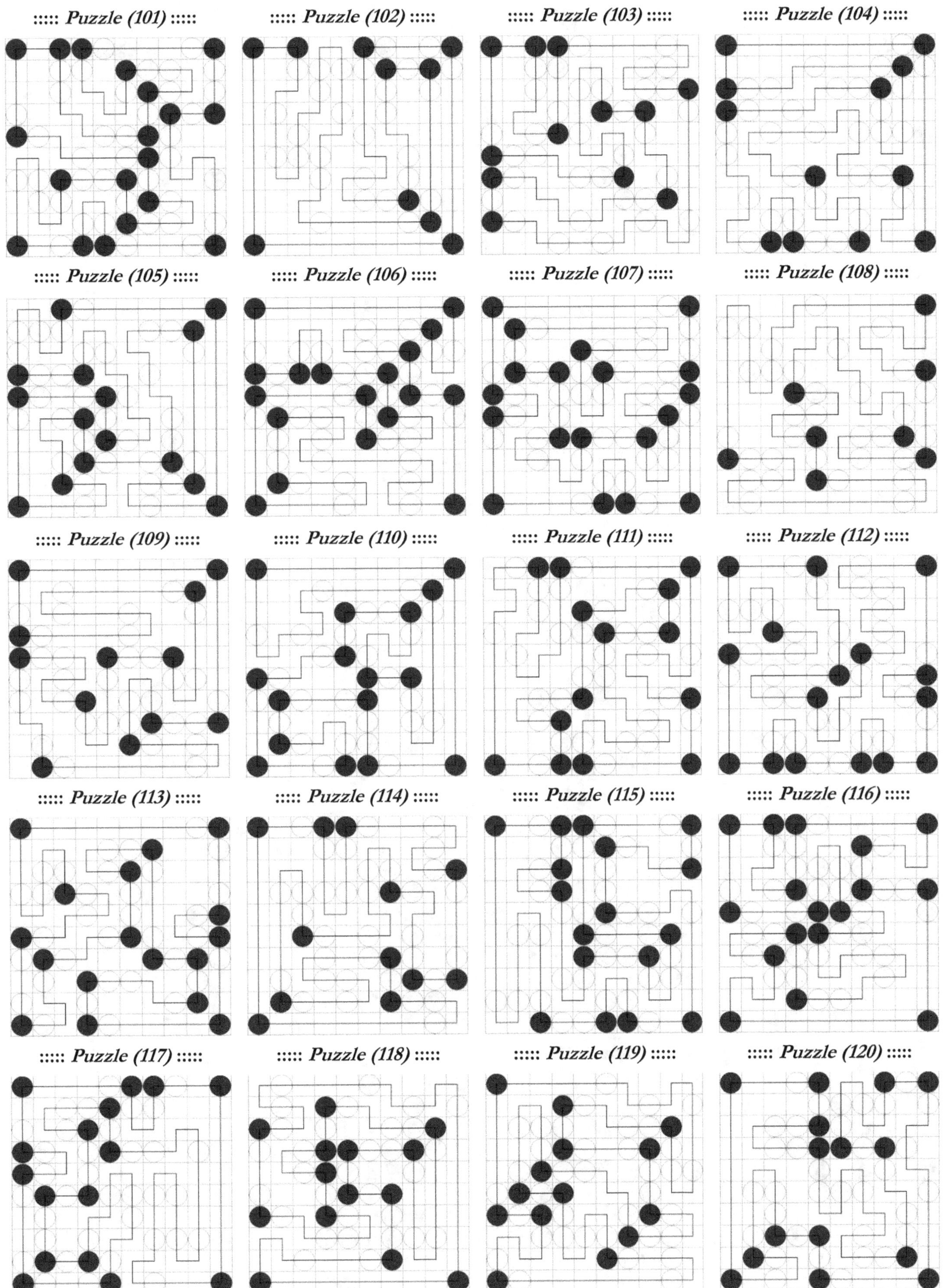

::::: Puzzle (101) :::::
::::: Puzzle (102) :::::
::::: Puzzle (103) :::::
::::: Puzzle (104) :::::
::::: Puzzle (105) :::::
::::: Puzzle (106) :::::
::::: Puzzle (107) :::::
::::: Puzzle (108) :::::
::::: Puzzle (109) :::::
::::: Puzzle (110) :::::
::::: Puzzle (111) :::::
::::: Puzzle (112) :::::
::::: Puzzle (113) :::::
::::: Puzzle (114) :::::
::::: Puzzle (115) :::::
::::: Puzzle (116) :::::
::::: Puzzle (117) :::::
::::: Puzzle (118) :::::
::::: Puzzle (119) :::::
::::: Puzzle (120) :::::

(175)

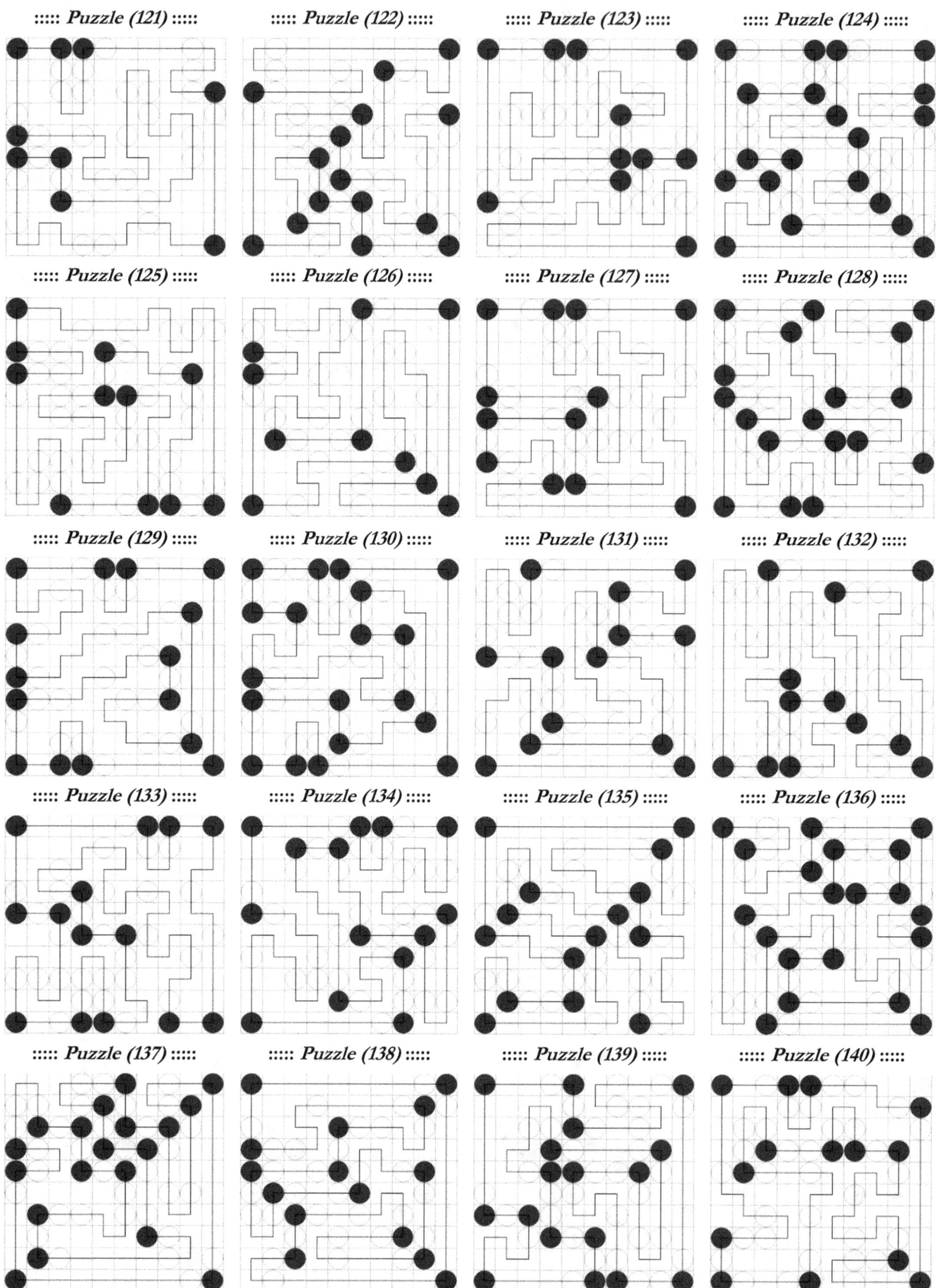

::::: Puzzle (121) :::::
::::: Puzzle (122) :::::
::::: Puzzle (123) :::::
::::: Puzzle (124) :::::
::::: Puzzle (125) :::::
::::: Puzzle (126) :::::
::::: Puzzle (127) :::::
::::: Puzzle (128) :::::
::::: Puzzle (129) :::::
::::: Puzzle (130) :::::
::::: Puzzle (131) :::::
::::: Puzzle (132) :::::
::::: Puzzle (133) :::::
::::: Puzzle (134) :::::
::::: Puzzle (135) :::::
::::: Puzzle (136) :::::
::::: Puzzle (137) :::::
::::: Puzzle (138) :::::
::::: Puzzle (139) :::::
::::: Puzzle (140) :::::

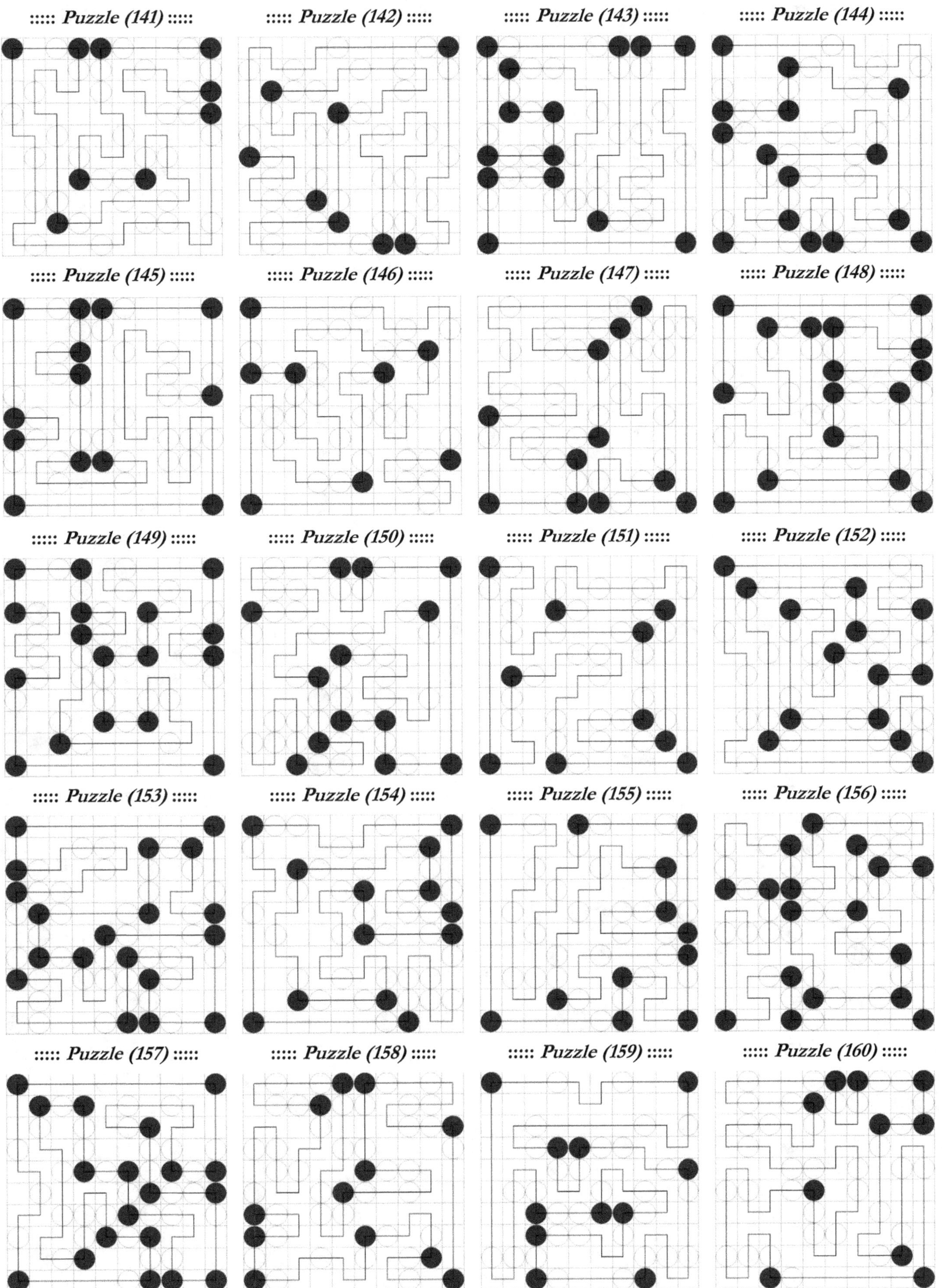

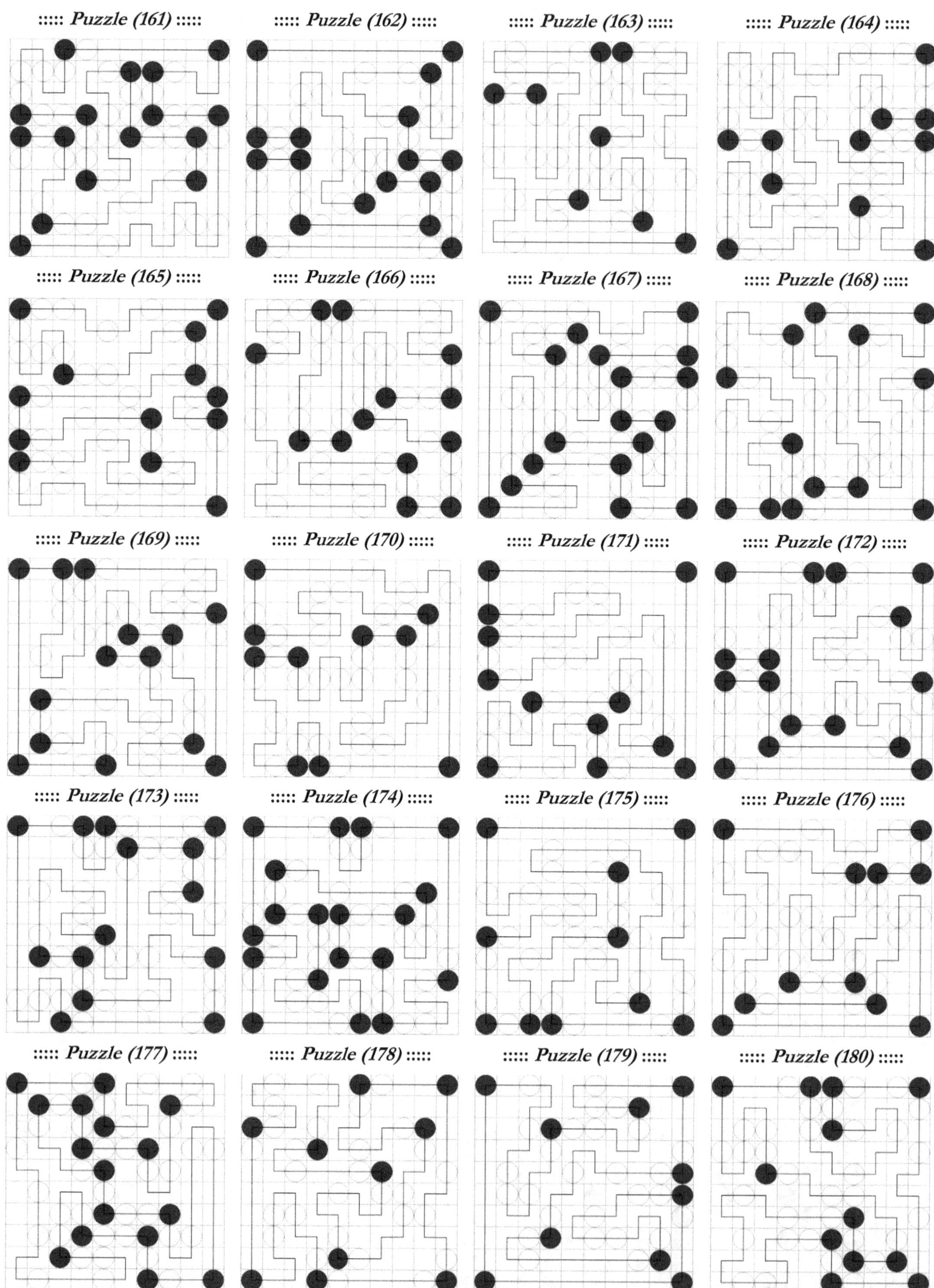

::::: Puzzle (161) :::::
::::: Puzzle (162) :::::
::::: Puzzle (163) :::::
::::: Puzzle (164) :::::
::::: Puzzle (165) :::::
::::: Puzzle (166) :::::
::::: Puzzle (167) :::::
::::: Puzzle (168) :::::
::::: Puzzle (169) :::::
::::: Puzzle (170) :::::
::::: Puzzle (171) :::::
::::: Puzzle (172) :::::
::::: Puzzle (173) :::::
::::: Puzzle (174) :::::
::::: Puzzle (175) :::::
::::: Puzzle (176) :::::
::::: Puzzle (177) :::::
::::: Puzzle (178) :::::
::::: Puzzle (179) :::::
::::: Puzzle (180) :::::

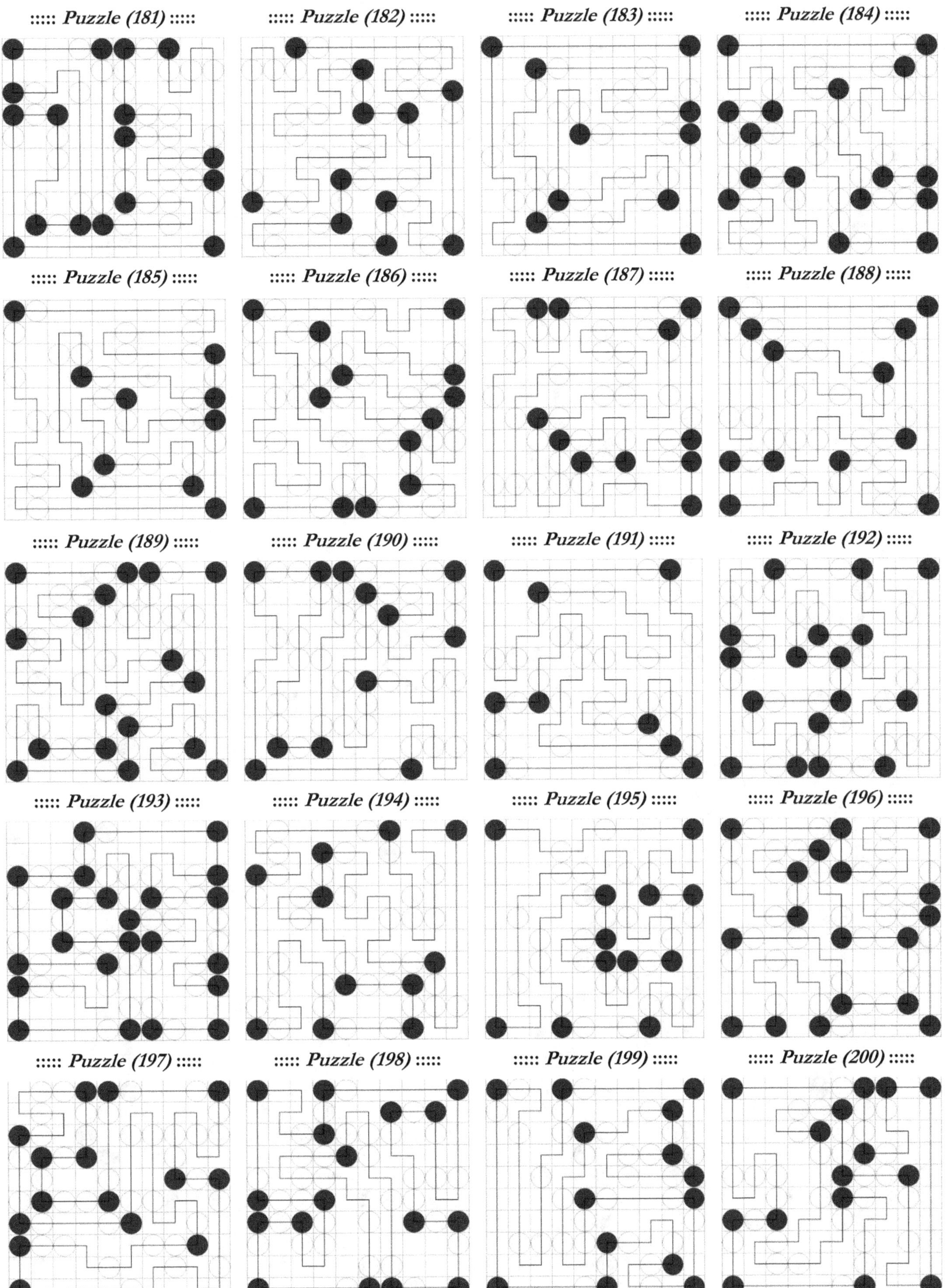

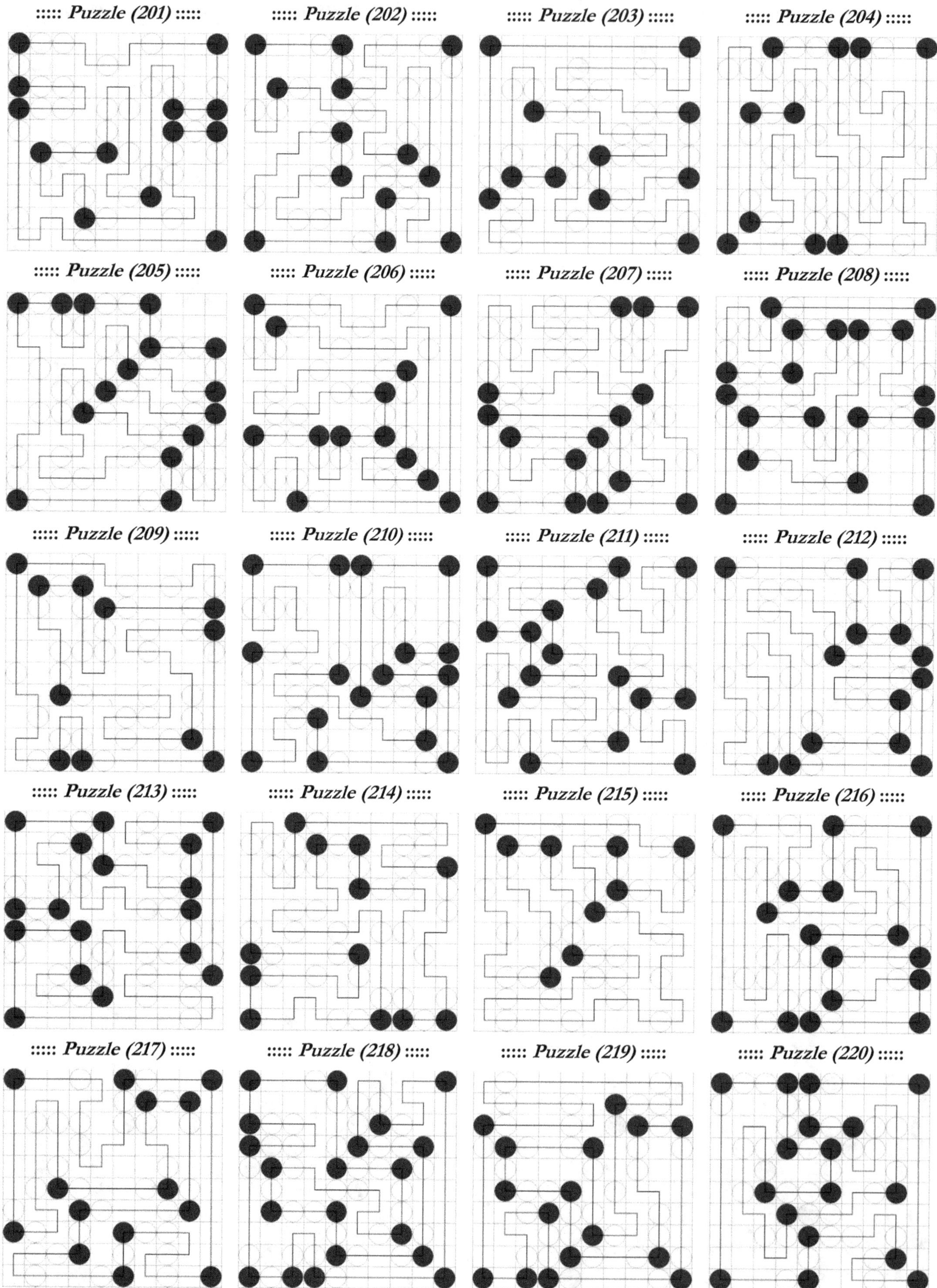

::::: *Puzzle (221)* :::::   ::::: *Puzzle (222)* :::::   ::::: *Puzzle (223)* :::::   ::::: *Puzzle (224)* :::::

::::: *Puzzle (225)* :::::   ::::: *Puzzle (226)* :::::   ::::: *Puzzle (227)* :::::   ::::: *Puzzle (228)* :::::

::::: *Puzzle (229)* :::::   ::::: *Puzzle (230)* :::::   ::::: *Puzzle (231)* :::::   ::::: *Puzzle (232)* :::::

::::: *Puzzle (233)* :::::   ::::: *Puzzle (234)* :::::   ::::: *Puzzle (235)* :::::   ::::: *Puzzle (236)* :::::

::::: *Puzzle (237)* :::::   ::::: *Puzzle (238)* :::::   ::::: *Puzzle (239)* :::::   ::::: *Puzzle (240)* :::::

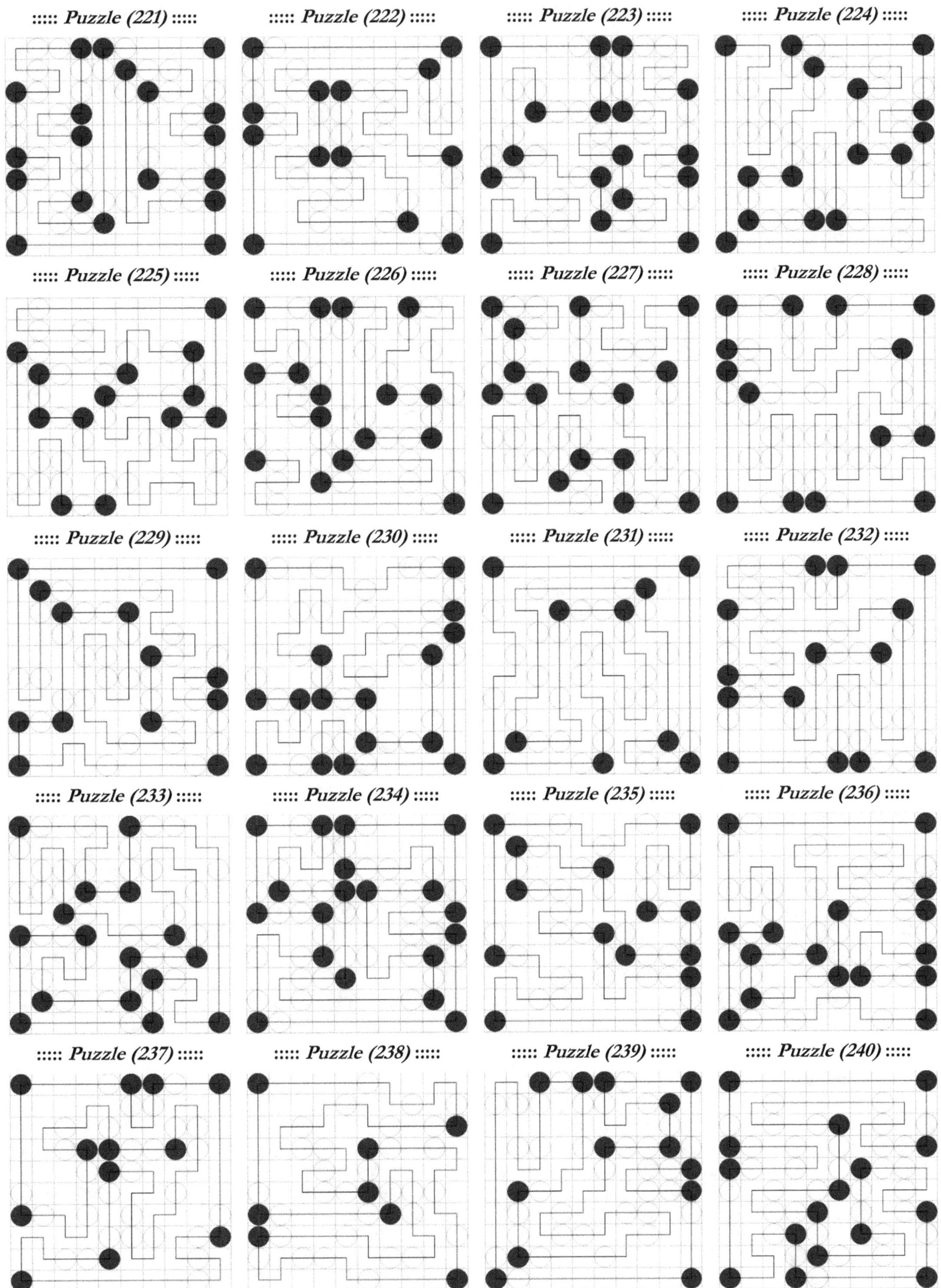

(181)

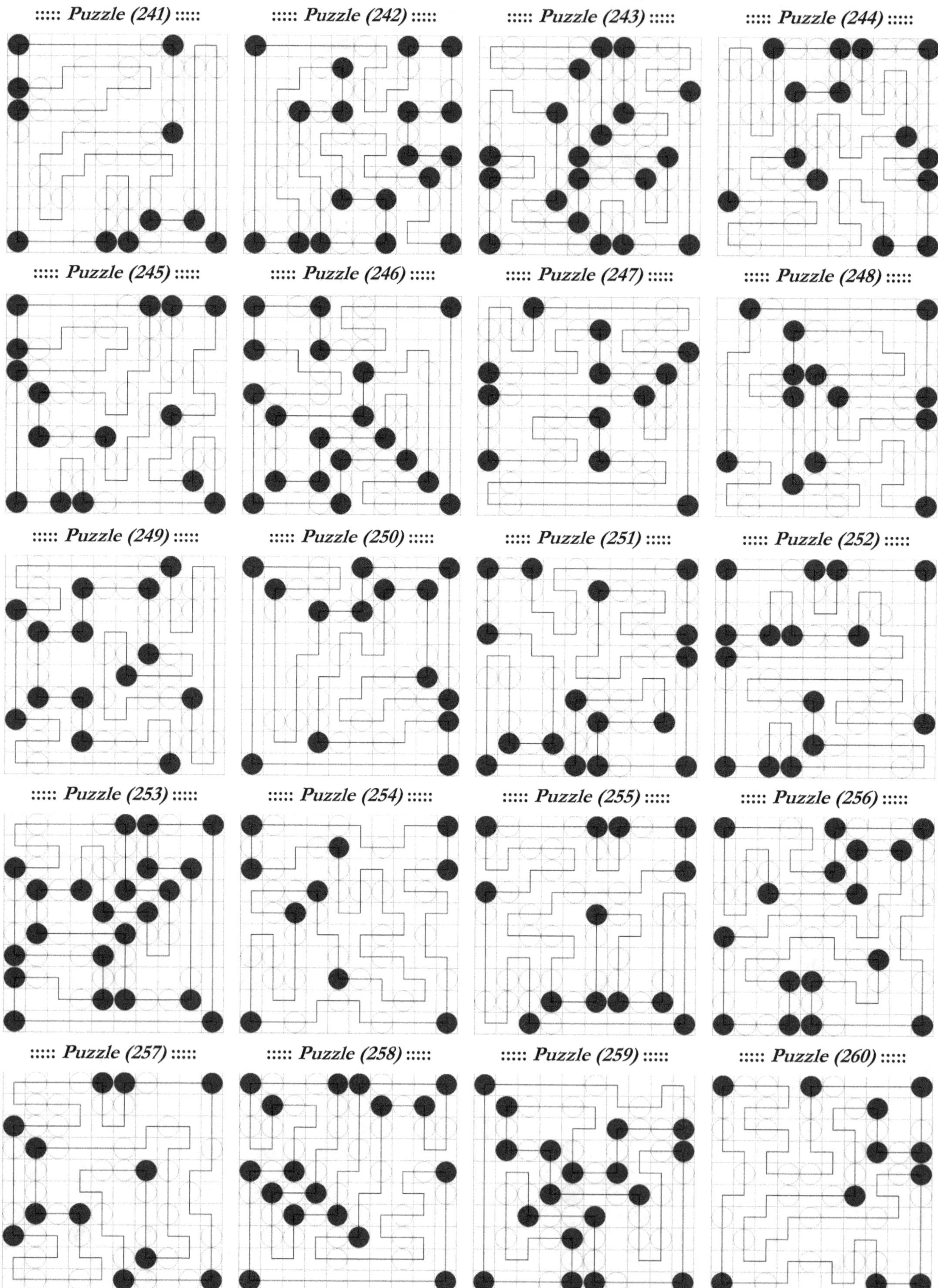

::::: Puzzle (241) :::::
::::: Puzzle (242) :::::
::::: Puzzle (243) :::::
::::: Puzzle (244) :::::
::::: Puzzle (245) :::::
::::: Puzzle (246) :::::
::::: Puzzle (247) :::::
::::: Puzzle (248) :::::
::::: Puzzle (249) :::::
::::: Puzzle (250) :::::
::::: Puzzle (251) :::::
::::: Puzzle (252) :::::
::::: Puzzle (253) :::::
::::: Puzzle (254) :::::
::::: Puzzle (255) :::::
::::: Puzzle (256) :::::
::::: Puzzle (257) :::::
::::: Puzzle (258) :::::
::::: Puzzle (259) :::::
::::: Puzzle (260) :::::

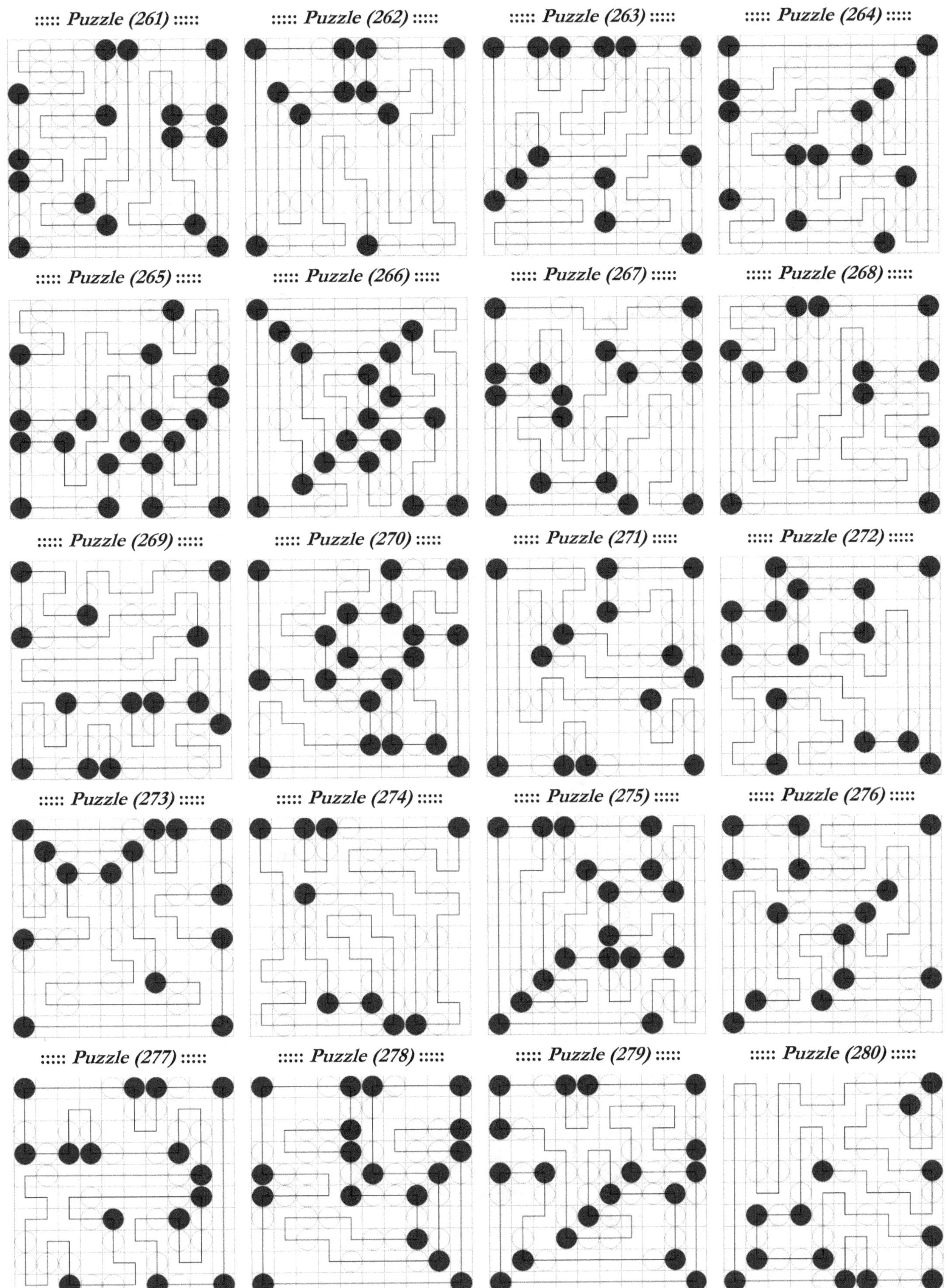

::::: Puzzle (261) :::::
::::: Puzzle (262) :::::
::::: Puzzle (263) :::::
::::: Puzzle (264) :::::
::::: Puzzle (265) :::::
::::: Puzzle (266) :::::
::::: Puzzle (267) :::::
::::: Puzzle (268) :::::
::::: Puzzle (269) :::::
::::: Puzzle (270) :::::
::::: Puzzle (271) :::::
::::: Puzzle (272) :::::
::::: Puzzle (273) :::::
::::: Puzzle (274) :::::
::::: Puzzle (275) :::::
::::: Puzzle (276) :::::
::::: Puzzle (277) :::::
::::: Puzzle (278) :::::
::::: Puzzle (279) :::::
::::: Puzzle (280) :::::

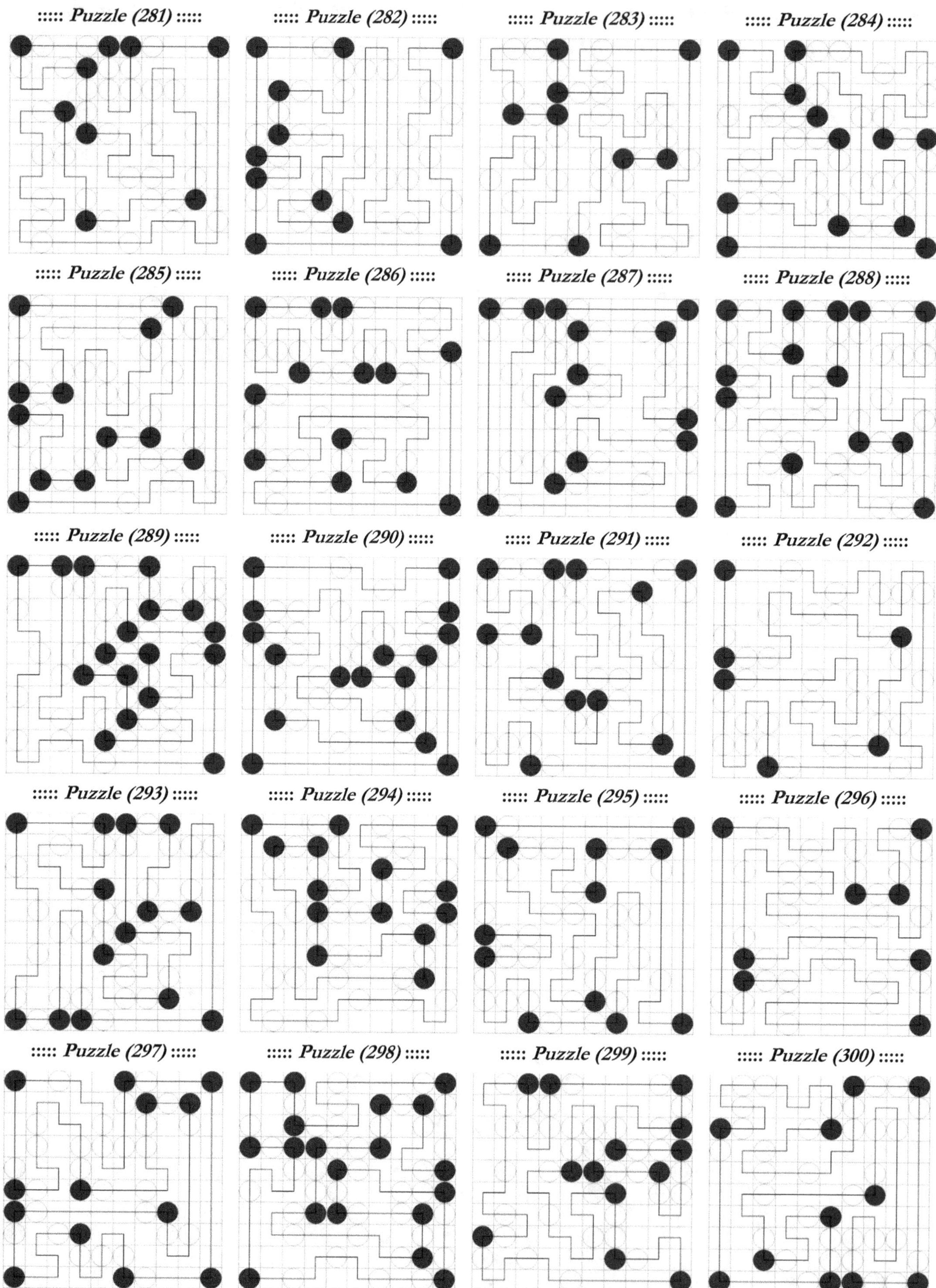

::::: Puzzle (281) :::::
::::: Puzzle (282) :::::
::::: Puzzle (283) :::::
::::: Puzzle (284) :::::
::::: Puzzle (285) :::::
::::: Puzzle (286) :::::
::::: Puzzle (287) :::::
::::: Puzzle (288) :::::
::::: Puzzle (289) :::::
::::: Puzzle (290) :::::
::::: Puzzle (291) :::::
::::: Puzzle (292) :::::
::::: Puzzle (293) :::::
::::: Puzzle (294) :::::
::::: Puzzle (295) :::::
::::: Puzzle (296) :::::
::::: Puzzle (297) :::::
::::: Puzzle (298) :::::
::::: Puzzle (299) :::::
::::: Puzzle (300) :::::

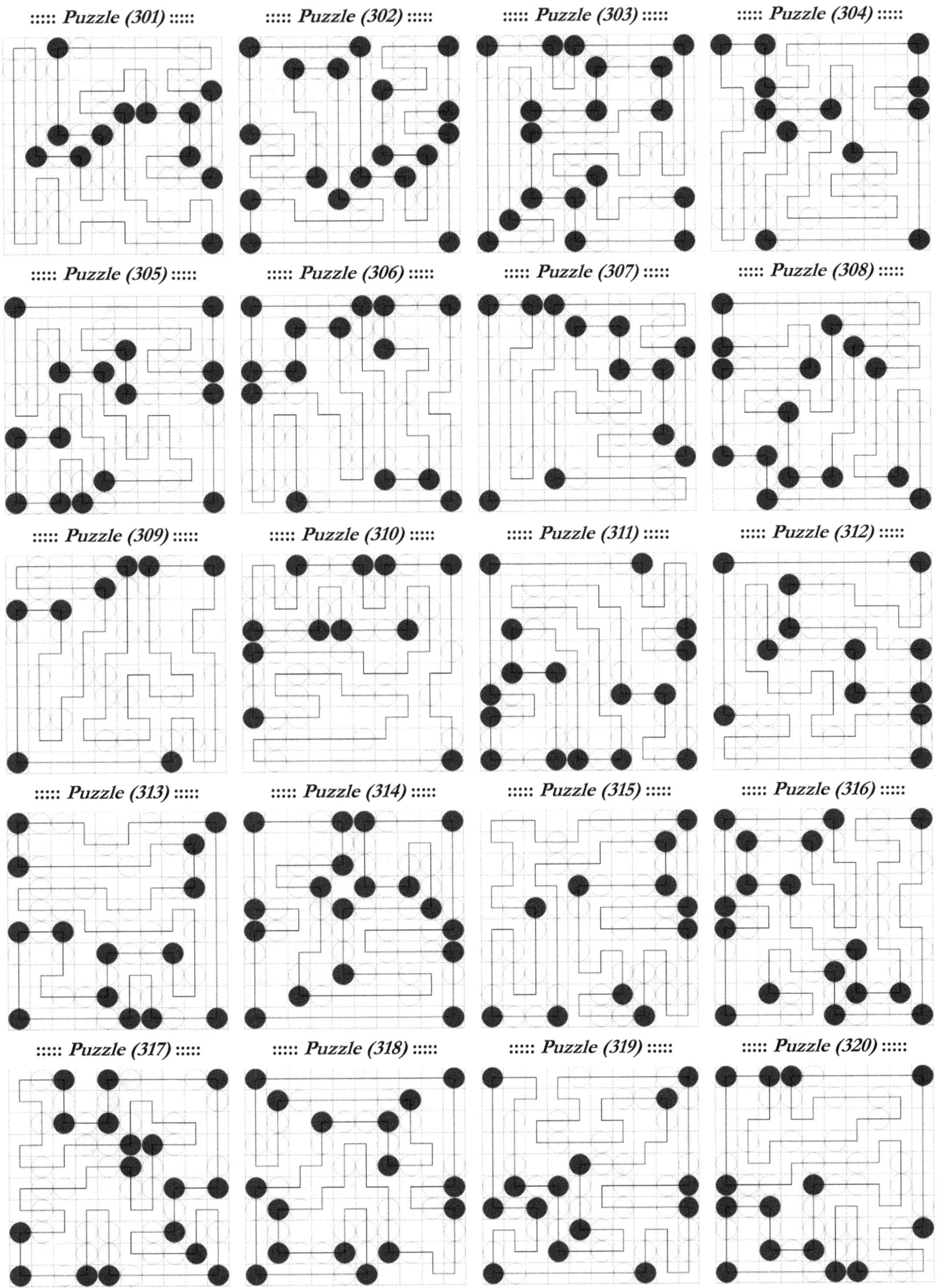

::::: Puzzle (301) :::::
::::: Puzzle (302) :::::
::::: Puzzle (303) :::::
::::: Puzzle (304) :::::
::::: Puzzle (305) :::::
::::: Puzzle (306) :::::
::::: Puzzle (307) :::::
::::: Puzzle (308) :::::
::::: Puzzle (309) :::::
::::: Puzzle (310) :::::
::::: Puzzle (311) :::::
::::: Puzzle (312) :::::
::::: Puzzle (313) :::::
::::: Puzzle (314) :::::
::::: Puzzle (315) :::::
::::: Puzzle (316) :::::
::::: Puzzle (317) :::::
::::: Puzzle (318) :::::
::::: Puzzle (319) :::::
::::: Puzzle (320) :::::

::::: *Puzzle (321)* ::::: ::::: *Puzzle (322)* ::::: ::::: *Puzzle (323)* ::::: ::::: *Puzzle (324)* :::::

::::: *Puzzle (325)* ::::: ::::: *Puzzle (326)* ::::: ::::: *Puzzle (327)* ::::: ::::: *Puzzle (328)* :::::

::::: *Puzzle (329)* ::::: ::::: *Puzzle (330)* ::::: ::::: *Puzzle (331)* ::::: ::::: *Puzzle (332)* :::::

::::: *Puzzle (333)* ::::: ::::: *Puzzle (334)* ::::: ::::: *Puzzle (335)* ::::: ::::: *Puzzle (336)* :::::

::::: *Puzzle (337)* ::::: ::::: *Puzzle (338)* ::::: ::::: *Puzzle (339)* ::::: ::::: *Puzzle (340)* :::::

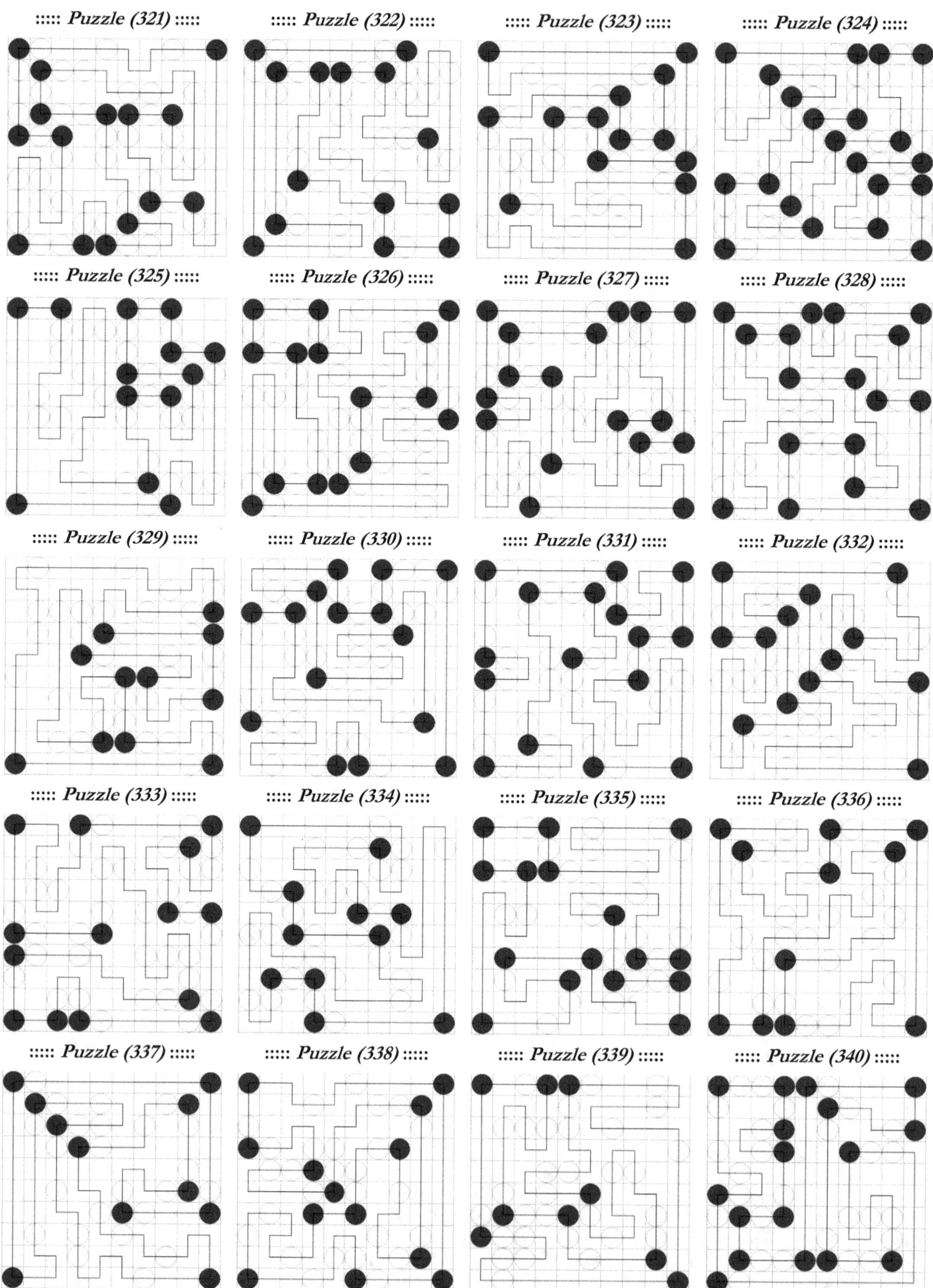

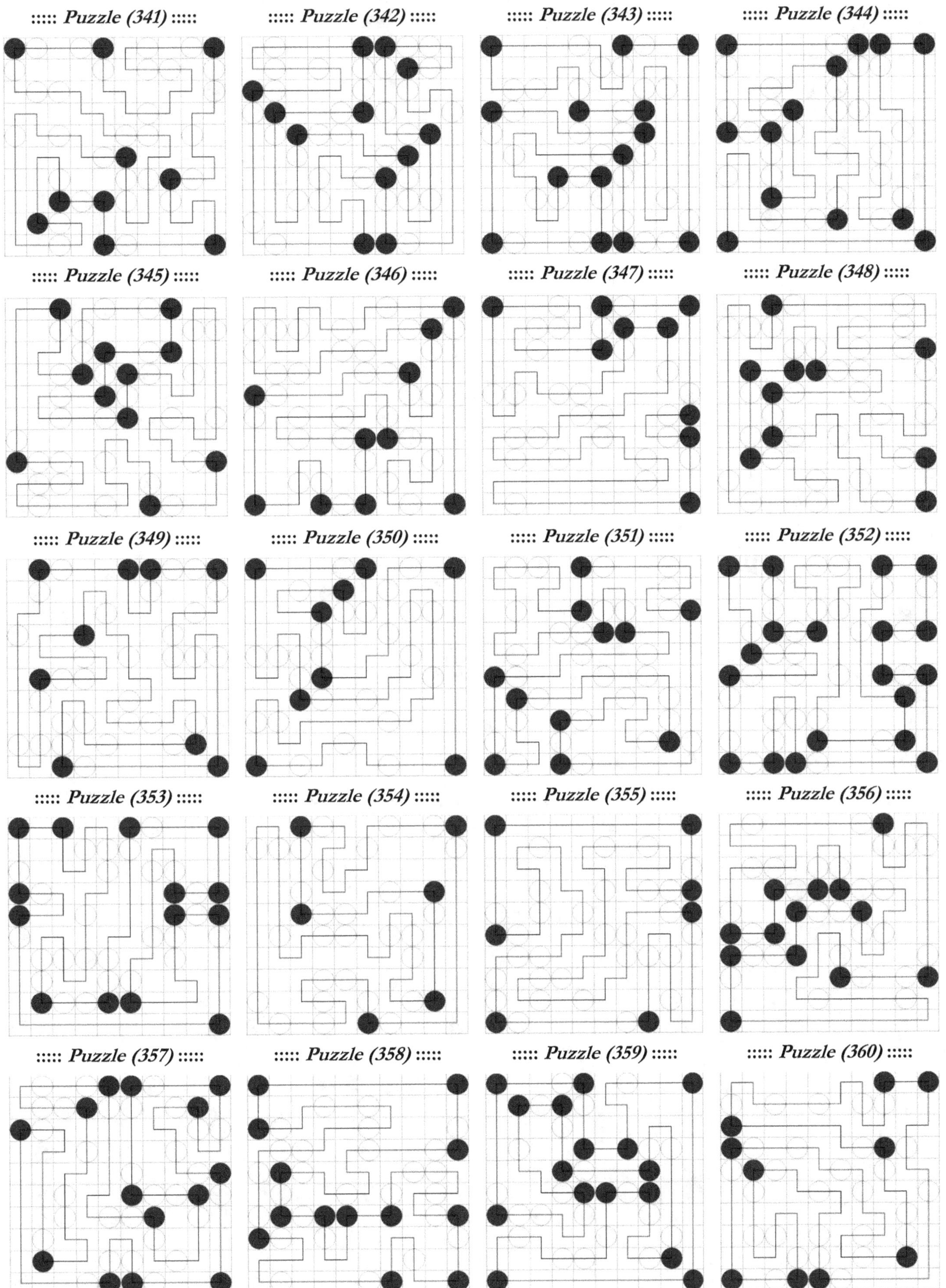

::::: Puzzle (341) :::::
::::: Puzzle (342) :::::
::::: Puzzle (343) :::::
::::: Puzzle (344) :::::
::::: Puzzle (345) :::::
::::: Puzzle (346) :::::
::::: Puzzle (347) :::::
::::: Puzzle (348) :::::
::::: Puzzle (349) :::::
::::: Puzzle (350) :::::
::::: Puzzle (351) :::::
::::: Puzzle (352) :::::
::::: Puzzle (353) :::::
::::: Puzzle (354) :::::
::::: Puzzle (355) :::::
::::: Puzzle (356) :::::
::::: Puzzle (357) :::::
::::: Puzzle (358) :::::
::::: Puzzle (359) :::::
::::: Puzzle (360) :::::

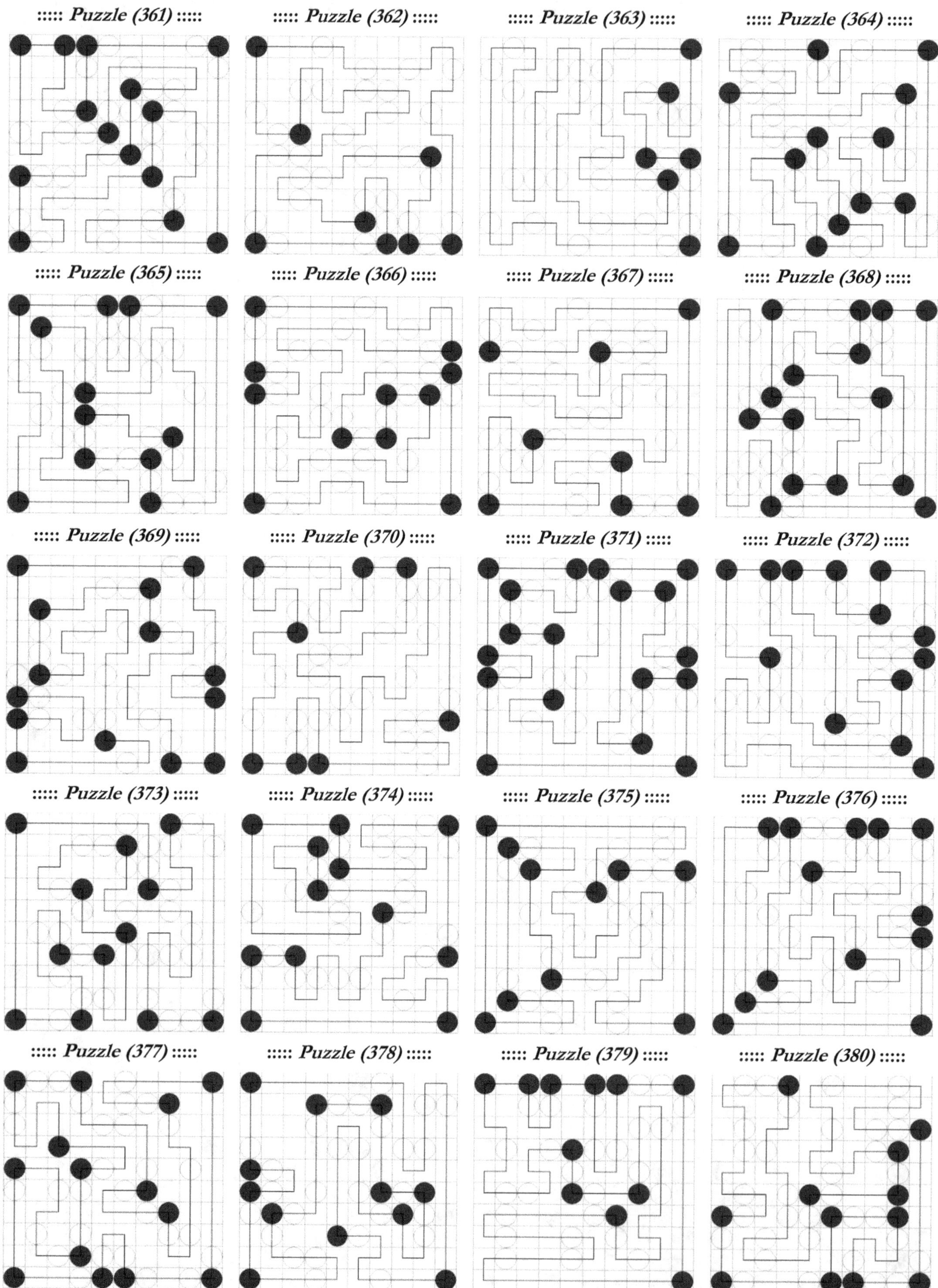

::::: Puzzle (361) :::::
::::: Puzzle (362) :::::
::::: Puzzle (363) :::::
::::: Puzzle (364) :::::
::::: Puzzle (365) :::::
::::: Puzzle (366) :::::
::::: Puzzle (367) :::::
::::: Puzzle (368) :::::
::::: Puzzle (369) :::::
::::: Puzzle (370) :::::
::::: Puzzle (371) :::::
::::: Puzzle (372) :::::
::::: Puzzle (373) :::::
::::: Puzzle (374) :::::
::::: Puzzle (375) :::::
::::: Puzzle (376) :::::
::::: Puzzle (377) :::::
::::: Puzzle (378) :::::
::::: Puzzle (379) :::::
::::: Puzzle (380) :::::

::::: *Puzzle (381)* ::::: ::::: *Puzzle (382)* ::::: ::::: *Puzzle (383)* ::::: ::::: *Puzzle (384)* :::::

::::: *Puzzle (385)* ::::: ::::: *Puzzle (386)* ::::: ::::: *Puzzle (387)* ::::: ::::: *Puzzle (388)* :::::

::::: *Puzzle (389)* ::::: ::::: *Puzzle (390)* ::::: ::::: *Puzzle (391)* ::::: ::::: *Puzzle (392)* :::::

::::: *Puzzle (393)* ::::: ::::: *Puzzle (394)* ::::: ::::: *Puzzle (395)* ::::: ::::: *Puzzle (396)* :::::

::::: *Puzzle (397)* ::::: ::::: *Puzzle (398)* ::::: ::::: *Puzzle (399)* ::::: ::::: *Puzzle (400)* :::::

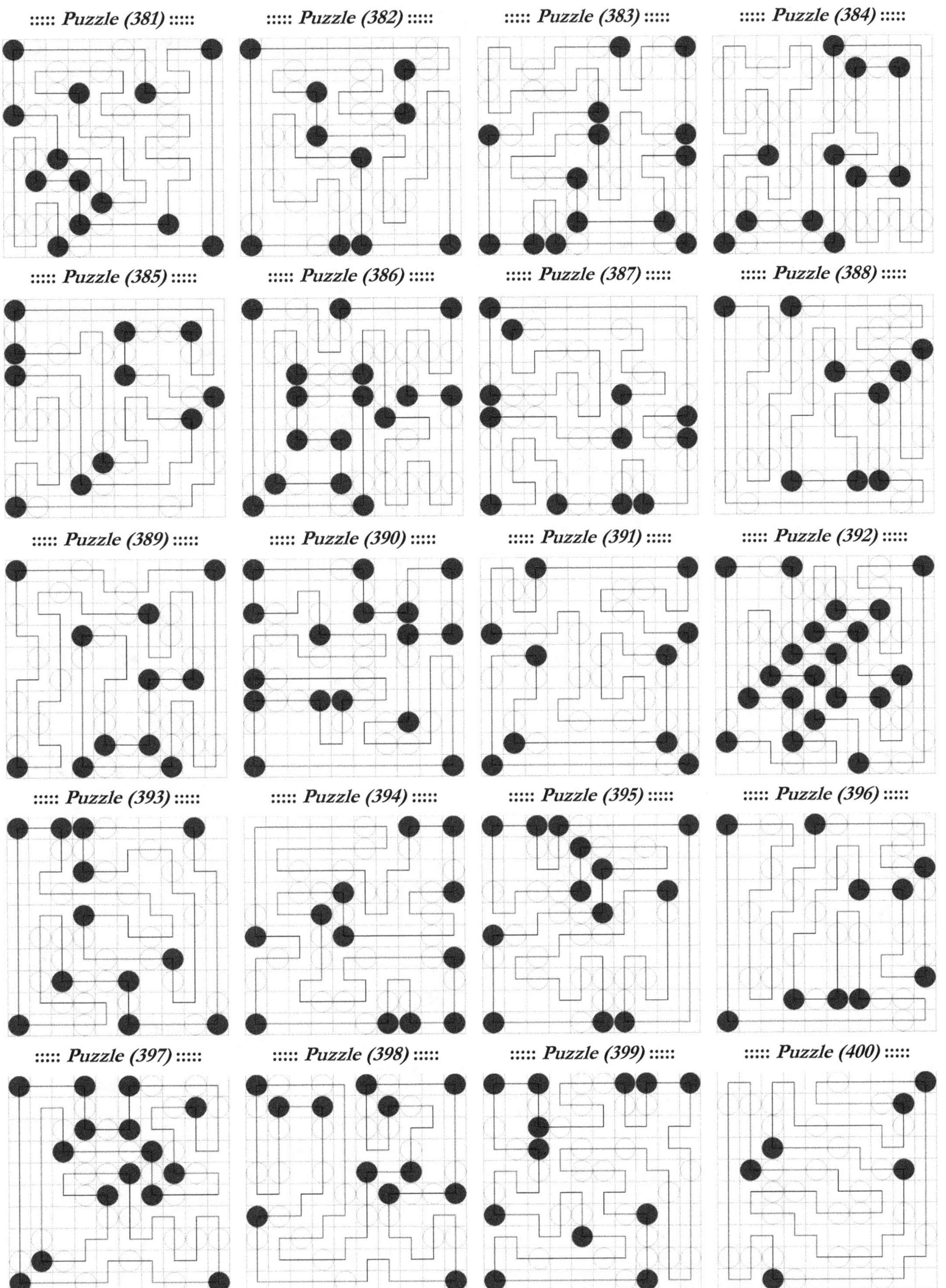

(189)

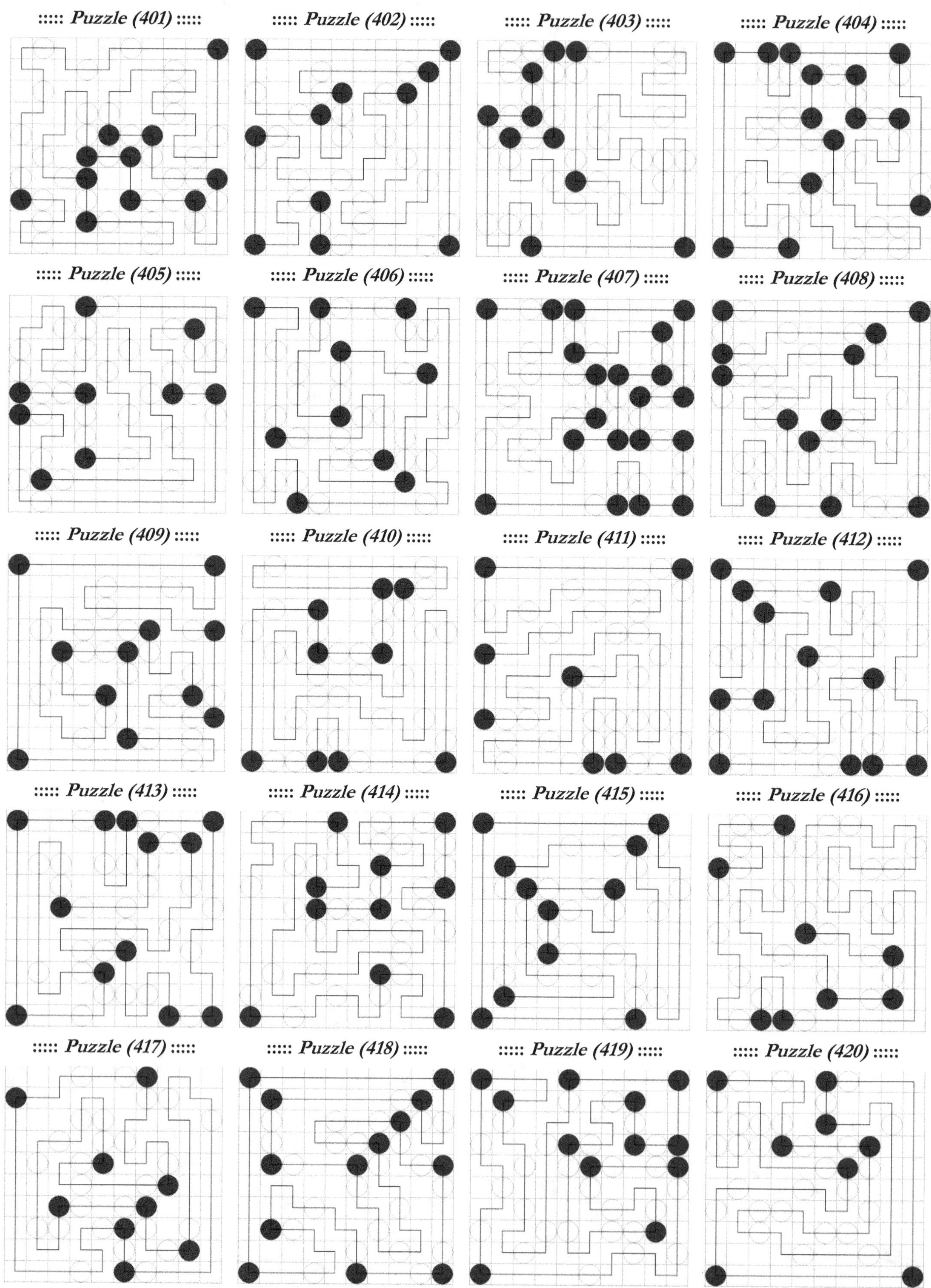

::::: Puzzle (401) :::::
::::: Puzzle (402) :::::
::::: Puzzle (403) :::::
::::: Puzzle (404) :::::
::::: Puzzle (405) :::::
::::: Puzzle (406) :::::
::::: Puzzle (407) :::::
::::: Puzzle (408) :::::
::::: Puzzle (409) :::::
::::: Puzzle (410) :::::
::::: Puzzle (411) :::::
::::: Puzzle (412) :::::
::::: Puzzle (413) :::::
::::: Puzzle (414) :::::
::::: Puzzle (415) :::::
::::: Puzzle (416) :::::
::::: Puzzle (417) :::::
::::: Puzzle (418) :::::
::::: Puzzle (419) :::::
::::: Puzzle (420) :::::

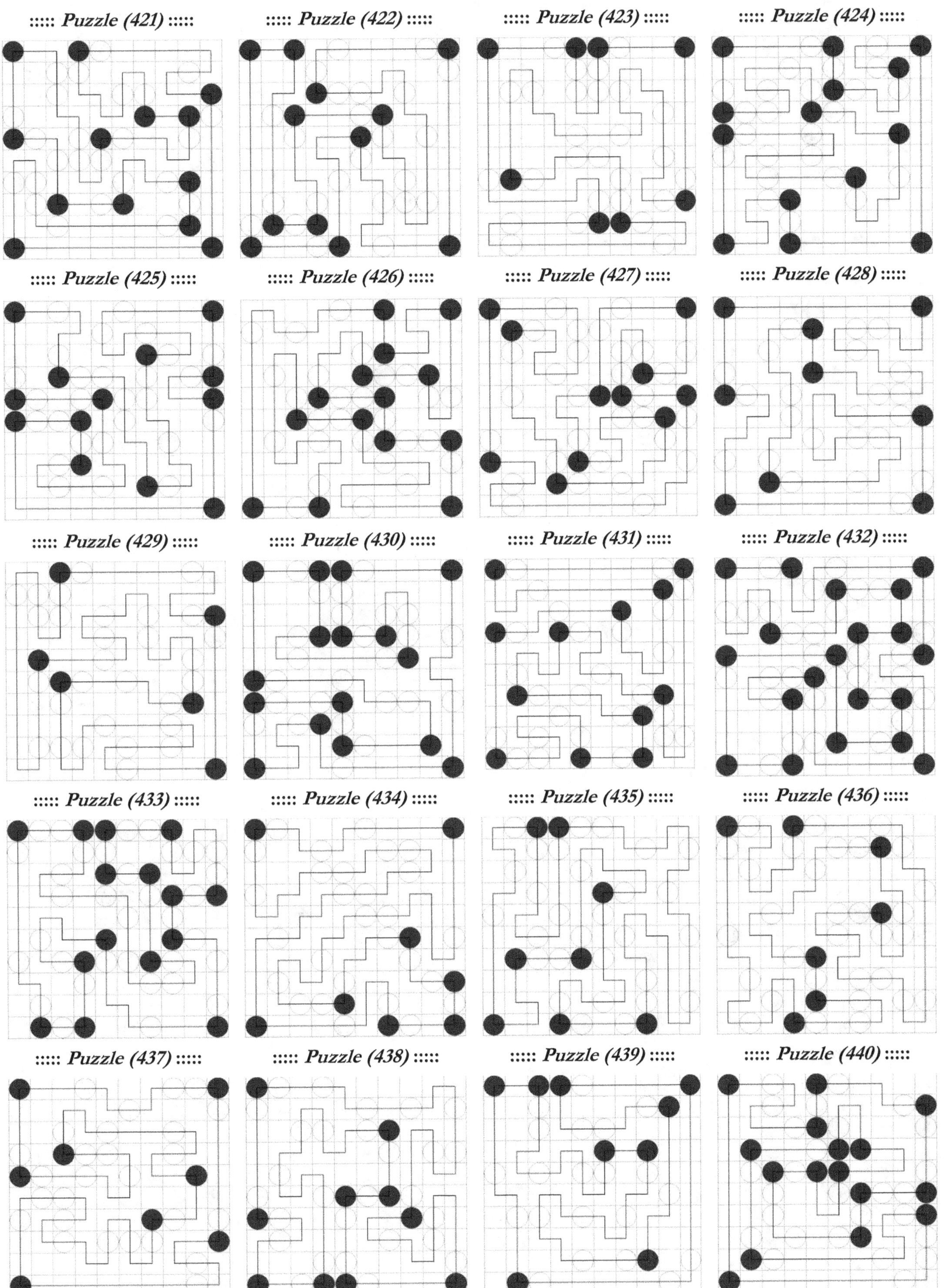

::::: Puzzle (421) :::::
::::: Puzzle (422) :::::
::::: Puzzle (423) :::::
::::: Puzzle (424) :::::
::::: Puzzle (425) :::::
::::: Puzzle (426) :::::
::::: Puzzle (427) :::::
::::: Puzzle (428) :::::
::::: Puzzle (429) :::::
::::: Puzzle (430) :::::
::::: Puzzle (431) :::::
::::: Puzzle (432) :::::
::::: Puzzle (433) :::::
::::: Puzzle (434) :::::
::::: Puzzle (435) :::::
::::: Puzzle (436) :::::
::::: Puzzle (437) :::::
::::: Puzzle (438) :::::
::::: Puzzle (439) :::::
::::: Puzzle (440) :::::

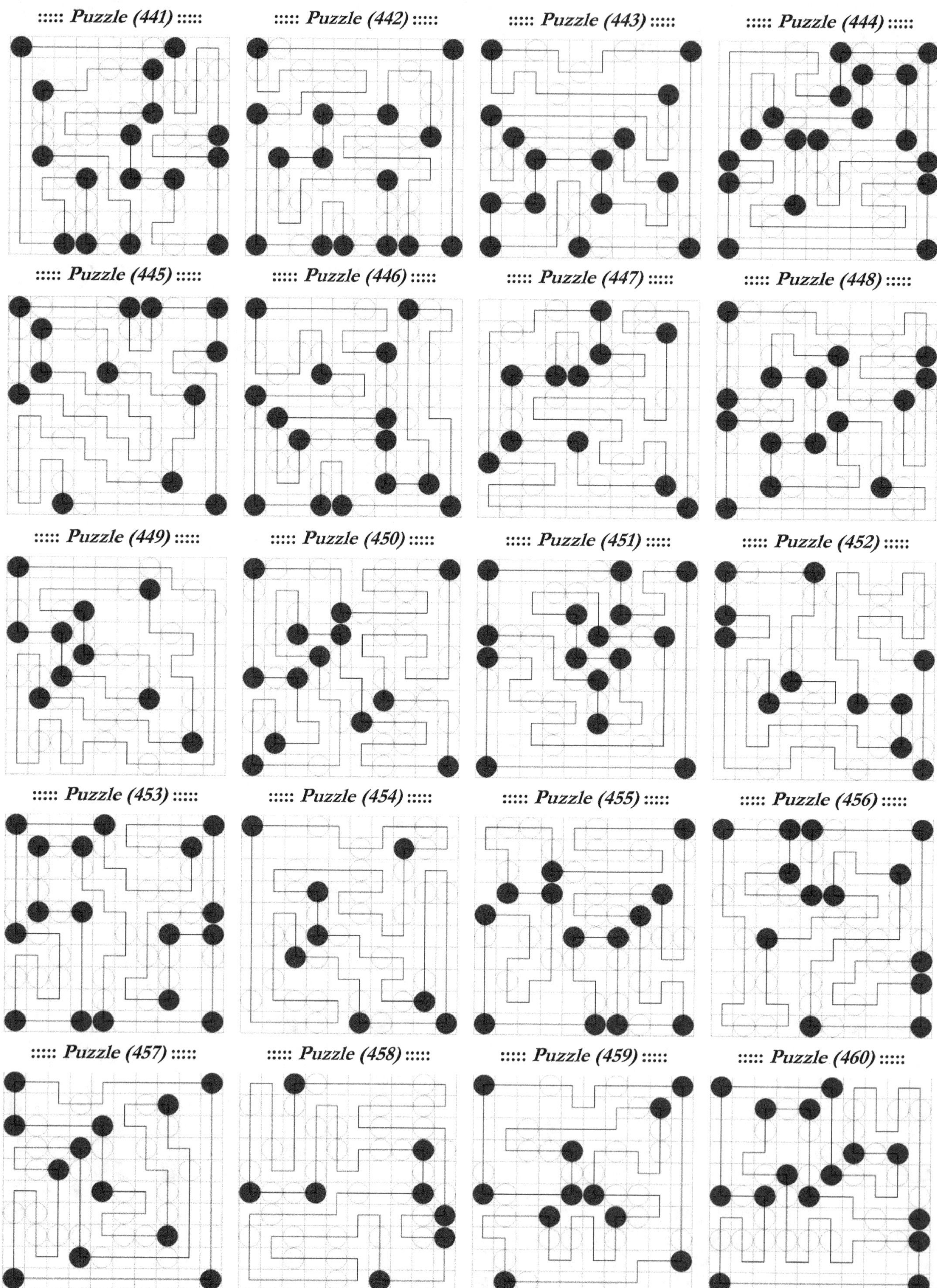

::::: Puzzle (441) :::::
::::: Puzzle (442) :::::
::::: Puzzle (443) :::::
::::: Puzzle (444) :::::
::::: Puzzle (445) :::::
::::: Puzzle (446) :::::
::::: Puzzle (447) :::::
::::: Puzzle (448) :::::
::::: Puzzle (449) :::::
::::: Puzzle (450) :::::
::::: Puzzle (451) :::::
::::: Puzzle (452) :::::
::::: Puzzle (453) :::::
::::: Puzzle (454) :::::
::::: Puzzle (455) :::::
::::: Puzzle (456) :::::
::::: Puzzle (457) :::::
::::: Puzzle (458) :::::
::::: Puzzle (459) :::::
::::: Puzzle (460) :::::

(192)

::::: *Puzzle (461)* :::::	::::: *Puzzle (462)* :::::	::::: *Puzzle (463)* :::::	::::: *Puzzle (464)* :::::
::::: *Puzzle (465)* :::::	::::: *Puzzle (466)* :::::	::::: *Puzzle (467)* :::::	::::: *Puzzle (468)* :::::
::::: *Puzzle (469)* :::::	::::: *Puzzle (470)* :::::	::::: *Puzzle (471)* :::::	::::: *Puzzle (472)* :::::
::::: *Puzzle (473)* :::::	::::: *Puzzle (474)* :::::	::::: *Puzzle (475)* :::::	::::: *Puzzle (476)* :::::
::::: *Puzzle (477)* :::::	::::: *Puzzle (478)* :::::	::::: *Puzzle (479)* :::::	::::: *Puzzle (480)* :::::

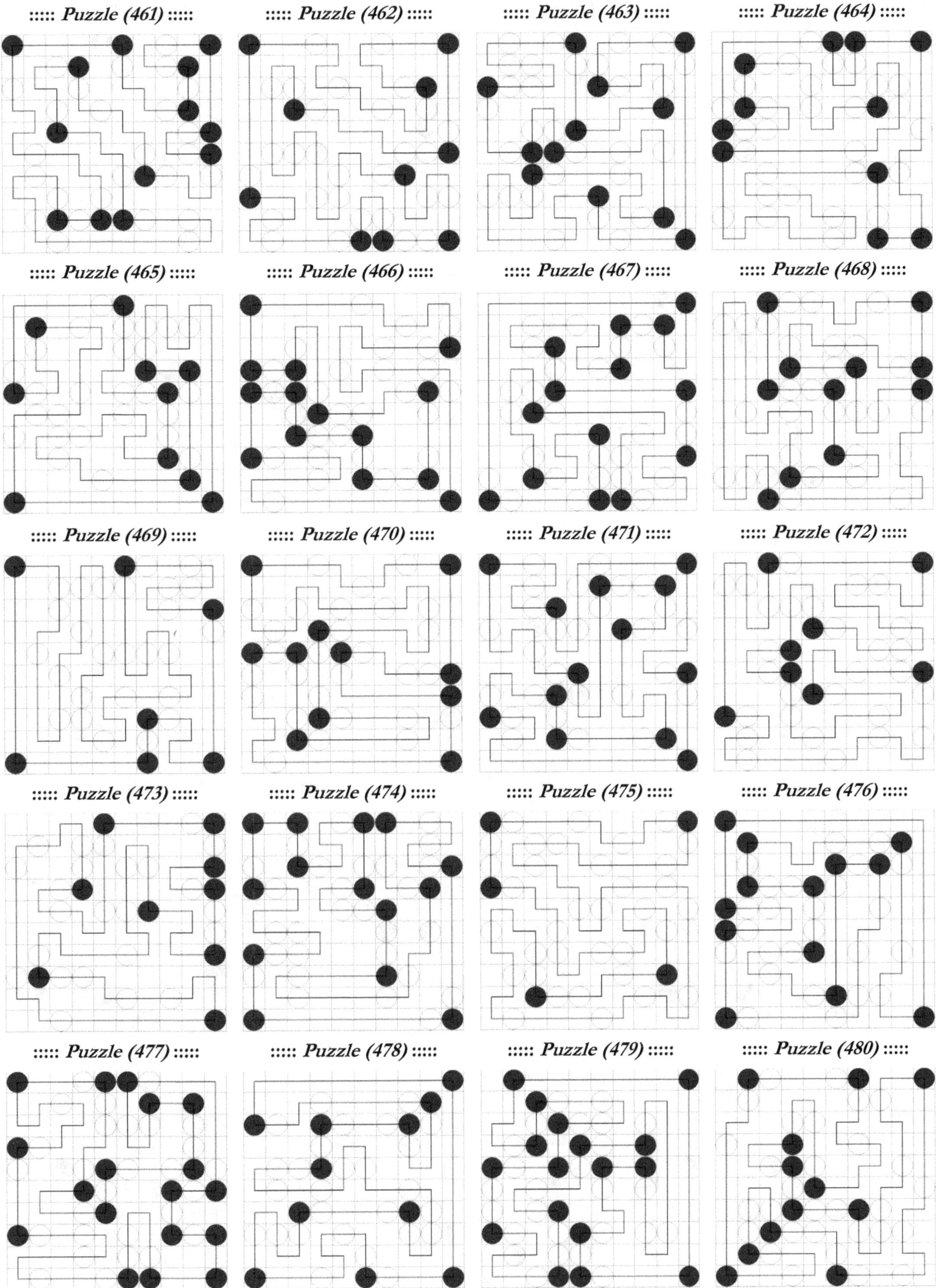

(193)

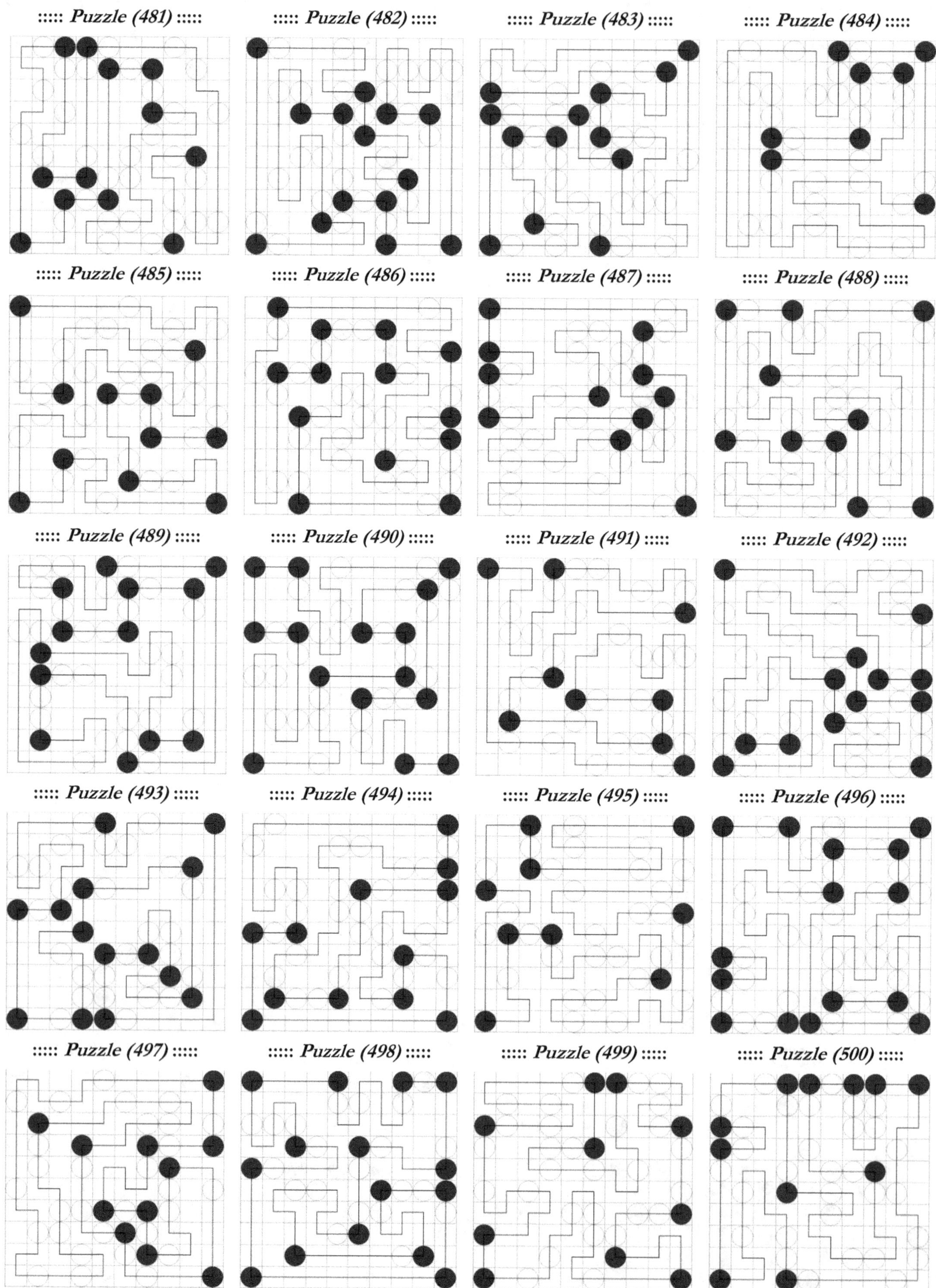

::::: Puzzle (481) :::::
::::: Puzzle (482) :::::
::::: Puzzle (483) :::::
::::: Puzzle (484) :::::
::::: Puzzle (485) :::::
::::: Puzzle (486) :::::
::::: Puzzle (487) :::::
::::: Puzzle (488) :::::
::::: Puzzle (489) :::::
::::: Puzzle (490) :::::
::::: Puzzle (491) :::::
::::: Puzzle (492) :::::
::::: Puzzle (493) :::::
::::: Puzzle (494) :::::
::::: Puzzle (495) :::::
::::: Puzzle (496) :::::
::::: Puzzle (497) :::::
::::: Puzzle (498) :::::
::::: Puzzle (499) :::::
::::: Puzzle (500) :::::

(194)

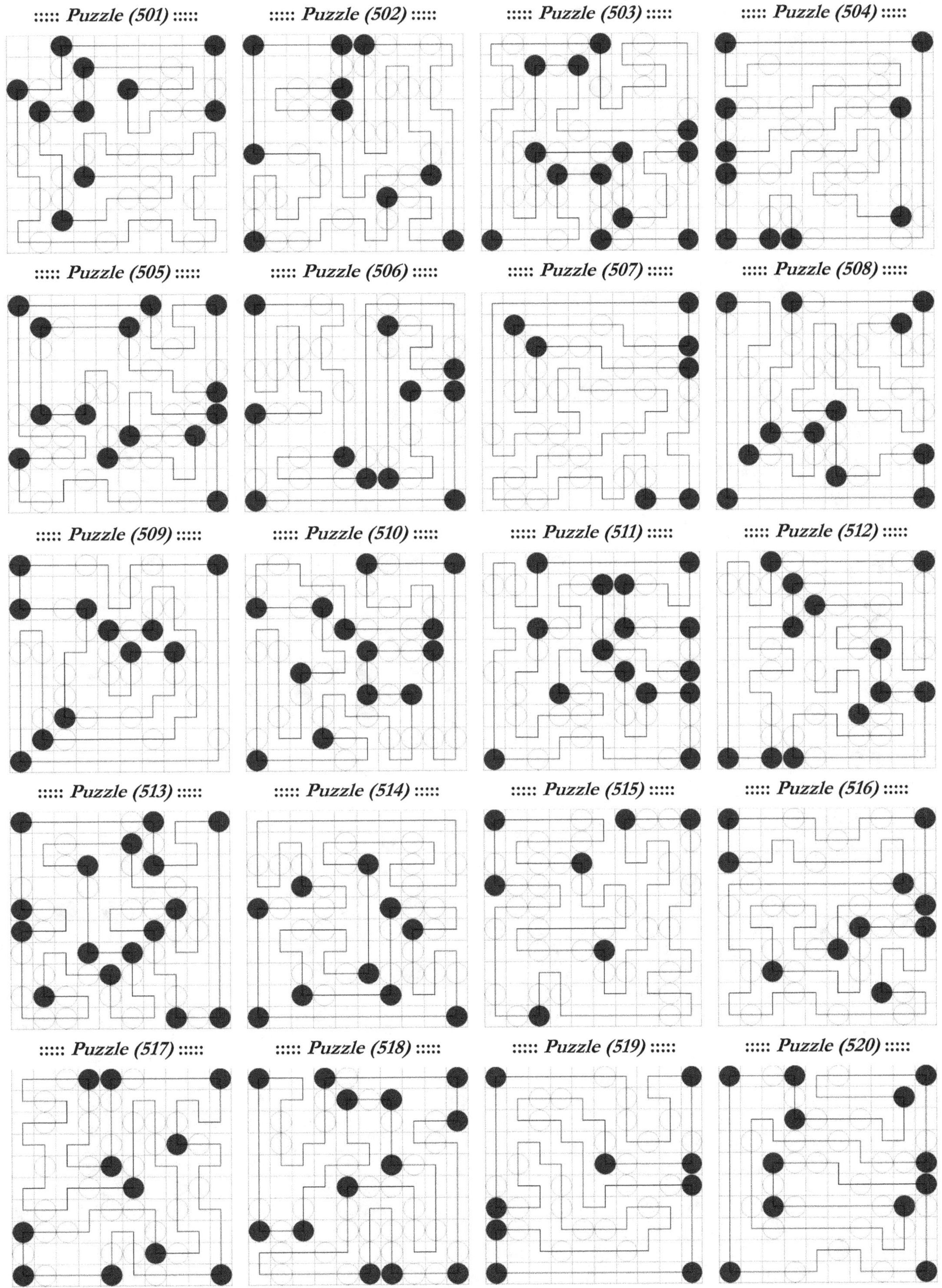

::::: Puzzle (501) :::::
::::: Puzzle (502) :::::
::::: Puzzle (503) :::::
::::: Puzzle (504) :::::
::::: Puzzle (505) :::::
::::: Puzzle (506) :::::
::::: Puzzle (507) :::::
::::: Puzzle (508) :::::
::::: Puzzle (509) :::::
::::: Puzzle (510) :::::
::::: Puzzle (511) :::::
::::: Puzzle (512) :::::
::::: Puzzle (513) :::::
::::: Puzzle (514) :::::
::::: Puzzle (515) :::::
::::: Puzzle (516) :::::
::::: Puzzle (517) :::::
::::: Puzzle (518) :::::
::::: Puzzle (519) :::::
::::: Puzzle (520) :::::

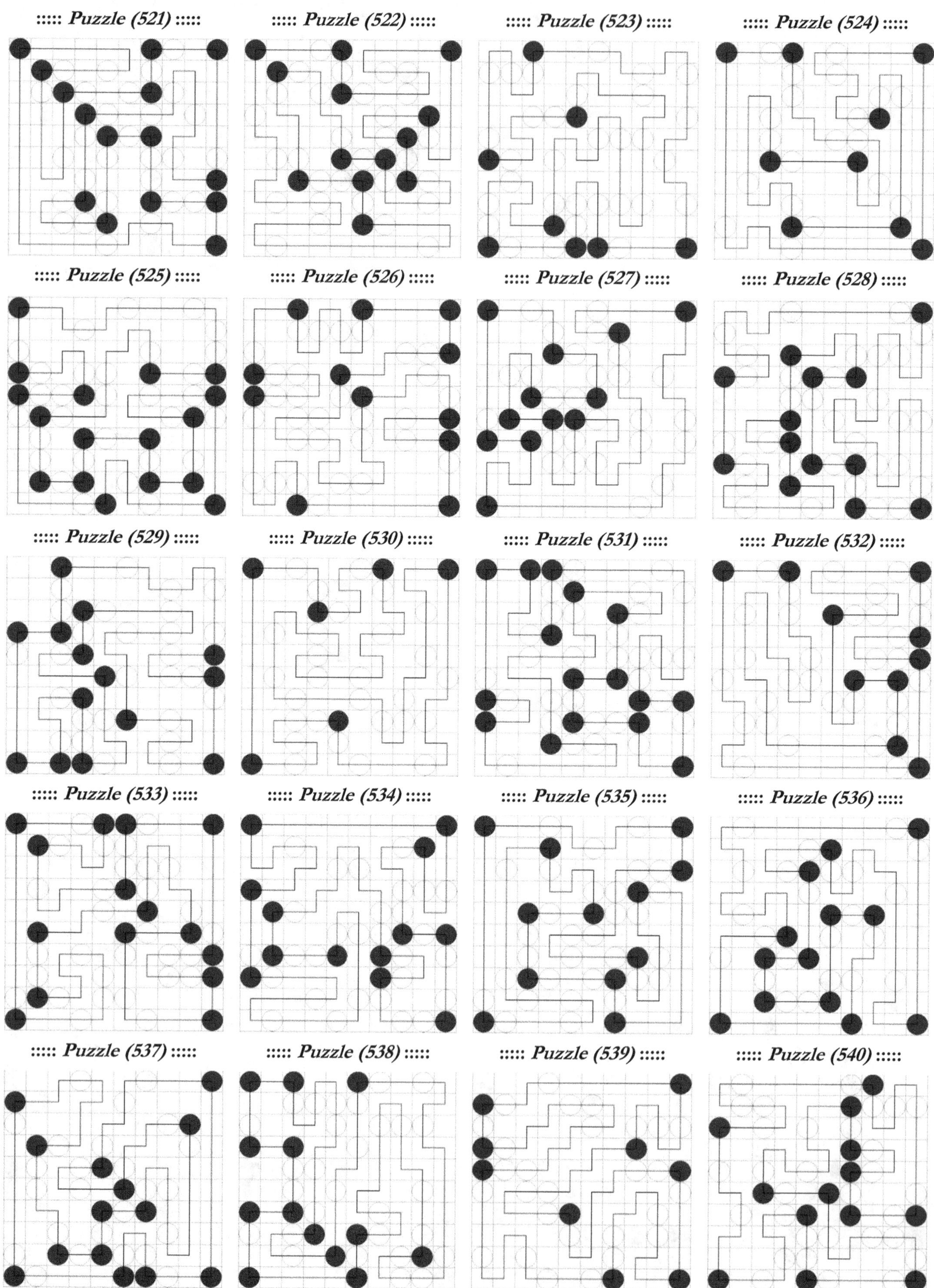

::::: Puzzle (521) :::::
::::: Puzzle (522) :::::
::::: Puzzle (523) :::::
::::: Puzzle (524) :::::
::::: Puzzle (525) :::::
::::: Puzzle (526) :::::
::::: Puzzle (527) :::::
::::: Puzzle (528) :::::
::::: Puzzle (529) :::::
::::: Puzzle (530) :::::
::::: Puzzle (531) :::::
::::: Puzzle (532) :::::
::::: Puzzle (533) :::::
::::: Puzzle (534) :::::
::::: Puzzle (535) :::::
::::: Puzzle (536) :::::
::::: Puzzle (537) :::::
::::: Puzzle (538) :::::
::::: Puzzle (539) :::::
::::: Puzzle (540) :::::

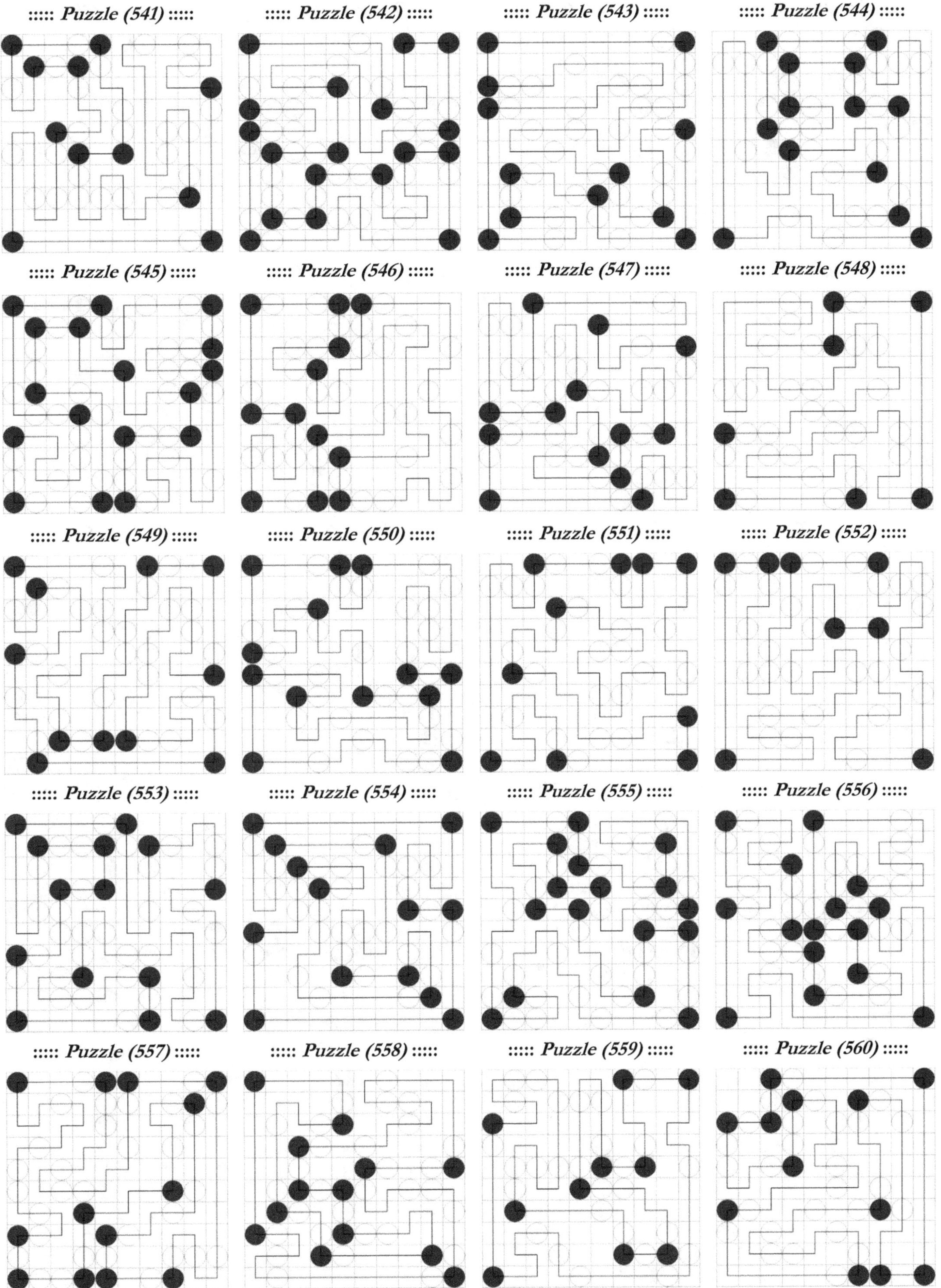

::::: Puzzle (541) :::::
::::: Puzzle (542) :::::
::::: Puzzle (543) :::::
::::: Puzzle (544) :::::
::::: Puzzle (545) :::::
::::: Puzzle (546) :::::
::::: Puzzle (547) :::::
::::: Puzzle (548) :::::
::::: Puzzle (549) :::::
::::: Puzzle (550) :::::
::::: Puzzle (551) :::::
::::: Puzzle (552) :::::
::::: Puzzle (553) :::::
::::: Puzzle (554) :::::
::::: Puzzle (555) :::::
::::: Puzzle (556) :::::
::::: Puzzle (557) :::::
::::: Puzzle (558) :::::
::::: Puzzle (559) :::::
::::: Puzzle (560) :::::

::::: Puzzle (561) :::::
::::: Puzzle (562) :::::
::::: Puzzle (563) :::::
::::: Puzzle (564) :::::
::::: Puzzle (565) :::::
::::: Puzzle (566) :::::
::::: Puzzle (567) :::::
::::: Puzzle (568) :::::
::::: Puzzle (569) :::::
::::: Puzzle (570) :::::
::::: Puzzle (571) :::::
::::: Puzzle (572) :::::
::::: Puzzle (573) :::::
::::: Puzzle (574) :::::
::::: Puzzle (575) :::::
::::: Puzzle (576) :::::
::::: Puzzle (577) :::::
::::: Puzzle (578) :::::
::::: Puzzle (579) :::::
::::: Puzzle (580) :::::

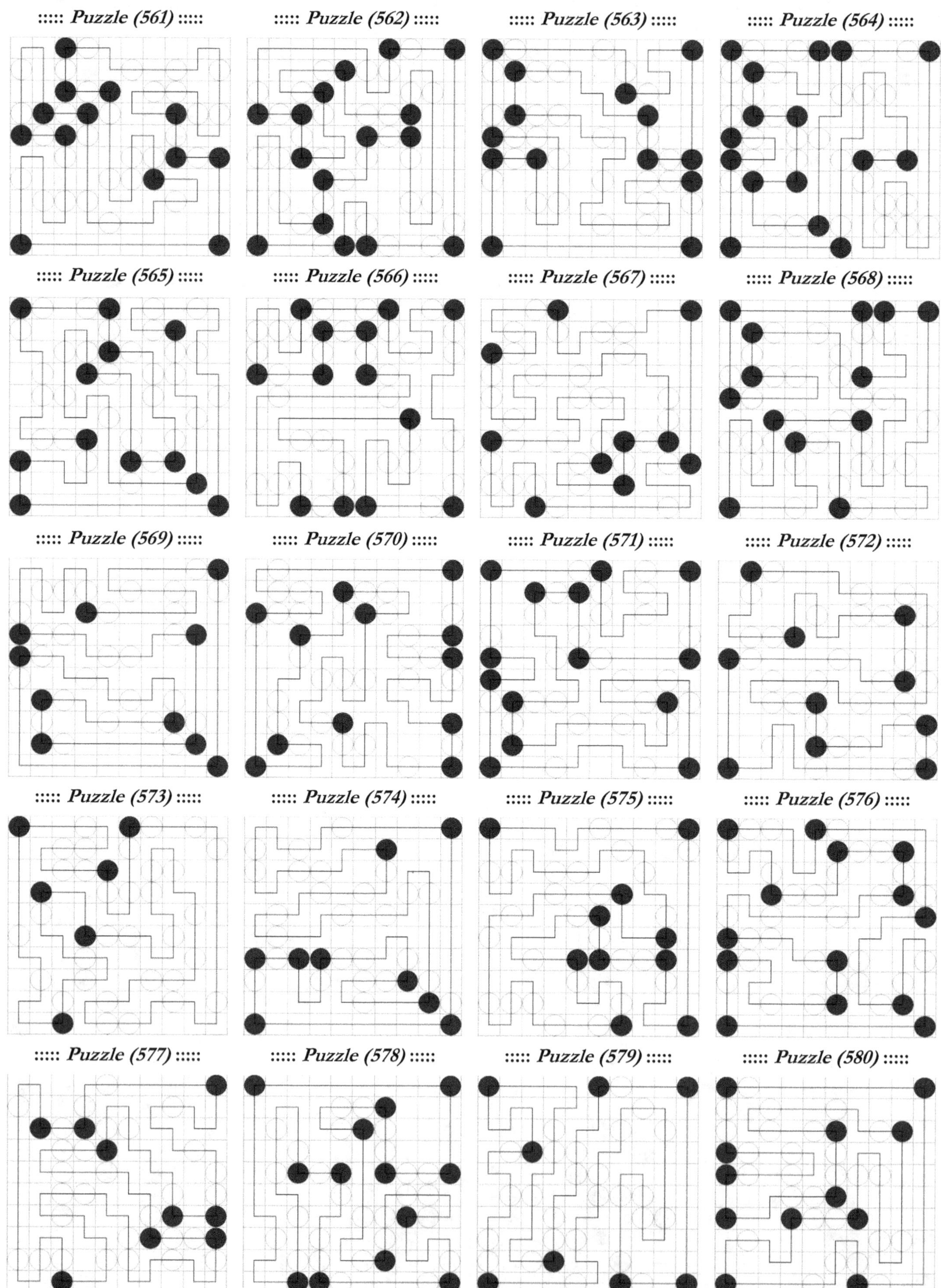

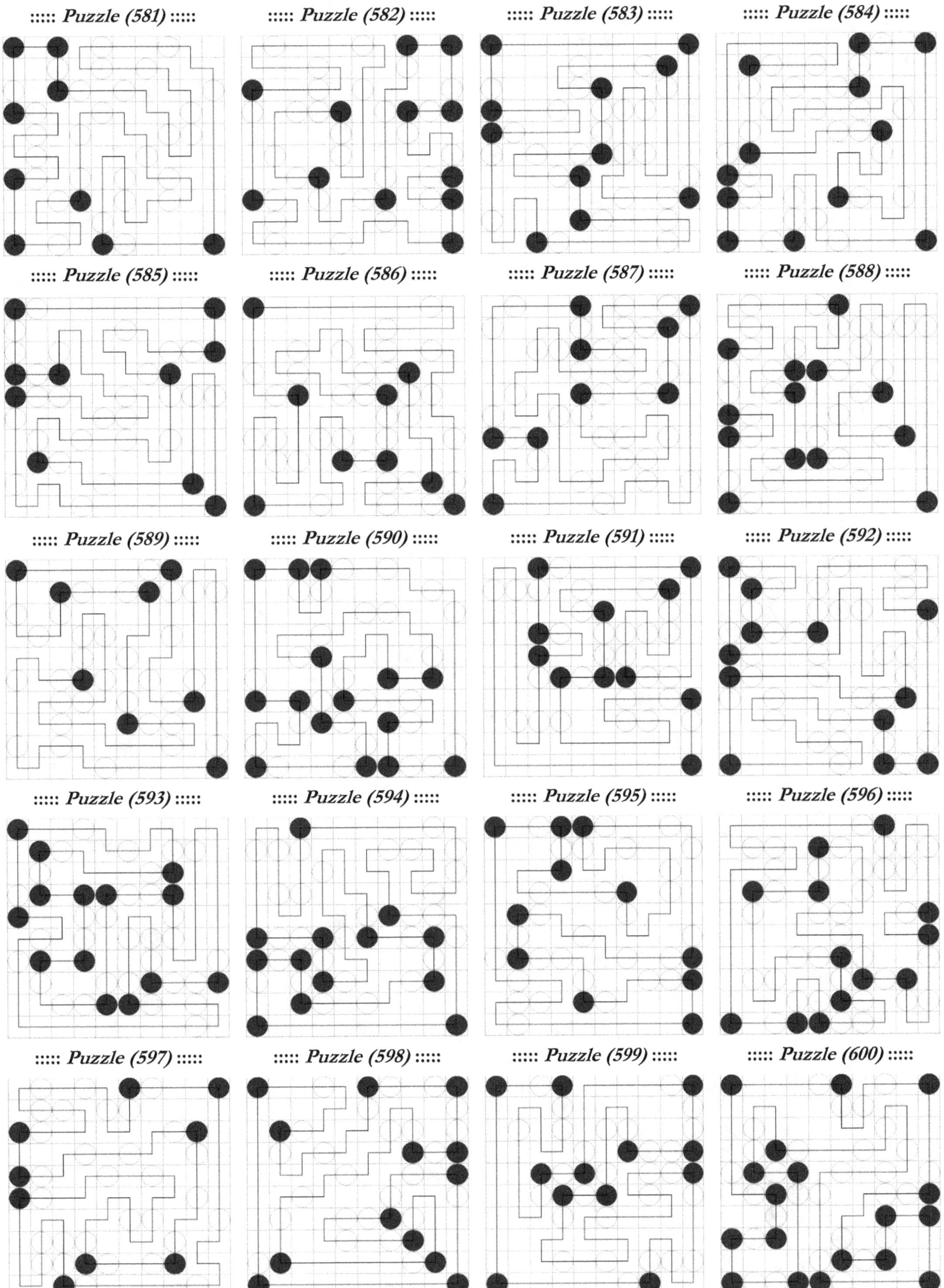

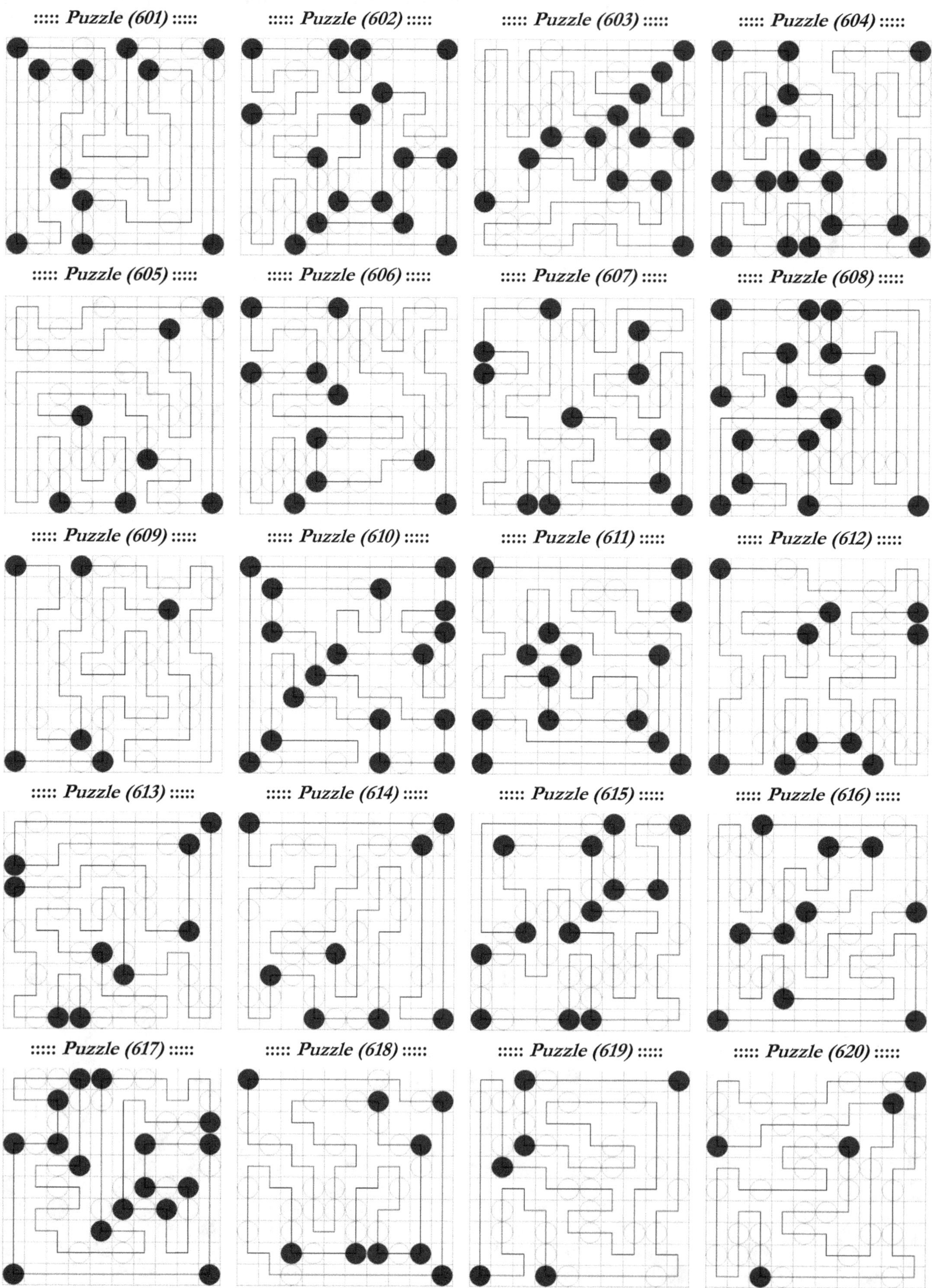

::::: Puzzle (601) :::::
::::: Puzzle (602) :::::
::::: Puzzle (603) :::::
::::: Puzzle (604) :::::
::::: Puzzle (605) :::::
::::: Puzzle (606) :::::
::::: Puzzle (607) :::::
::::: Puzzle (608) :::::
::::: Puzzle (609) :::::
::::: Puzzle (610) :::::
::::: Puzzle (611) :::::
::::: Puzzle (612) :::::
::::: Puzzle (613) :::::
::::: Puzzle (614) :::::
::::: Puzzle (615) :::::
::::: Puzzle (616) :::::
::::: Puzzle (617) :::::
::::: Puzzle (618) :::::
::::: Puzzle (619) :::::
::::: Puzzle (620) :::::

::::: *Puzzle (621)* :::::　::::: *Puzzle (622)* :::::　::::: *Puzzle (623)* :::::　::::: *Puzzle (624)* :::::

::::: *Puzzle (625)* :::::　::::: *Puzzle (626)* :::::　::::: *Puzzle (627)* :::::　::::: *Puzzle (628)* :::::

::::: *Puzzle (629)* :::::　::::: *Puzzle (630)* :::::　::::: *Puzzle (631)* :::::　::::: *Puzzle (632)* :::::

::::: *Puzzle (633)* :::::　::::: *Puzzle (634)* :::::　::::: *Puzzle (635)* :::::　::::: *Puzzle (636)* :::::

::::: *Puzzle (637)* :::::　::::: *Puzzle (638)* :::::　::::: *Puzzle (639)* :::::　::::: *Puzzle (640)* :::::

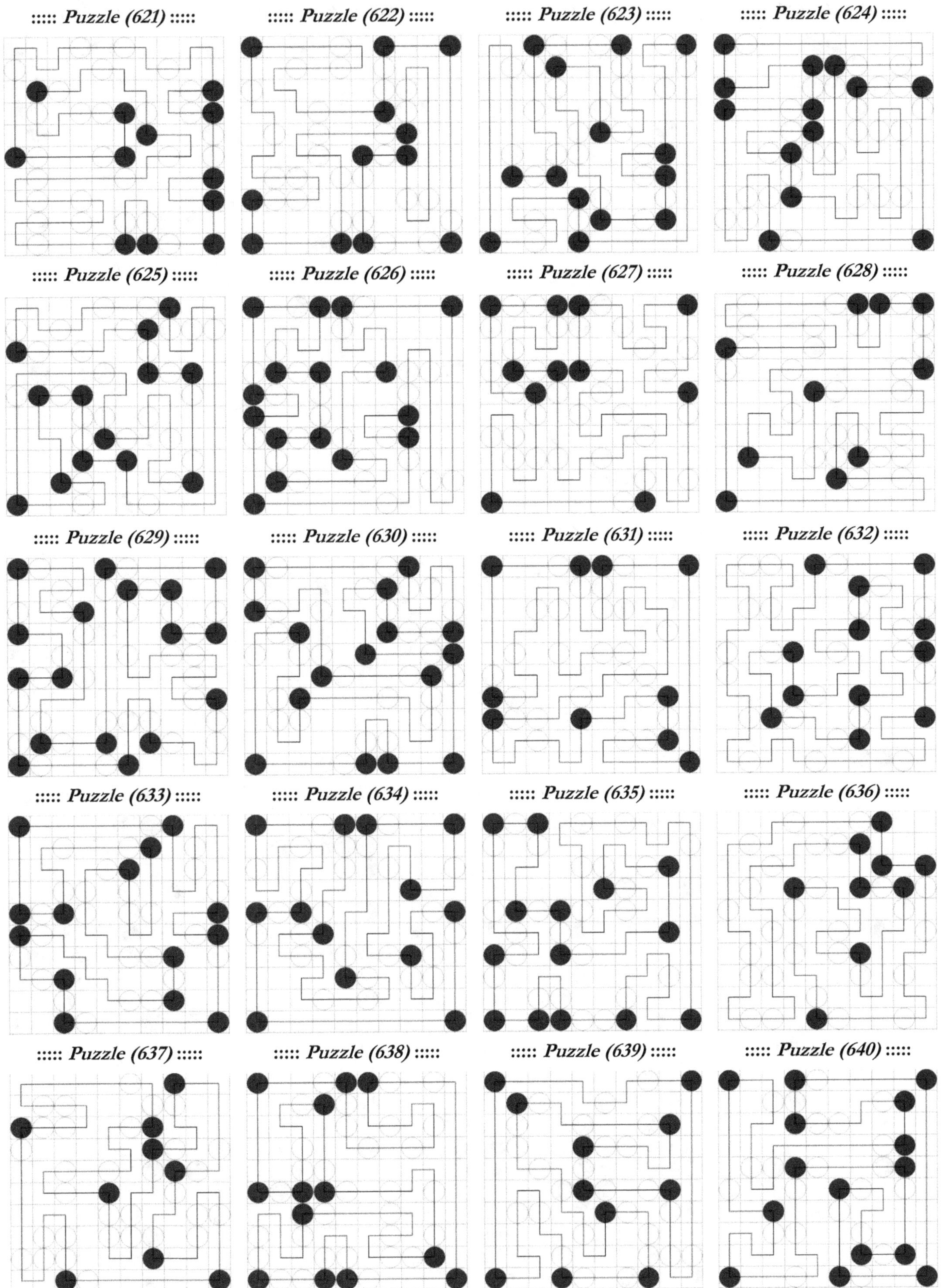

::::: *Puzzle (641)* :::::   ::::: *Puzzle (642)* :::::   ::::: *Puzzle (643)* :::::   ::::: *Puzzle (644)* :::::

::::: *Puzzle (645)* :::::   ::::: *Puzzle (646)* :::::   ::::: *Puzzle (647)* :::::   ::::: *Puzzle (648)* :::::

::::: *Puzzle (649)* :::::   ::::: *Puzzle (650)* :::::   ::::: *Puzzle (651)* :::::   ::::: *Puzzle (652)* :::::

::::: *Puzzle (653)* :::::   ::::: *Puzzle (654)* :::::   ::::: *Puzzle (655)* :::::   ::::: *Puzzle (656)* :::::

::::: *Puzzle (657)* :::::   ::::: *Puzzle (658)* :::::   ::::: *Puzzle (659)* :::::   ::::: *Puzzle (660)* :::::

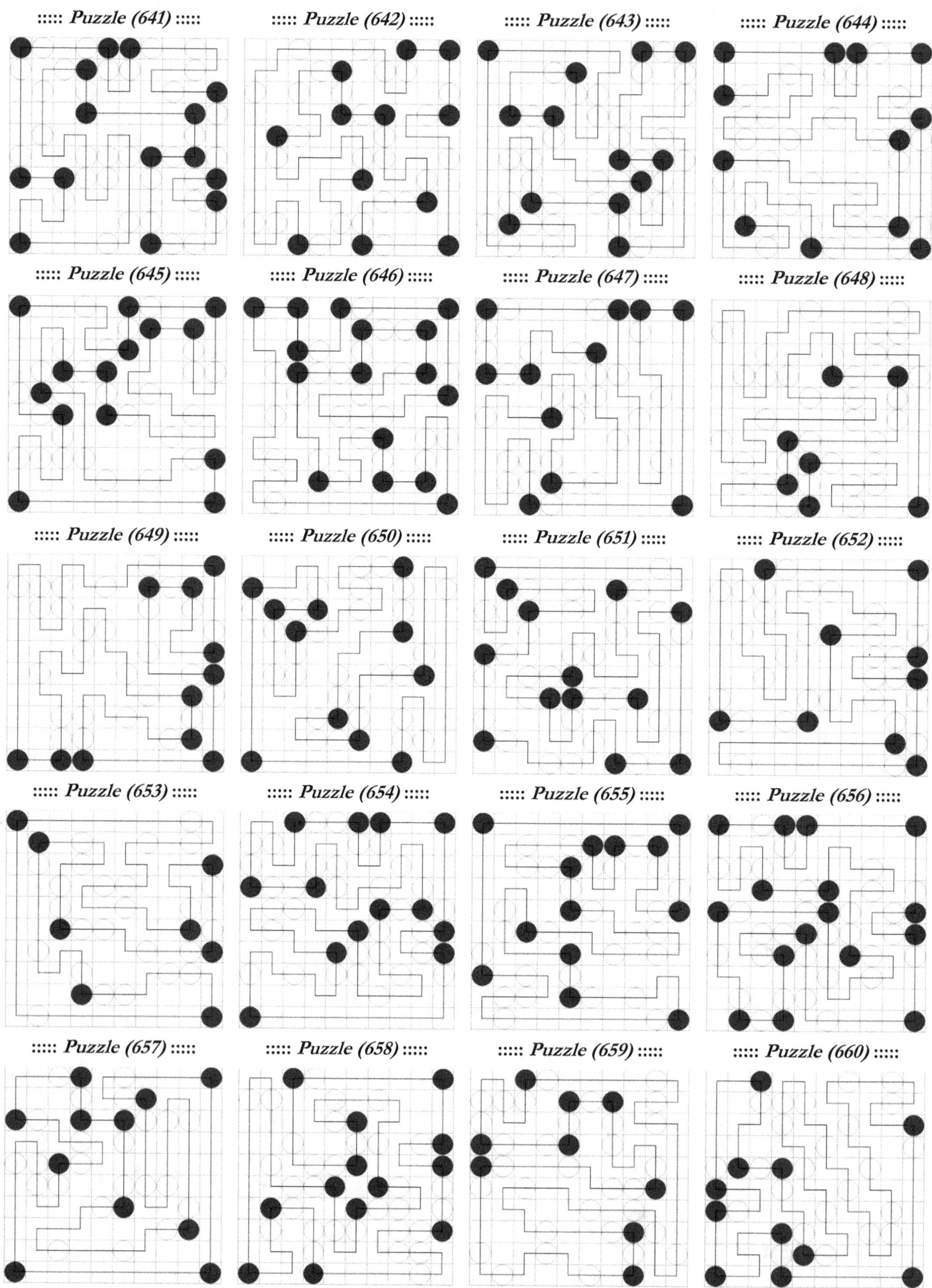

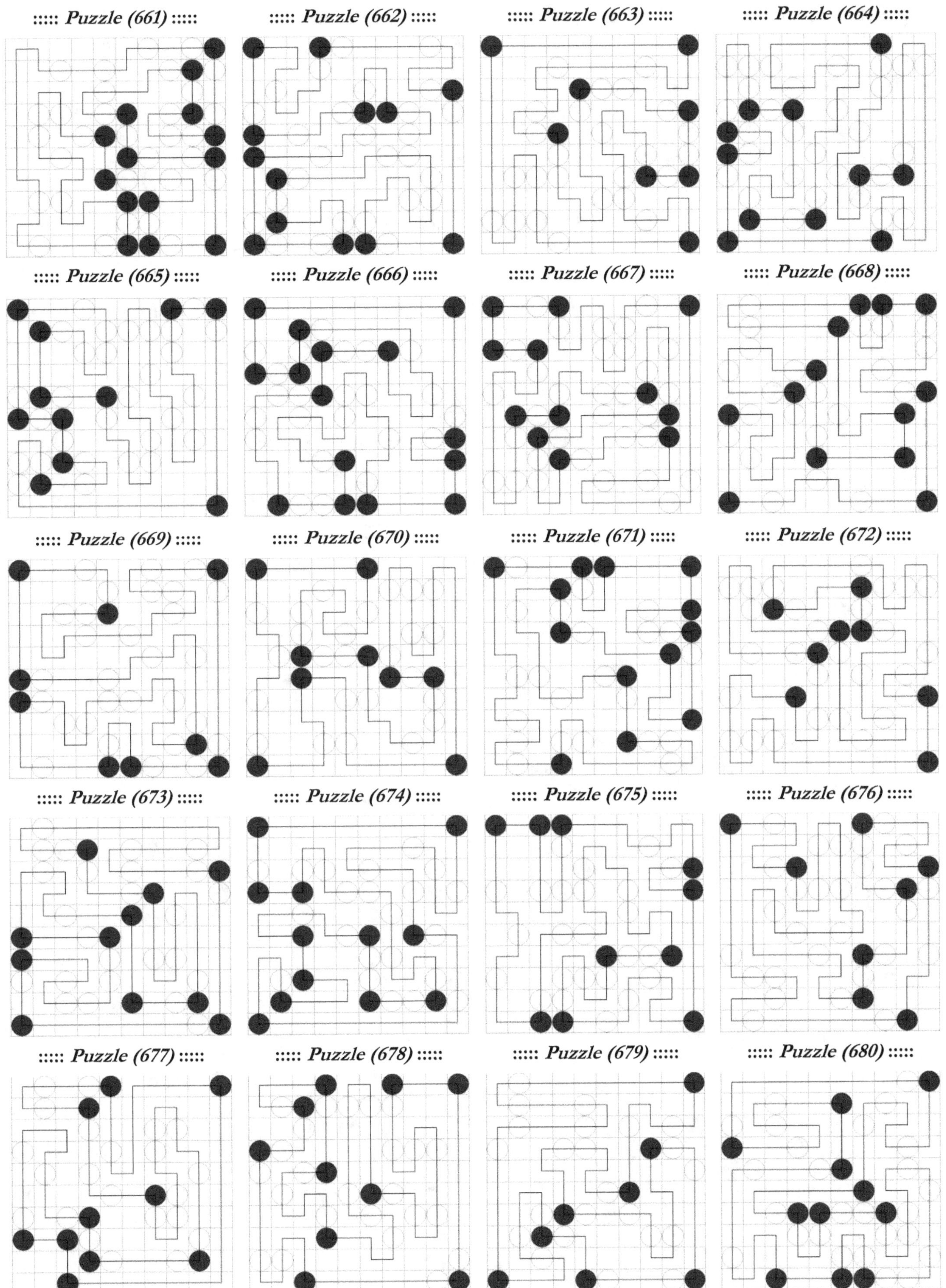

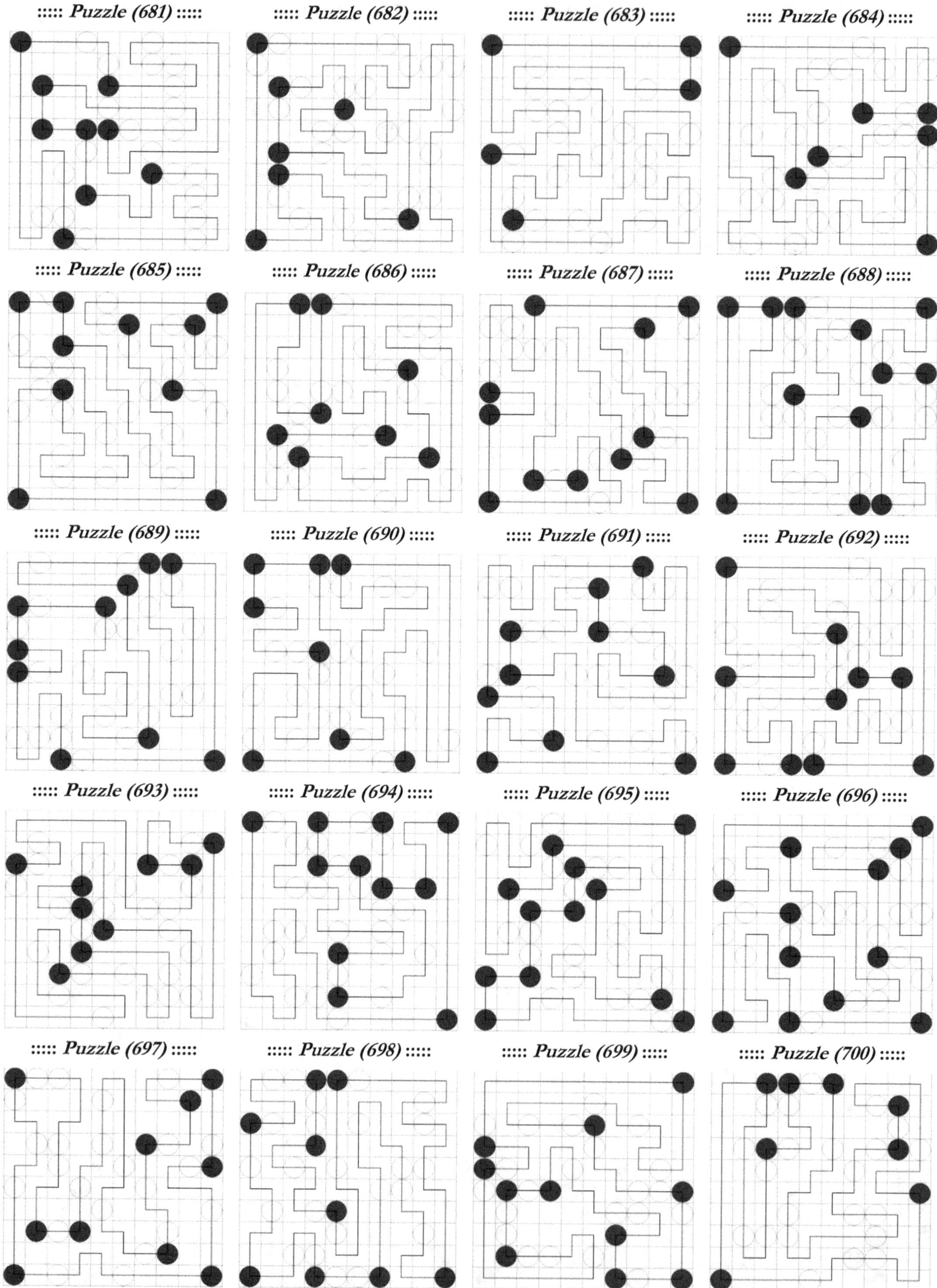

::::: Puzzle (681) :::::
::::: Puzzle (682) :::::
::::: Puzzle (683) :::::
::::: Puzzle (684) :::::
::::: Puzzle (685) :::::
::::: Puzzle (686) :::::
::::: Puzzle (687) :::::
::::: Puzzle (688) :::::
::::: Puzzle (689) :::::
::::: Puzzle (690) :::::
::::: Puzzle (691) :::::
::::: Puzzle (692) :::::
::::: Puzzle (693) :::::
::::: Puzzle (694) :::::
::::: Puzzle (695) :::::
::::: Puzzle (696) :::::
::::: Puzzle (697) :::::
::::: Puzzle (698) :::::
::::: Puzzle (699) :::::
::::: Puzzle (700) :::::

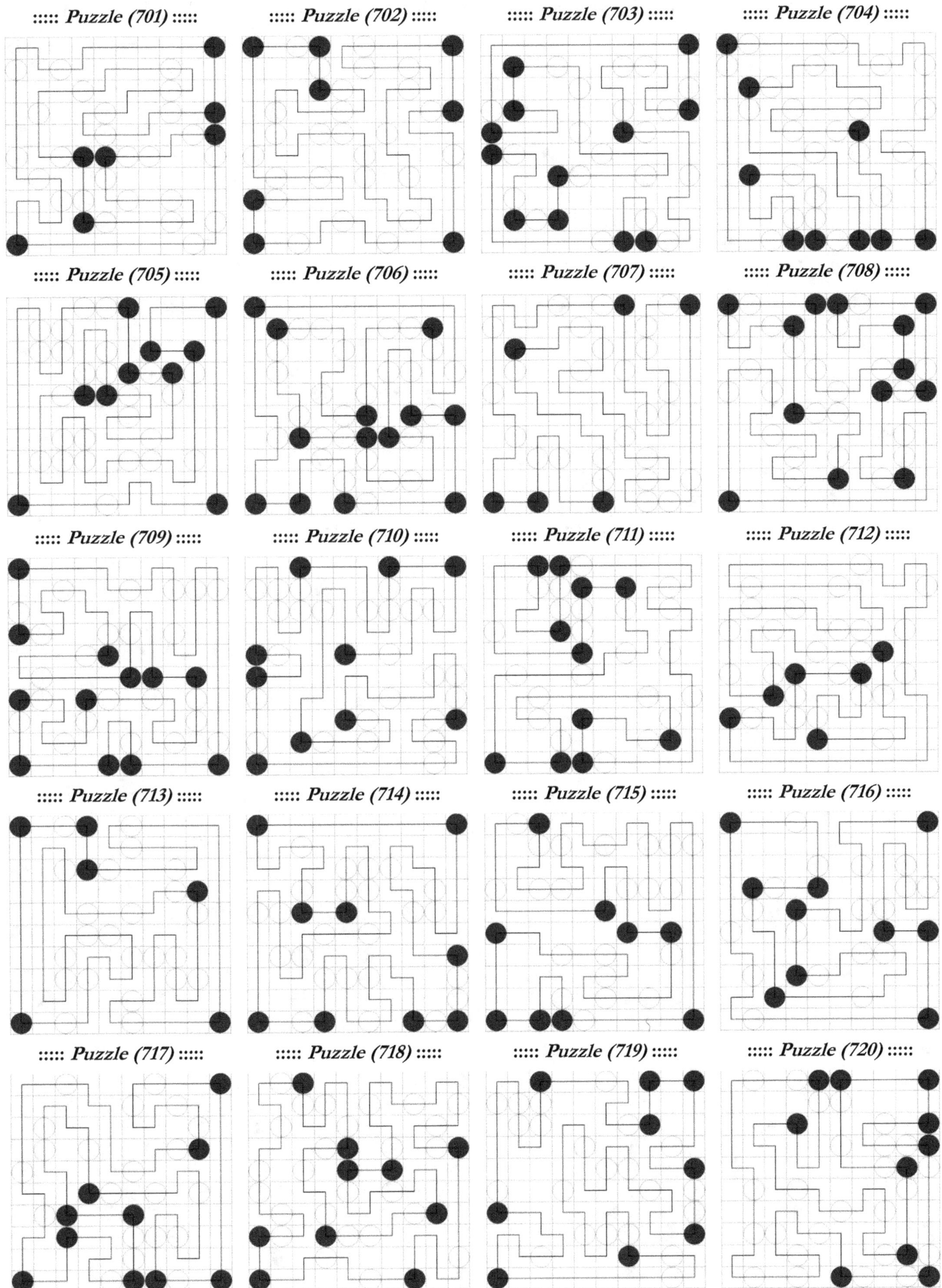

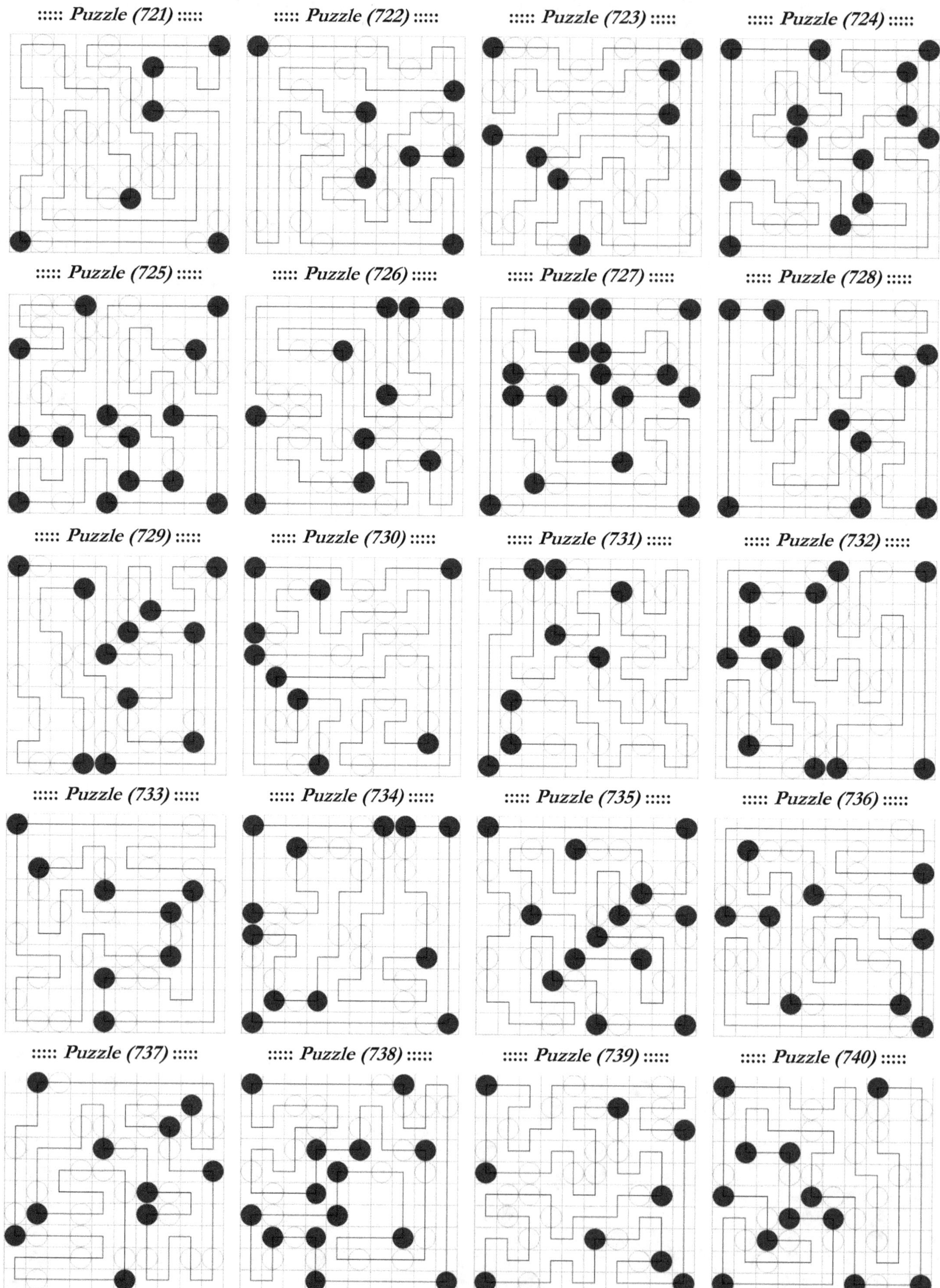

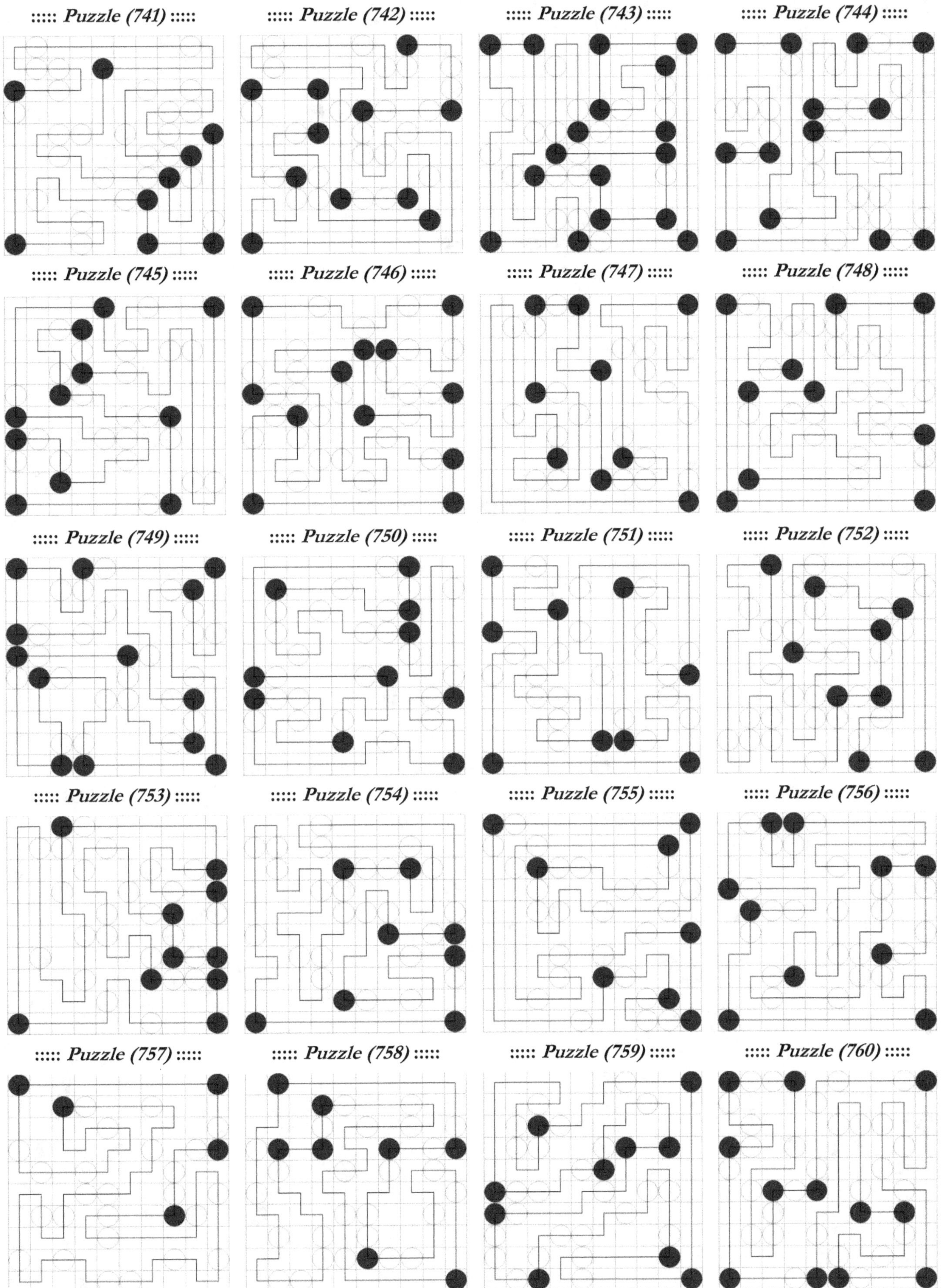

::::: Puzzle (741) :::::
::::: Puzzle (742) :::::
::::: Puzzle (743) :::::
::::: Puzzle (744) :::::
::::: Puzzle (745) :::::
::::: Puzzle (746) :::::
::::: Puzzle (747) :::::
::::: Puzzle (748) :::::
::::: Puzzle (749) :::::
::::: Puzzle (750) :::::
::::: Puzzle (751) :::::
::::: Puzzle (752) :::::
::::: Puzzle (753) :::::
::::: Puzzle (754) :::::
::::: Puzzle (755) :::::
::::: Puzzle (756) :::::
::::: Puzzle (757) :::::
::::: Puzzle (758) :::::
::::: Puzzle (759) :::::
::::: Puzzle (760) :::::

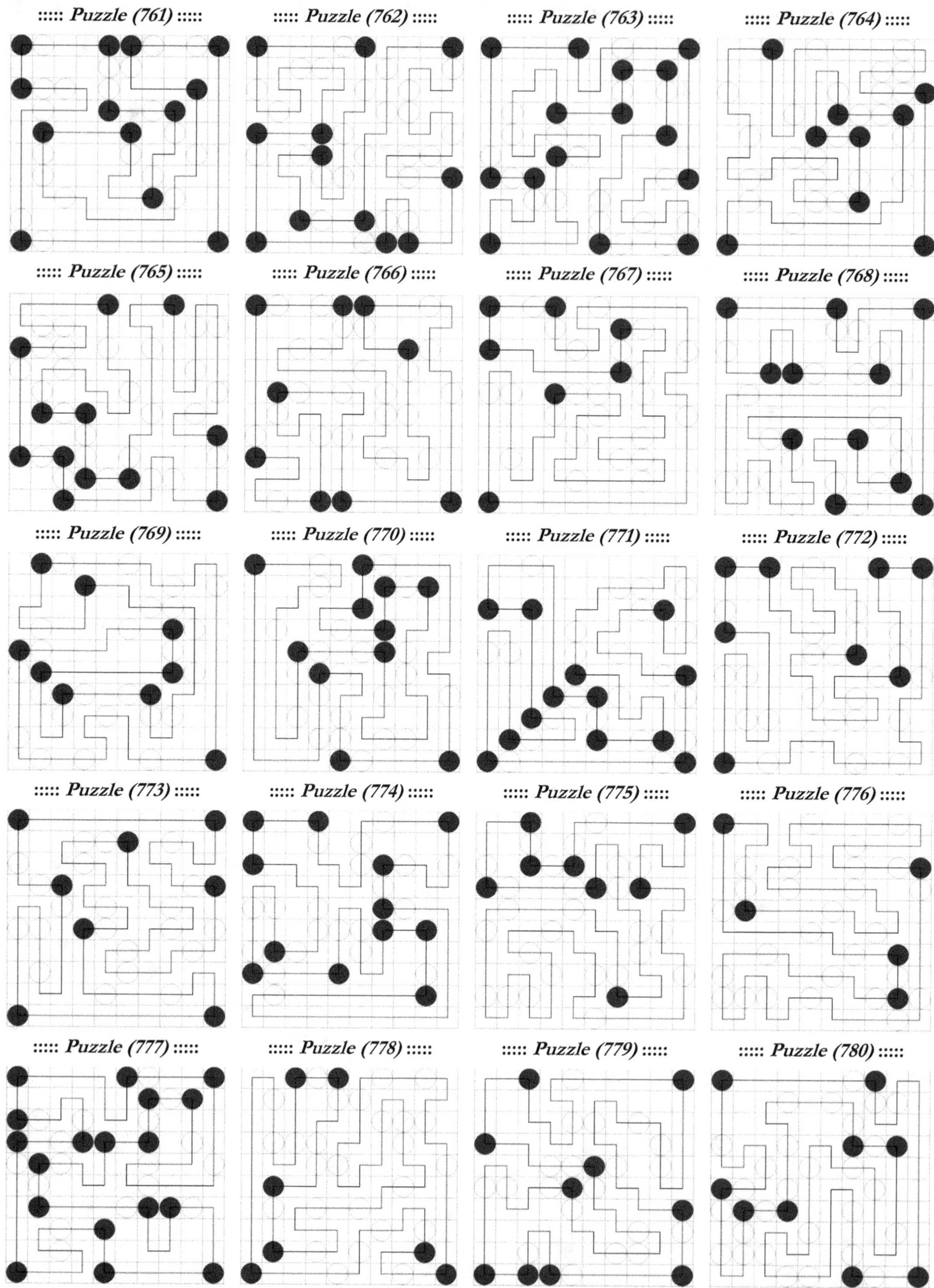

::::: Puzzle (761) :::::
::::: Puzzle (762) :::::
::::: Puzzle (763) :::::
::::: Puzzle (764) :::::
::::: Puzzle (765) :::::
::::: Puzzle (766) :::::
::::: Puzzle (767) :::::
::::: Puzzle (768) :::::
::::: Puzzle (769) :::::
::::: Puzzle (770) :::::
::::: Puzzle (771) :::::
::::: Puzzle (772) :::::
::::: Puzzle (773) :::::
::::: Puzzle (774) :::::
::::: Puzzle (775) :::::
::::: Puzzle (776) :::::
::::: Puzzle (777) :::::
::::: Puzzle (778) :::::
::::: Puzzle (779) :::::
::::: Puzzle (780) :::::

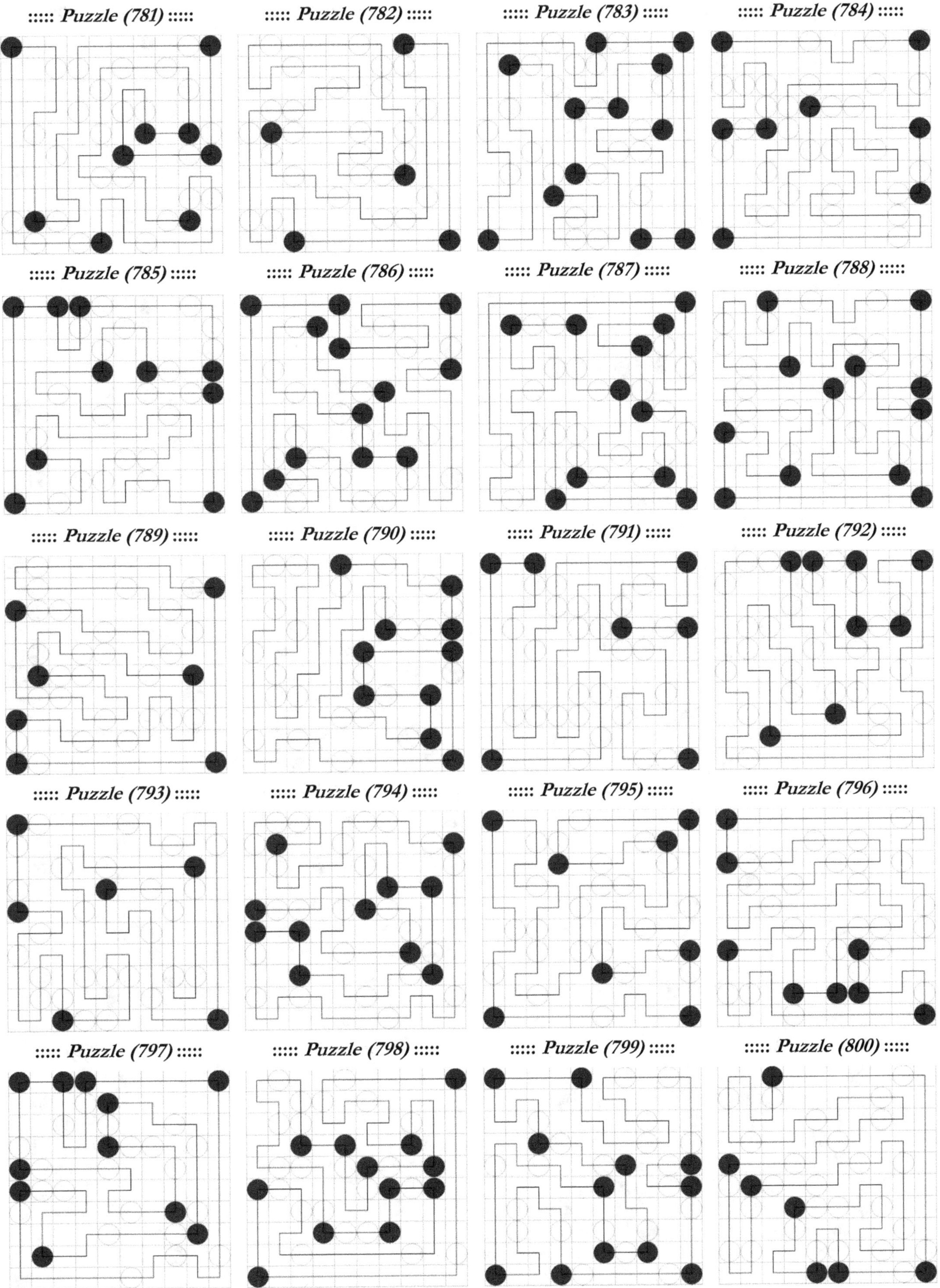

::::: Puzzle (781) :::::
::::: Puzzle (782) :::::
::::: Puzzle (783) :::::
::::: Puzzle (784) :::::
::::: Puzzle (785) :::::
::::: Puzzle (786) :::::
::::: Puzzle (787) :::::
::::: Puzzle (788) :::::
::::: Puzzle (789) :::::
::::: Puzzle (790) :::::
::::: Puzzle (791) :::::
::::: Puzzle (792) :::::
::::: Puzzle (793) :::::
::::: Puzzle (794) :::::
::::: Puzzle (795) :::::
::::: Puzzle (796) :::::
::::: Puzzle (797) :::::
::::: Puzzle (798) :::::
::::: Puzzle (799) :::::
::::: Puzzle (800) :::::

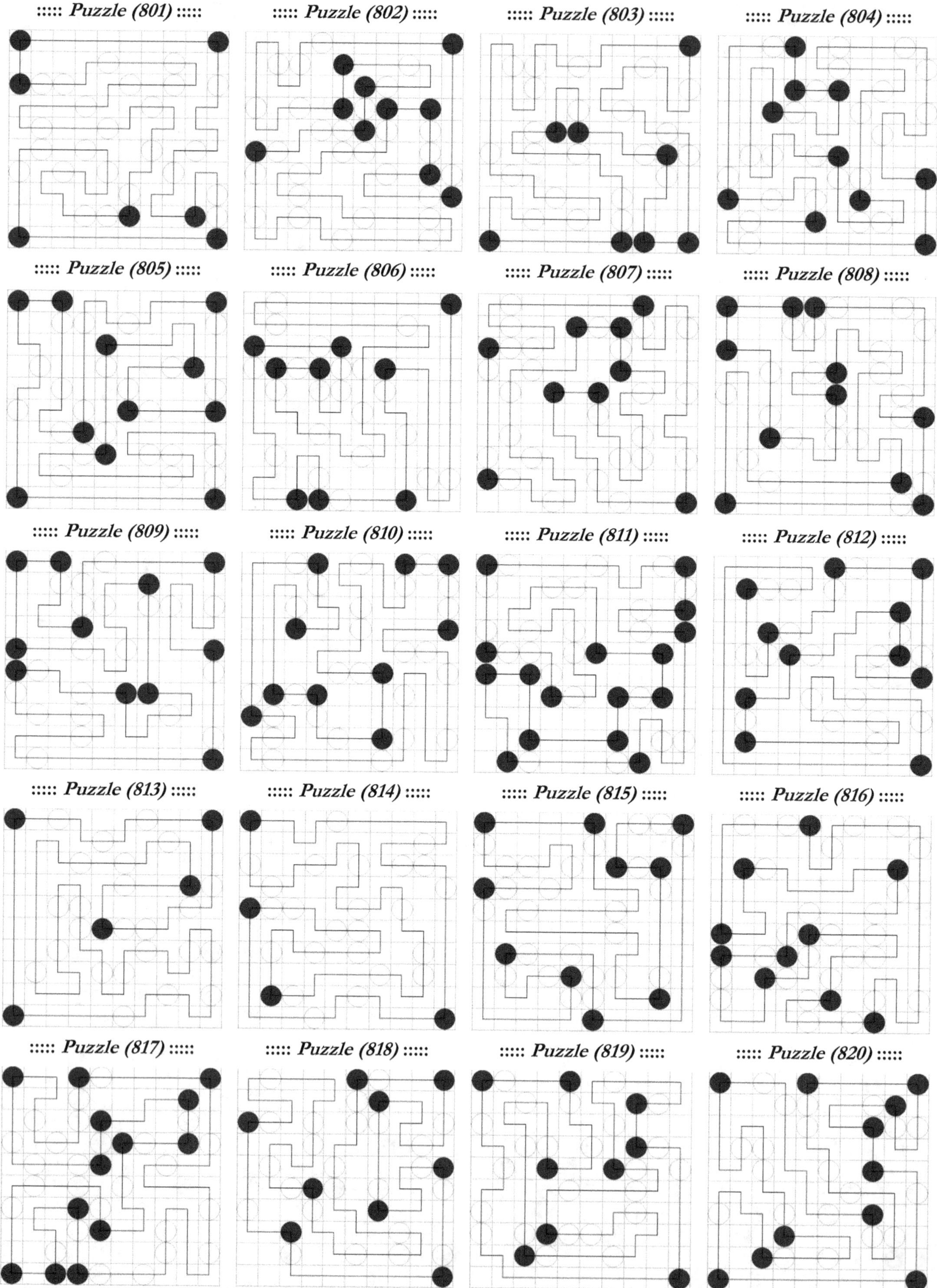

::::: Puzzle (801) :::::
::::: Puzzle (802) :::::
::::: Puzzle (803) :::::
::::: Puzzle (804) :::::
::::: Puzzle (805) :::::
::::: Puzzle (806) :::::
::::: Puzzle (807) :::::
::::: Puzzle (808) :::::
::::: Puzzle (809) :::::
::::: Puzzle (810) :::::
::::: Puzzle (811) :::::
::::: Puzzle (812) :::::
::::: Puzzle (813) :::::
::::: Puzzle (814) :::::
::::: Puzzle (815) :::::
::::: Puzzle (816) :::::
::::: Puzzle (817) :::::
::::: Puzzle (818) :::::
::::: Puzzle (819) :::::
::::: Puzzle (820) :::::

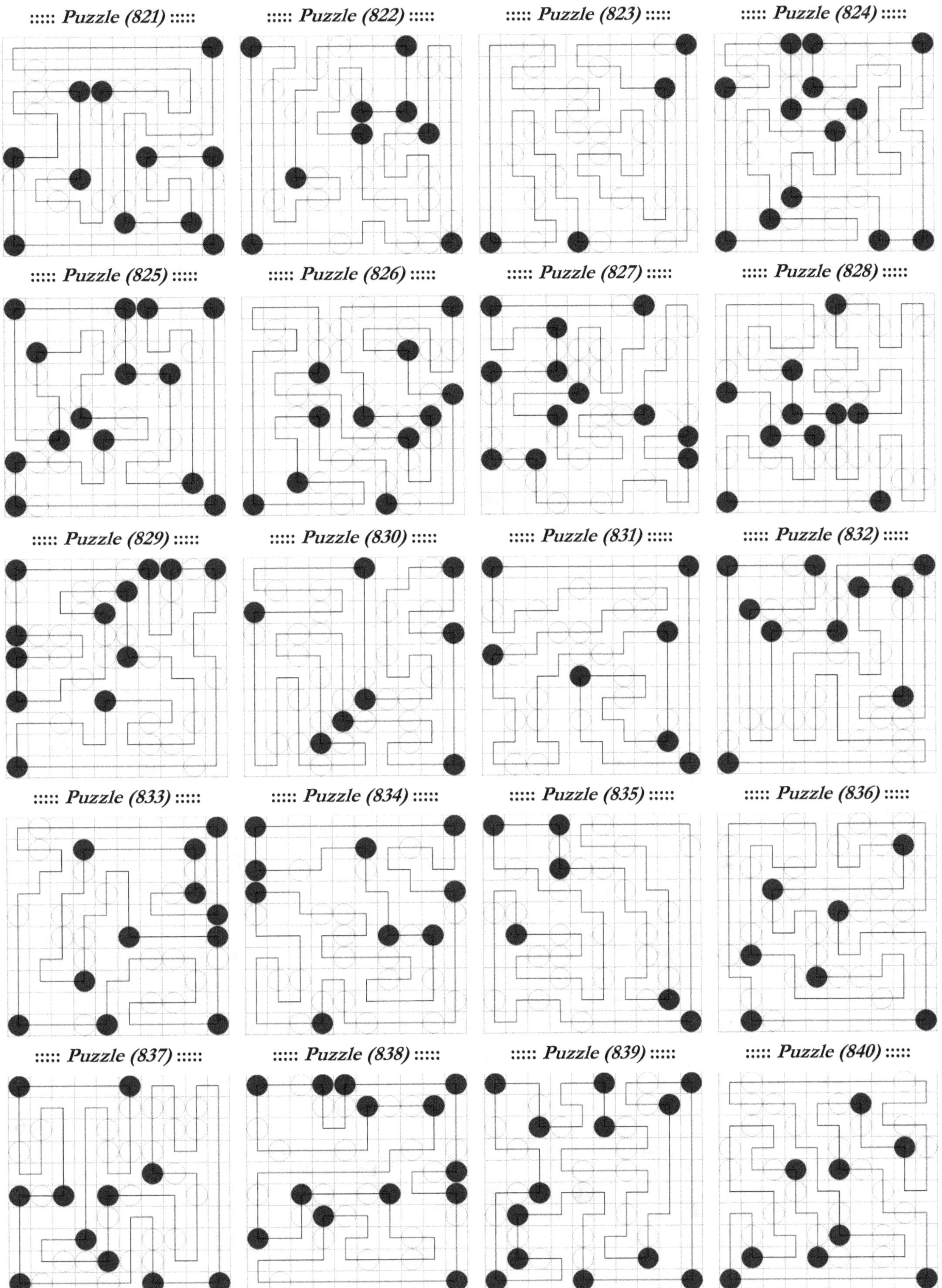

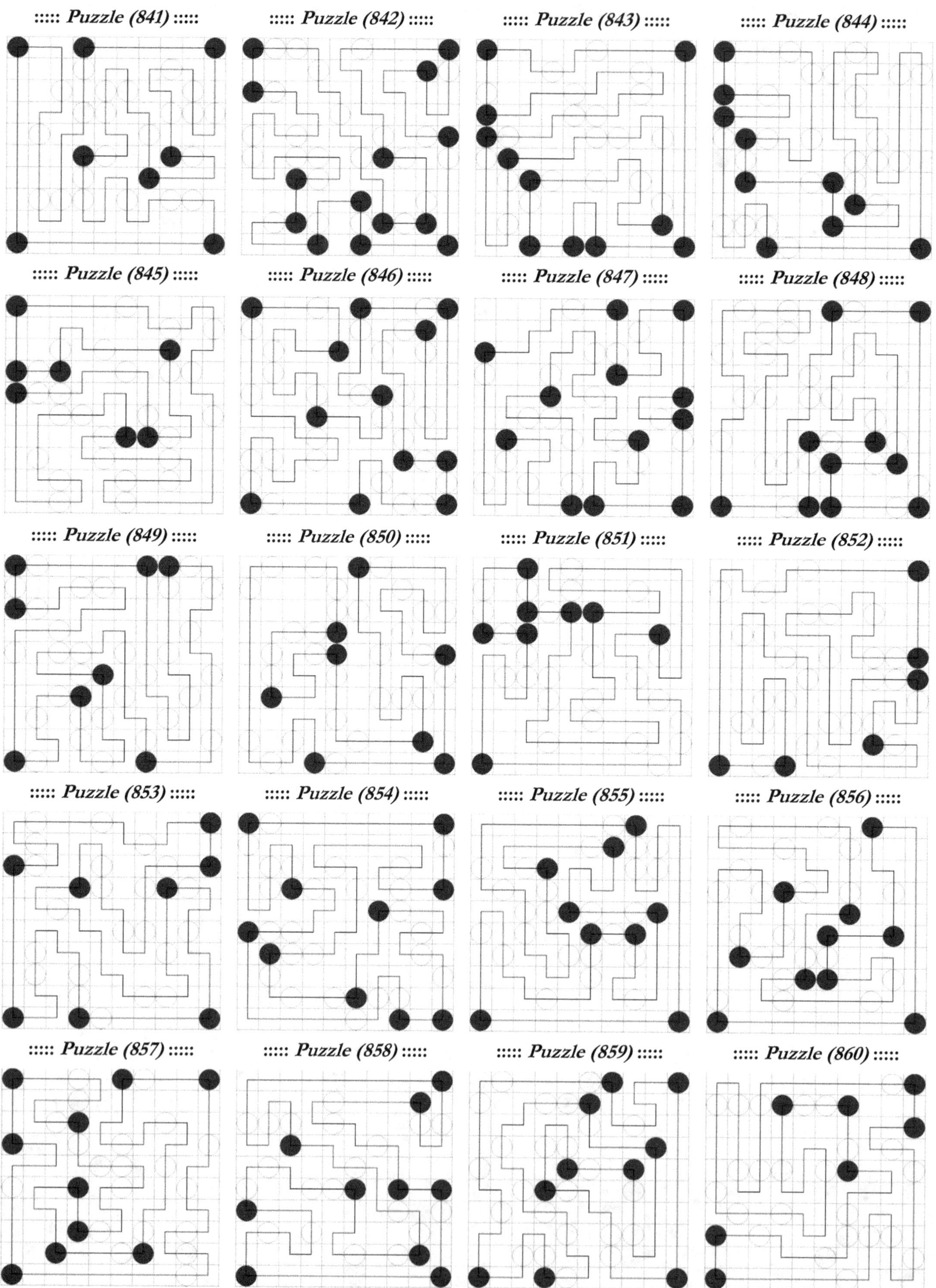

::::: *Puzzle (861)* ::::: ::::: *Puzzle (862)* ::::: ::::: *Puzzle (863)* ::::: ::::: *Puzzle (864)* :::::

::::: *Puzzle (865)* ::::: ::::: *Puzzle (866)* ::::: ::::: *Puzzle (867)* ::::: ::::: *Puzzle (868)* :::::

::::: *Puzzle (869)* ::::: ::::: *Puzzle (870)* ::::: ::::: *Puzzle (871)* ::::: ::::: *Puzzle (872)* :::::

::::: *Puzzle (873)* ::::: ::::: *Puzzle (874)* ::::: ::::: *Puzzle (875)* ::::: ::::: *Puzzle (876)* :::::

::::: *Puzzle (877)* ::::: ::::: *Puzzle (878)* ::::: ::::: *Puzzle (879)* ::::: ::::: *Puzzle (880)* :::::

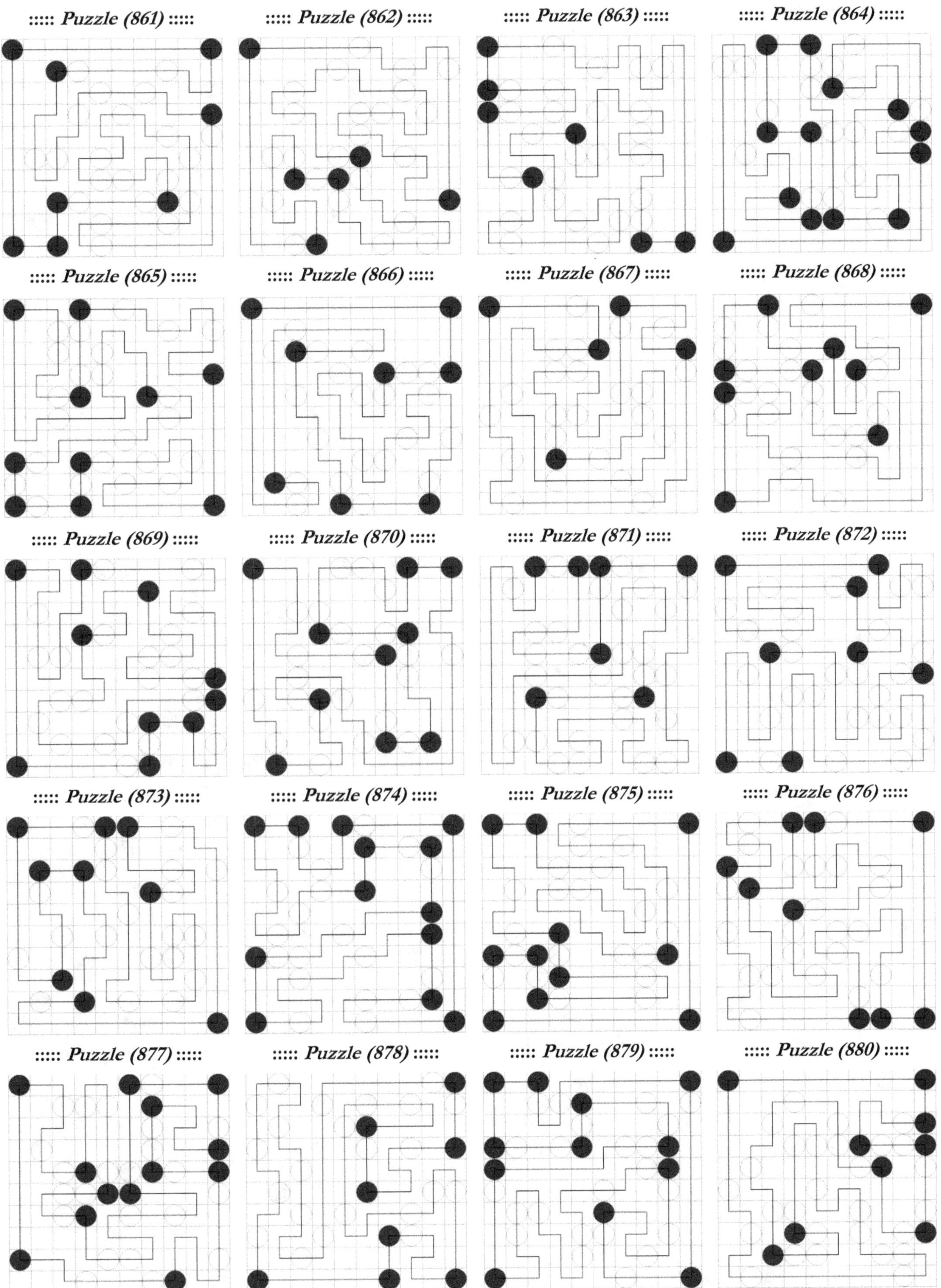

(213)

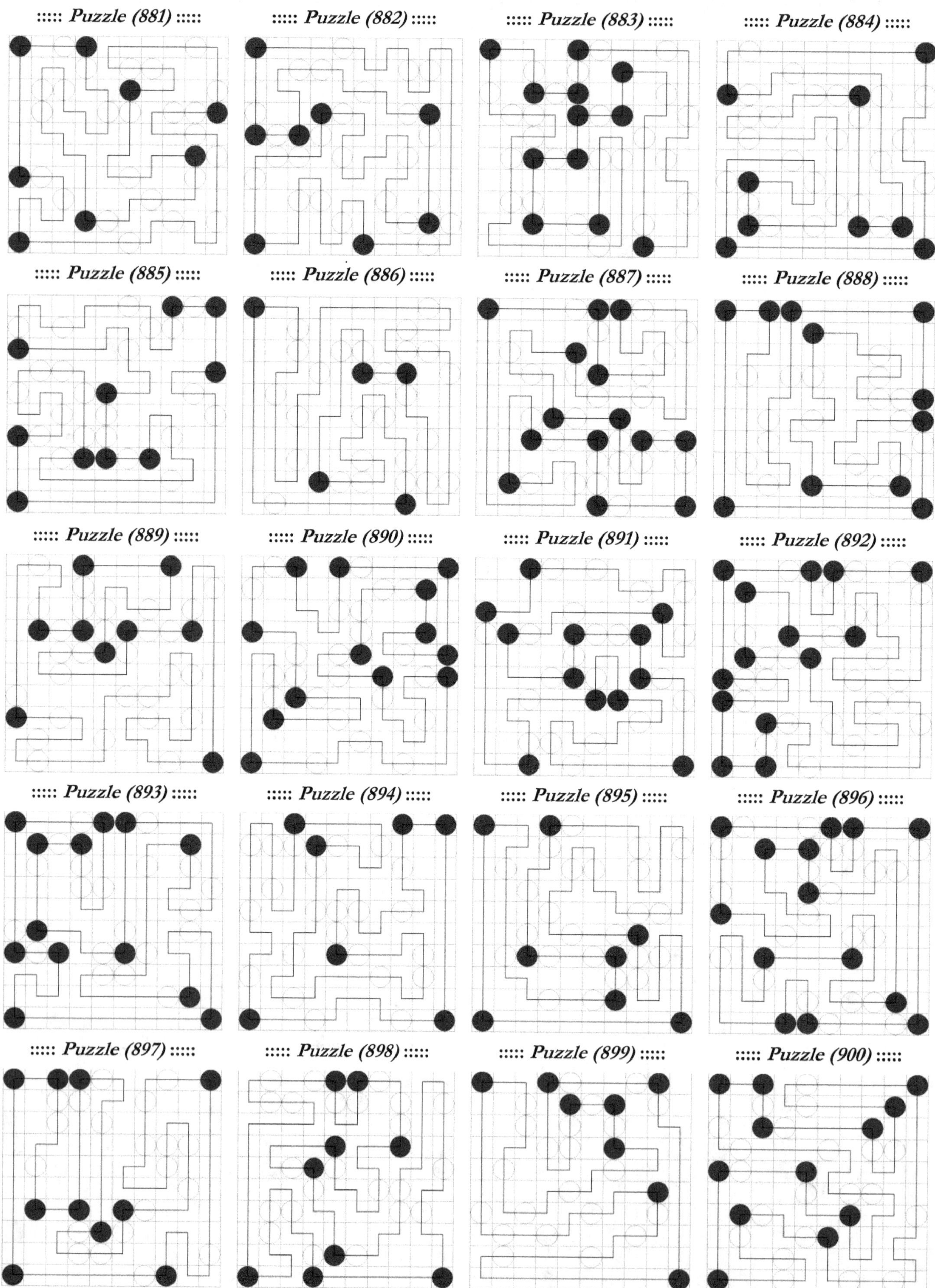

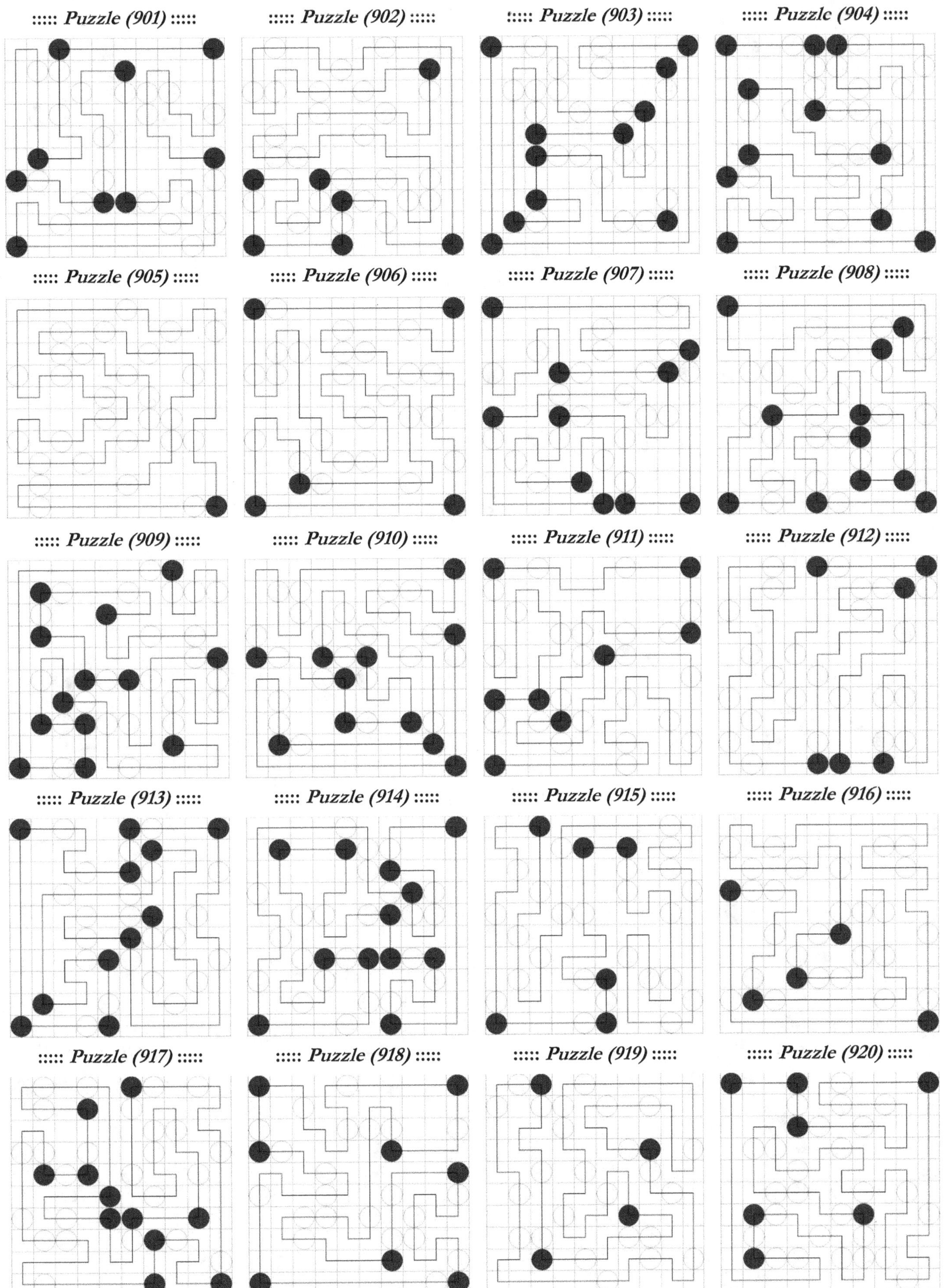

::::: Puzzle (901) :::::
::::: Puzzle (902) :::::
::::: Puzzle (903) :::::
::::: Puzzle (904) :::::
::::: Puzzle (905) :::::
::::: Puzzle (906) :::::
::::: Puzzle (907) :::::
::::: Puzzle (908) :::::
::::: Puzzle (909) :::::
::::: Puzzle (910) :::::
::::: Puzzle (911) :::::
::::: Puzzle (912) :::::
::::: Puzzle (913) :::::
::::: Puzzle (914) :::::
::::: Puzzle (915) :::::
::::: Puzzle (916) :::::
::::: Puzzle (917) :::::
::::: Puzzle (918) :::::
::::: Puzzle (919) :::::
::::: Puzzle (920) :::::

(215)

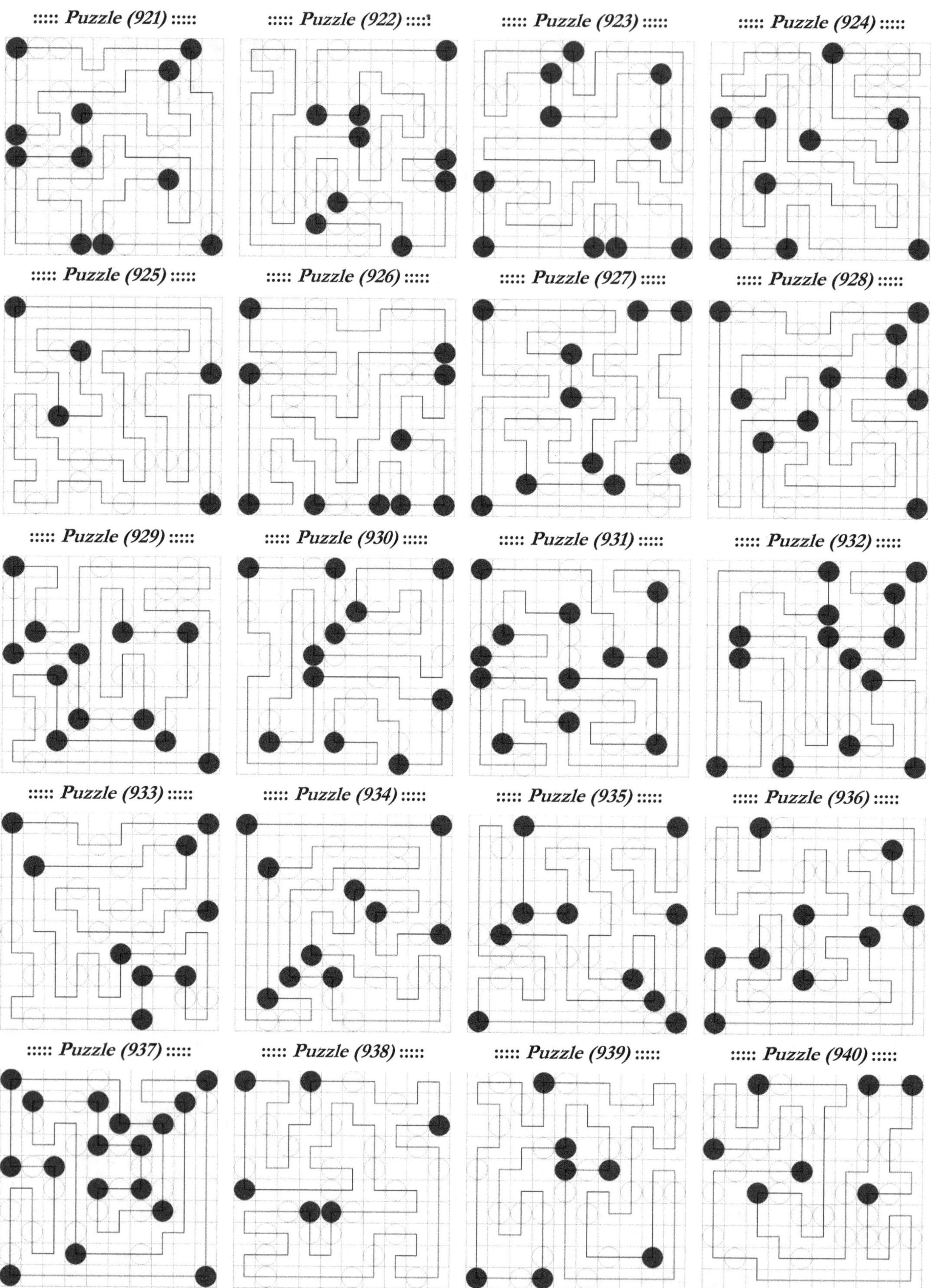

::::: Puzzle (921) :::::
::::: Puzzle (922) :::::
::::: Puzzle (923) :::::
::::: Puzzle (924) :::::
::::: Puzzle (925) :::::
::::: Puzzle (926) :::::
::::: Puzzle (927) :::::
::::: Puzzle (928) :::::
::::: Puzzle (929) :::::
::::: Puzzle (930) :::::
::::: Puzzle (931) :::::
::::: Puzzle (932) :::::
::::: Puzzle (933) :::::
::::: Puzzle (934) :::::
::::: Puzzle (935) :::::
::::: Puzzle (936) :::::
::::: Puzzle (937) :::::
::::: Puzzle (938) :::::
::::: Puzzle (939) :::::
::::: Puzzle (940) :::::

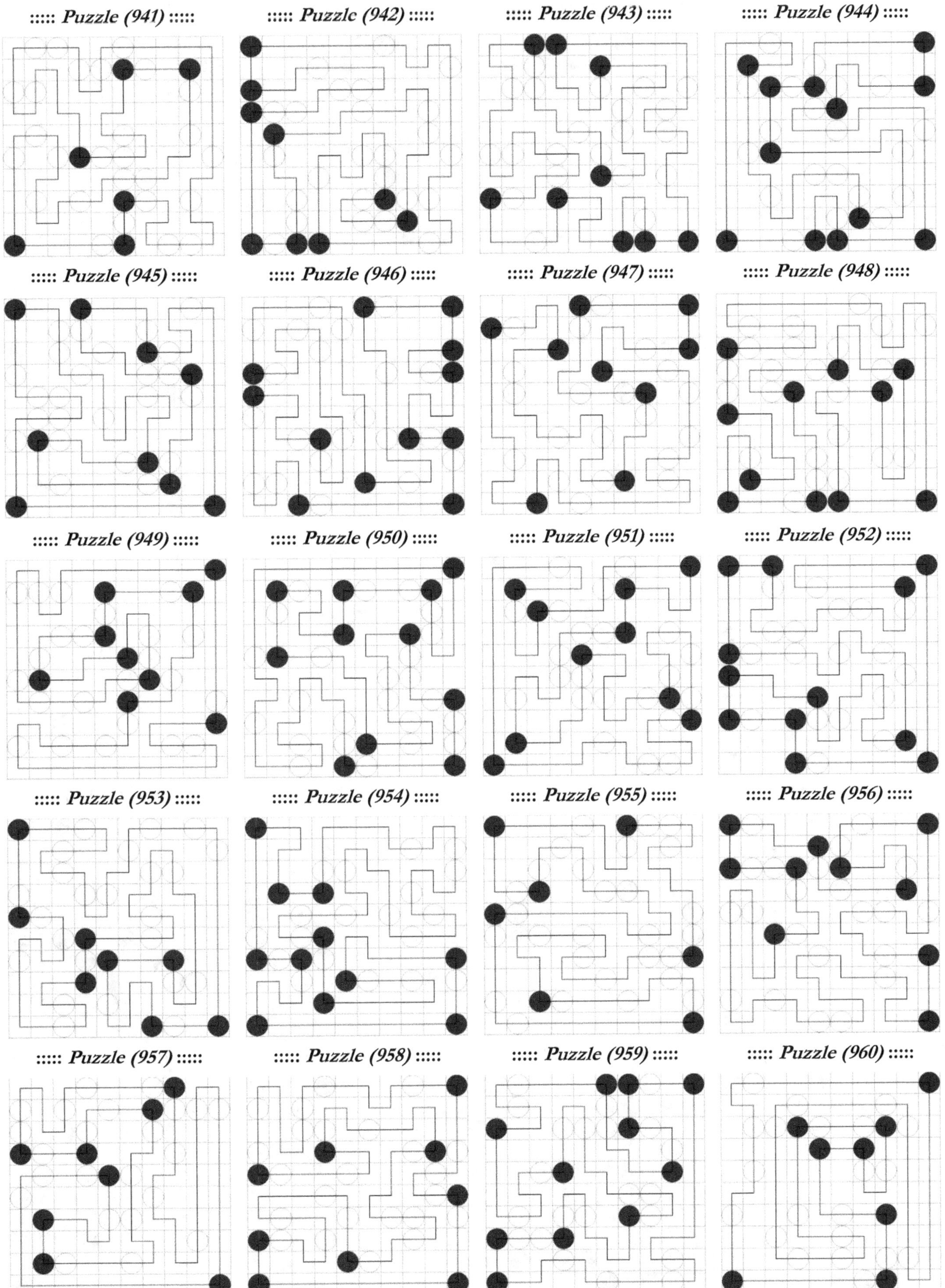

::::: *Puzzle (941)* ::::: ::::: *Puzzle (942)* ::::: ::::: *Puzzle (943)* ::::: ::::: *Puzzle (944)* :::::

::::: *Puzzle (945)* ::::: ::::: *Puzzle (946)* ::::: ::::: *Puzzle (947)* ::::: ::::: *Puzzle (948)* :::::

::::: *Puzzle (949)* ::::: ::::: *Puzzle (950)* ::::: ::::: *Puzzle (951)* ::::: ::::: *Puzzle (952)* :::::

::::: *Puzzle (953)* ::::: ::::: *Puzzle (954)* ::::: ::::: *Puzzle (955)* ::::: ::::: *Puzzle (956)* :::::

::::: *Puzzle (957)* ::::: ::::: *Puzzle (958)* ::::: ::::: *Puzzle (959)* ::::: ::::: *Puzzle (960)* :::::

(217)

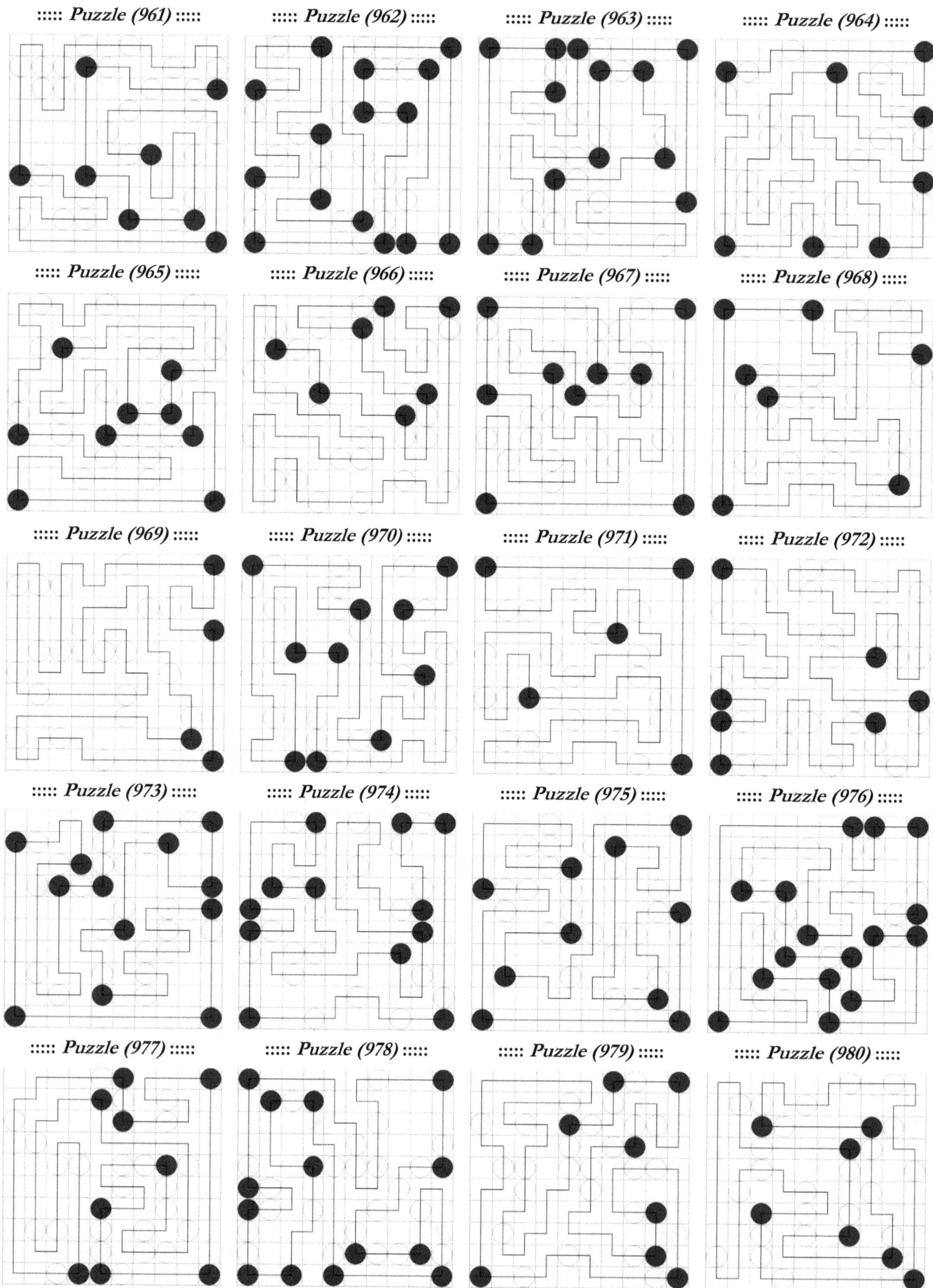

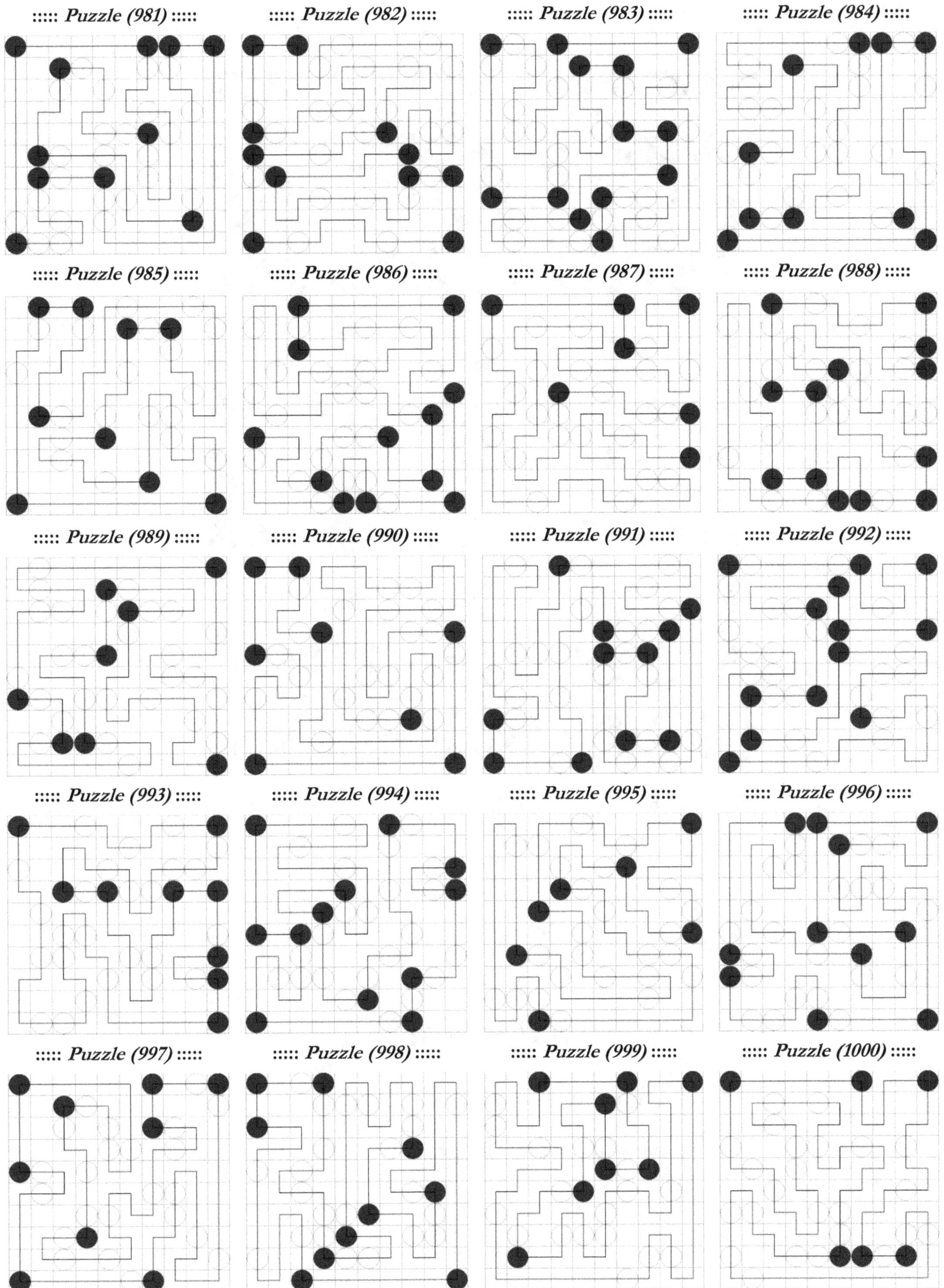

Please go to the link below to download and print this 500 Easy to Hard logic puzzles and start having fun.

https://bit.ly/2BM2LSC

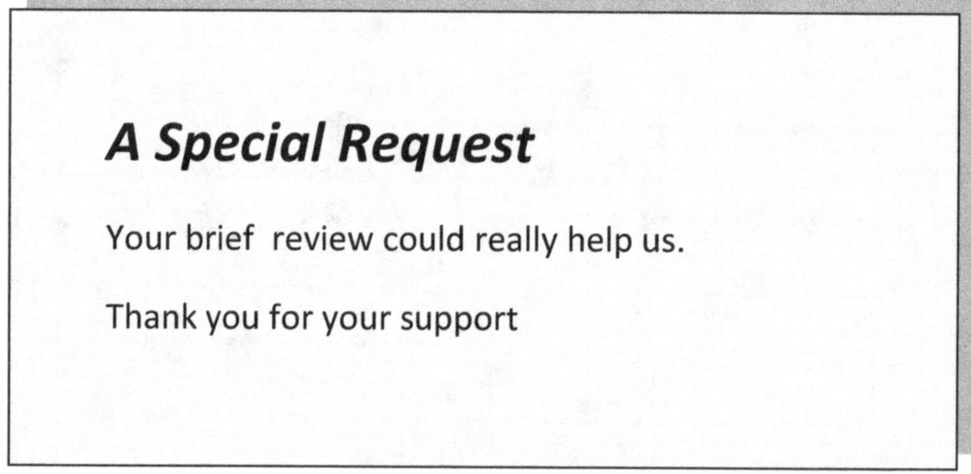

A Special Request

Your brief review could really help us.

Thank you for your support